LOST Opportunities

EXPLORATIONS OF EDUCATIONAL PURPOSE

Volume 23

Series Scope

In today's dominant modes of pedagogy, questions about issues of race, class, gender, sexuality, colonialism, religion, and other social dynamics are rarely asked. Questions about the social spaces where pedagogy takes place – in schools, media, and corporate think tanks – are not raised. And they need to be.

The *Explorations of Educational Purpose* book series can help establish a renewed interest in such questions and their centrality in the larger study of education and the preparation of teachers and other educational professionals. The editors of this series feel that education matters and that the world is in need of a rethinking of education and educational purpose.

Coming from a critical pedagogical orientation, *Explorations of Educational Purpose* aims to have the study of education transcend the trivialization that often degrades it. Rather than be content with the frivolous, scholarly lax forms of teacher education and weak teaching prevailing in the world today, we should work towards education that truly takes the unattained potential of human beings as its starting point. The series will present studies of all dimensions of education and offer alternatives. The ultimate aim of the series is to create new possibilities for people around the world who suffer under the current design of socio-political and educational institutions.

For further volumes:
http://www.springer.com/series/7472

Bronwyn Bevan • Philip Bell
Reed Stevens • Aria Razfar
Editors

LOST Opportunities

Learning in Out-of-School Time

Springer

Editors
Bronwyn Bevan
Exploratorium, San Francisco
CA, USA

Philip Bell
Learning Sciences and Human Development
University of Washington
Seattle, WA, USA

Reed Stevens
Northwestern University
Chicago, IL, USA

Aria Razfar
College of Education
University of Illinois at Chicago
Chicago, IL, USA

ISBN 978-94-007-4303-8 ISBN 978-94-007-4304-5 (eBook)
DOI 10.1007/978-94-007-4304-5
Springer Dordrecht Heidelberg New York London

Library of Congress Control Number: 2012943270

Printed on acid-free paper

Springer is part of Springer Science+Business Media (www.springer.com)

Preface

This volume focuses on learning outside of schools, a largely neglected topic at least in comparison to learning in schools. The chapters included here span studies of everyday and situated science and mathematical practices; the development of new theoretical and empirical frameworks for studying learning trajectories that span both in and out-of-school-time (OST) and settings; and the structured, organized OST programs where oftentimes everyday and situated knowledge and practices are leveraged to engage with more formalized practices and conceptions of science, technology, engineering, and mathematics.

While our title refers to OST learning opportunities as *lost,* in fact we know that in some ways they are just being *found*. Amid decades of educational research focusing almost entirely on learning in the classroom, quietly there has been developing a small but robust literature examining how scientific and mathematical practices occur across a wide array of everyday settings (Bransford et al., 2006; Gonzalez et al., 2005; Rogoff & Lave, 1984), in settings that may be less high-stakes than today's classrooms (Cole, 2006; Gutiérrez, 2008), and in ways that children, youth, and adults take up science and mathematics in order to achieve local and meaningful purposes (Fusco, 2001; Nasir, 2002). Today, a growing number of scholars in the learning sciences are working to understand how everyday experiences and opportunities to engage in scientific or mathematically rich practices support the emergence of interest, fluency, and mastery in different disciplinary areas. New departments and faculty positions are being established at universities across the country, and consensus volumes and policy documents are being issued by governing bodies (National Research Council, 2009; President's Council of Advisors on Science and Technology, 2010). Underpinning much of this work is a strong emphasis on educational equity and making available and accessible more socially supportive and intellectually engaging opportunities for children from nondominant communities to engage with and pursue learning experiences in science, technology, engineering, and mathematics (STEM).

A New Way of Looking

Conducting research on learning in OST settings requires broadening and deepening conceptions of learning, better specifying the nature of different learning settings, and more appropriately conceptualizing how learning in such contexts relates to school-based ways of knowing. Many scholars working in the domain of OST learning seek to develop knowledge that can be used not only to understand the science of learning but also to help to create, structure, and support new opportunities for learning both in and out of school.

Indeed, although the conventional discussion is frequently cast as "how can OST learning be structured to support school learning?" many scholars investigating the OST space and time are more interested in asking how learning across settings *relate* to one another, seeing the learning that happens outside of the school day as being oftentimes quite different in scope, meaning, and utility but also equally legitimate in terms of how it supports children's emergent interests, identities, and disciplinary understanding and experience. Many of the chapters in this volume illuminate the powerful learning that can happen in OST settings, both in less structured home or community settings and in designed spaces such as in after-school programs, and what learning scientists themselves can learn by looking more closely at this space.

This in part is the opportunity of learning about OST learning.

This Volume

The chapters in this book address a diverse set of issues related to current research in the OST space. The scholars who have contributed to this book include leading thinkers in the field of education as well as many emerging scholars, some still in graduate school at the time of writing, who are pushing the boundaries on how to think about and research learning. The issues addressed in this book range from how mathematical practices are taken up in the home to how after-school programs can serve as sites for teacher development. Proposals for new research frameworks that can account for learning as it develops across settings and over time are included. Ways of conceptualizing different learning settings and indeed what counts as science or what counts as mathematics in different settings are also addressed.

The diversity of subjects in this volume reflects the burgeoning and diverse field of research on learning in informal environments and settings. This field spans studies that look closely at cultural and developmental issues related to learners and communities of learners, as well as studies that consider institutional and policy dimensions of supporting learning in OST.

This volume is the result of a series of meetings that was organized by the Exploratorium's Center for Informal Learning and Schools to support collaboration and communication across four research centers funded by the National Science Foundation (NSF) under its Centers for Learning and Teaching (CLT) program and its Science of Learning Centers (SLC) program. The four contributing centers were the following:

Center for the Mathematics Education of Latinos/as (*CEMELA*), which focuses on the research and practice of the teaching and learning of mathematics with Latino students in the USA and involves the University of Arizona Tucson, the University of Illinois at Chicago, and the University of California Santa Cruz.

Center for Informal Learning and Schools (*CILS*), which supports research, scholarship, and professional development for informal educators to strengthen collaborations between, and learning across, formal and informal science education settings and involves the Exploratorium, King's College London, and the University of California Santa Cruz.

The Learning in Informal and Formal Environments (LIFE) Center, which develops and tests principles about the social foundations of human learning in informal and formal environments, including how people learn to innovate in contemporary society, with the goal of enhancing human learning from infancy to adulthood, and involves the University of Washington, Stanford University, SRI International Northwestern University, and University of California, Berkeley.

MetroMath, the Center for Mathematics in America's Cities, which focuses on improving mathematics teaching and learning in urban communities and schools and involves Rutgers University, the University of Pennsylvania, and City University of New York.

The CLT program was established by NSF to support the development of a new generation of scholars and leaders in key domains of science and mathematics education. CLTs primarily supported graduate studentships and postdoctoral research, as well as professional development and research activities supervised by center faculty. The SLCs supported research agendas that created the intellectual, organizational, and physical infrastructure needed for the long-term advancement of Science of Learning research.

Faculty and graduate students from CEMELA, CILS, LIFE, and MetroMath first met together in 2007 at the Exploratorium in San Francisco to share our work and discuss what we were learning about science; mathematics and learning in the OST space. The focus of CILS was explicitly on learning in OST settings. LIFE's work partially focused on OST. CEMELA and MetroMath primarily focused on learning in schools, but both were beginning to examine the OST setting as a way of testing particular ideas or tools. When these disparate groups came together, we quickly discovered that we were all grappling with foundational issues of (1) what counts as science/mathematics in the OST setting and (2) what counts as learning in the OST setting. These two issues were at the core of our interests and challenges relating to research methods and methodologies, to program design, professional development, and even policy analysis.

We worked together to host two series of small video seminars—at the 2007 CILS Bay Area Institute and at the University of Pennsylvania Ethnography Conference in 2008—where we engaged small groups of scholars and informal educators in viewing and considering the question of "What counts as math and science?" in relation to its appearance in everyday practices—from people fixing their cars, to walking their dogs, to visiting museums. We found that responses among participants varied tremendously, depending on their training and institutional perspectives. Out of these sessions, and after a poster session we collaboratively organized at the 2008 American Educational Research Association conference, we identified the need for a volume that specifically addressed research issues related to studying STEM in OST settings.

This volume is divided into three parts representing three of the different trends in the work of the contributing centers.

Part I raises the fundamental question of what counts as science and mathematics in everyday settings. The question seems easy to answer when we look inside the school at the subjects, textbooks, and teachers that go by the names "mathematics" and "science." But once we step outside of schools, the question about what counts becomes complex and important. The chapters in this part take us inside some of this complexity, often relating what counts as and is experienced as STEM in OST settings to trends and representations of STEM in school settings. A commentary by Ray McDermott of Stanford University notes how narrow conceptions of what counts as math or science have operated to close doors for many children. In order to develop a theoretical understanding of the life-course outcomes of particular individuals, we need to better understand how people move and learn across a varied set of cultural niches with variable practices, materials, and evaluation systems that are used to gauge human behavior (Gutiérrez and Rogoff, 2003; Lee, 2008).

Part II contains four chapters describing emerging research frameworks for studying learning as it develops over time and across settings. These frameworks include the use of methods, such as technobiographies, discourse analysis, and ethnographic work including longitudinal ethnographic studies. A commentary by Kris D. Gutiérrez of the University of Colorado Boulder highlights how cross-setting accounts of learning could promote equity and transformative outcomes for youth from nondominant communities, including suggesting how educational systems might provide coordinated, redundant supports for learning across multiple settings over longitudinal time.

Part III contains five chapters that address teaching and learning in organized OST settings, primarily after-school programs. This section problematizes studies of the after-school setting, showing how the after-school space operates as a unique in-between space, adopting many of the norms of schooling as well as the norms of home or community time. A commentary by Mike Cole, from the University of California San Diego, closes this section with an appeal to policymakers to broaden conceptions of learning in order to understand and leverage the potential of the structured OST space. Without broadening conceptions of learning, without developing more appropriate ways to evaluate programs and assess learning in these settings, the opportunity of learning in OST settings is largely lost.

Thank Yous

Schools are essential institutions in our democratic society, and all of the authors in this book are products of and contributors to schooling. We have the utmost respect for the teachers and educators who spend their days and lives trying to make schools a powerful experience for the children who spend so much of their formative years in classrooms or preparing for time in classrooms. The work in this book is meant to broaden conceptions of learning and education to encompass but go beyond schools, and is in no way meant to devalue the contribution of schools. We believe that learning about learning in OST settings can strengthen teaching and learning across the educational landscape.

The editors of the volume wish to thank all of the chapter authors for their dedication, patience, and collaboration. We also want to thank our program officers at the National Science Foundation, particularly Janice Earle, who supported this effort through a supplemental grant to the Center for Informal Learning and Schools (ESI-0119787). Thanks also to Fan Kong, who helped to produce and complete the project. Finally, we would like to thank the teachers in our lives, both in and out of school.

Exploratorium	Bronwyn Bevan
University of Washington	Philip Bell
Northwestern University	Reed Stevens
University of Illinois at Chicago	Aria Razfar

References

Bransford, J. D., Barron, B., Pea, R. D., Meltzoff, A., Kuhl, P., Bell, P., et al. (2006). Foundations and opportunities for an interdisciplinary science of learning. In R. K. Sawyer (Ed.), *The Cambridge handbook of the learning sciences* (pp. 19–34). New York: Cambridge University Press.

Cole, M. (2006). *The fifth dimension: An after-school program built on diversity*. New York: Russell Sage Foundation.

Fusco, D. (2001). Creating relevant science through urban planning and gardening. *Journal of Research in Science Teaching, 38*(8), 860–877.

Gonzalez, N., Moll, L. C., & Amanti, C. (2005). *Funds of knowledge: Theorizing practices in households and classrooms*. Mahwah, NJ: Lawrence Erlbaum Associates.

Gutiérrez, K. D. (2008). Developing a sociocritical literacy in the third space. *Reading Research Quarterly, 43*(2), 148–164.

Gutiérrez, K. D., & Rogoff, B. (2003). Cultural ways of learning: Individual traits or repertoires of practice. *Educational Researcher, 32*(5), 19–25.

Lee, C. D. (2008). The centrality of culture to the scientific study of learning and development: How an ecological framework in education research facilitates civic responsibility. *Educational Researcher, 37*(5), 267–279.

Nasir, N. S. (2002). Identity, goals, and learning: Mathematics in cultural practice. *Mathematical Thinking and Learning, 4*(2&3), 213–247.

National Research Council. (2009). *Learning science in informal environments: People, places, and pursuits*. Washington, DC: The National Academy Press.

President's Council of Advisors on Science and Technology. (2010). *Report to the President: Prepare and inspire: K-12 education in Science, Technology, Engineering, and Math (STEM) for America's future*. Washington, DC: Executive Office of the President.

Rogoff, B., & Lave, J. (1984). *Everyday cognition: Development in social context.* New York: Harvard University Press.

Contents

Part I
What Counts as Math and Science?

Chapter 1
Introduction: What Counts as Math and Science?

Reed Stevens

Each of the four chapters in this section addresses, in different ways, the question of what counts as math or science. Three of the chapters are about math and one is about science. At first glance, this "what counts" question may seem a strange one because what counts as math or science would seem to be a fairly straightforward question to answer. The answer one might give is that math and science are things you learn in school and if you stay in school long enough, you may eventually do these things as a professional mathematician or scientist. This suggests a pretty straight line through school into professional life. It might seem to settle the question of what counts as math or science.

However, of course people spend most of their time outside of school as Fig. 1.1 shows. This immediately complicates the issue, because about these activities that people are doing outside of school we may ask, "Do these activities count as math or science?" How and in what ways does it matter if we count people's outside of school activities as math or science? One way it matters is that it bears on the practical utility we attribute to school math or science. If we counted very few of the activities that people engage in out of school as math or science, then we might rightly conclude that school math and school science really do not "transfer out" (Schwartz & Nasir, 2003) to people's other activities. That would make school math and sciences two species of activity whose relevance begins and ends at the schoolroom door. And that, of course, would run counter to time-honored if not time-worn rhetoric about the central importance of school math and science to people's well-being, their ability to participate as citizens, and nothing less than our nation's economic survival. Alternatively, we might find that school math or science are to be

R. Stevens (✉)
Learning Sciences, School of Education and Social Policy,
Northwestern University, Evanston, IL, USA
e-mail: reed-stevens@northwestern.edu

B. Bevan et al. (eds.), *LOST Opportunities: Learning in Out-of-School Time*,
Explorations of Educational Purpose 23, DOI 10.1007/978-94-007-4304-5_1,

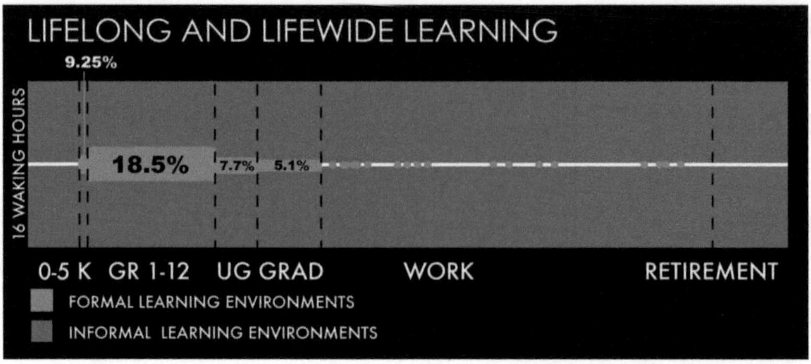

Fig. 1.1 This diagram shows the amount of time, averaged over a year, that people spend in formal or academic subject matter school experience as compared to the time they spend in other waking activities. This diagram was originally conceived by Reed Stevens and John Bransford to represent the range of learning environments being studied at the NSF Learning in Informal and Formal Environments (LIFE) Center (http://life-slc.org). Design, documentation, and calculations were conducted by Reed Stevens, with key assistance from Anne Stevens (graphic design) and Nathan Parham (calculations). Licensed under a Creative Commons Attribution-Share Alike 3.0 United States License

counted among many of people's activities in everyday life, which would affirm that school math and science do indeed have great utility outside of school.

The "what counts" question is important because there is symbolic capital (Bourdieu, 1991) attached to both "math" and "science." People who are counted as doing these things are more often than not seen as smart and allocated social opportunities, usually through the educational system, on that basis. Of course, being counted as doing math and science can also attach a person to a negative cultural stereotype like "geek" or "nerd." (For some, of course, these labels are worn proudly.) The point is that what counts as math or science can cut both ways in terms of attendant social meanings. Moreover, what counts as math or science depends on how the culture represents them, and school is but one setting where math and science are represented. They are represented also in the activities of professionals, which may or may not be seen as continuous with school math or science. Some mathematicians and scientists with a stake in education have made quite clear that they count little of school math or science, as it is typically practiced, as math or science as they understand it. Math and science are also, like most prominent cultural practices, featured in a wide range of public media. There are TV shows about scientists and mathematicians and they are usually solving crimes. There are representations of science and math in video games, in magazines, and in books and movies. All of these representations must play a role in what counts as math or science.

Clearly, the sites that can matter for what counts as math or science are multiple. School is clearly an important site but it is not the only one. Homes are probably among the more important sites, especially among families who understand the symbolic capital and social opportunities that may present to their school-aged children if they are successful with math and science in school. For these families, there is probably a strong pedagogical impetus to recognize and connect much of a child's interests with their growing capacities to do math or science, since this sets them on a positive trajectory with these school subjects. Equally plausible, families may be sites for negotiating disjunctions between what they see as legitimate math or science in their children's capacities and interests, at the same time that those capacities and interests go unrecognized in school.

The chapters in this section engage these issues of what counts as math or science in a number of ways and are spread across a range of societal contexts. In Chap. 2, entitled "Math I Am: What we learn from stories that people tell about math in their lives," Esmonde and colleagues from the "Family Math" project report on a study that involved home-based interviews with families about how they use math in their daily lives. They find families do count a good deal of their activity as math and only some of what they count do they relate directly to school math. For these families, more than school math counts as math. Like the other chapters in this section, the authors identify some tensions between school math and "family math" but also some opportunities for connections. The authors make analytic use of some of these tensions and some of the qualities that families report about their mathematical activity to recommend some different elements for school math.

In Chap. 3, "What counts as science in everyday and family interactions?" Callanan and colleagues take a different approach to the "what counts" question. In their chapter they are not studying what counts as science to their research participants, who are children and their parents. They are deliberately trying to identify qualities in children's everyday conversations with parents that they as social scientists count as science but the children and adults do not, at least in the sense that they do not label or otherwise identify them as such. The authors choose not to define science writ large but rather in terms of four qualities that they see as central to science: "(1) Are they using and discussing scientific terms? (2) Are they generalizing beyond the specific event to broader classes of events either through analogy or generalizing language? (3) Are they discussing causal explanations for events? (4) Are they seeking evidence to test hypotheses or experimenting?" (Chap. 3, Callanan et al., this volume, p. 30). The authors find these qualities mostly present in the moments they have selected for the paper and, therefore, they argue that many everyday moments contribute to children's development in math and science, whether they are recognized as such or not.

In Chap. 4, Esmonde addresses the neglected question of what counts as math in public media representations, in this case in the television show *NUMB3RS*. In this show, professional mathematicians help FBI agents with criminal cases they cannot solve on their own. (Based on shows like this and the *CSI* franchise, a whole generation of young people might grow up to understand that the primary practical function of math and science is solving grizzly crimes.) Esmonde's chapter involves a critical analysis of the way the show uses mathematical language and the ways it positions the characters' overall authority and intelligence in relation to their attributed mathematical authority. The analysis points to an irony in what is counted by this show as math. Despite a tagline, "we all do math every day," Esmonde's analysis makes clear that only the mathematicians, and particularly the star mathematician, really legitimately do anything that counts as math. In quoting from the character, the chapter shows that the authority for what counts as math is proportional to a character's status as a professional mathematician. Math is thus depicted as the province of mathematicians rather than something we all do every day. Her analysis also includes an interesting empirical component in which she asked a set of viewers to identify moments in one of the shows they counted as math. Based on this analysis, she affirms that the show's elitist cultural representation of what counts as math is being reinforced for viewers.

In Chap. 5, "What counts too much and too little as math," Reed Stevens draws on math teaching and ethnographic field research in and out of school to show how multifarious is what counts as math. Like the other papers in this section, Stevens' paper makes clear that across society "math" is no unitary thing. This paper goes further to show that for particular individual people, multiple coexisting senses of what counts as math can create very different possible futures and very different social identifications of self and by others. The primary boundaries across which he casts a comparative lens are school and occupational contexts, but the arguments would seem to generalize to the boundaries between school math and other out-of-school contexts of mathematical practice. What Stevens' paper shows is that sometimes math counts for reasons it should, because it does work and helps people advance their life projects, be they practical or those conventionally understood as more "purely intellectual" or disciplinary. However, this sense of math has an evil twin that is used to make moral judgments about people and label them for the purposes of the allocation of social opportunities. This chapter is framed by extended anchoring narratives about two people caught in the competing senses of what counts. The paper closes with a critical dialogue with a collection of rhetorical positions, insufficiently examined, that continue to rationalize school math vis-à-vis out-of-school mathematical practices.

Taken together, the chapters in this section invite us to think about what each of us counts as math and science and when and why it matters. Each chapter invites us to see mathematics and science as activities that are something more than just what happens in school and to think about the relationships between math and science in school and math and science beyond school's doors. As Ray McDermott notes in his commentary on this section, a failure to recognize and account for how children capably engage with and use math and science in everyday life creates a *lost opportunity* for teachers seeking to support students' science and math learning trajectories.

References

Bourdieu, P. (1991). *Language & symbolic power*. Cambridge: Harvard University Press.

Schwartz, D. L., & Nasir, N. (2003). Transfer of learning. In W. Guthrie (Ed.), *Encyclopedia of education* (2nd ed., pp. 1449–1452). New York: Macmillan.

Chapter 2
Math I Am: What We Learn from Stories That People Tell About Math in Their Lives

Indigo Esmonde, Kristen Pilner Blair, Shelley Goldman, Lee Martin, Osvaldo Jimenez, and Roy Pea

Introduction

What is mathematics? And what counts as mathematics in people's activities at home, work, and school in daily life? One might try to answer this question by consulting expert mathematicians and philosophers, or by examining the historical role that mathematics has played in shaping major scientific and technical advances. We take a far less lofty approach, and try to find answers to this question in the everyday experiences of adults and children. Without denying the possibility of a universal mathematics, we assume that the question *What is mathematics?* may garner markedly different answers from person to person and, therefore, the meaning of mathematics may vary from person to person and from context to context.

We might expect many families to deny any mathematical involvement, except in school and certain professional contexts. Or families might report engaging in a very narrow slice of the mathematical world, such as counting change at the grocery store. Or they might report broad participation in mathematical activities across varied contexts. In order to find out how different families perceived mathematics in their lives, we asked family members to tell us stories about their mathematical experiences.

Through stories gathered from interviews with 20 families reflecting the ethnic, racial, and economic diversity of the San Francisco Bay Area, we investigate the diverse contexts and activities in which families engage. We are especially interested

I. Esmonde (✉)
Ontario Institute for Studies in Education, University of Toronto, Toronto, ON, Canada
e-mail: indigo.esmonde@utoronto.ca

K.P. Blair • S. Goldman • O. Jimenez • R. Pea
Stanford University, Stanford, CA, USA

L. Martin
School of Education, University of California, Davis, Davis, CA, USA

B. Bevan et al. (eds.), *LOST Opportunities: Learning in Out-of-School Time*,
Explorations of Educational Purpose 23, DOI 10.1007/978-94-007-4304-5_2,

in understanding how and when mathematics plays a part in these contexts. We seek to characterize the structure of mathematical activities, describe the resources that are brought to bear, and analyze the contributing social conditions and arrangements. Ultimately, we hope to understand the ways that family life is replete with mathematically relevant thinking and problem-solving and to identify possible intersections between mathematics in the home and mathematics in the school (Goldman, 2005).

We base our analysis on a set of "Math in a Minute" (MIAM) stories we collected near the beginning of our interviews (described in detail below). We asked each family member to tell us a story from his or her life involving mathematics. These stories offer a glimpse into people's views of mathematics – what it is, what role it plays in their lives, and what is a reasonable way to tell a story about it. We find that families described mathematical experiences across a wide range of activities. The stories they told reflected their conceptions of what mathematics was, and their conceptions of who they were as individuals and as a family.

The stories are, as we had hoped, both personal and mathematical. As such, we found that when our interview participants reflected on mathematics, they also considered how their mathematics reflected on them. The two questions, "What is mathematics?" and "Who am I, (and who are we, as a family) in relation to mathematics?" are the two foci around which we form our analysis in this chapter. In tackling these two questions, we make specific for mathematics the general observation for learning theory that "learning to know" and "learning to be" are intertwined: "What people learn about, then, is always refracted through who they are and what they are learning to be" (Brown & Duguid, 1996, p. 138; also see Lave & Wenger, 1991). Because we asked people to tell personal stories, their observations about mathematics and about themselves were tightly interconnected.

Methods

The chapter is based on interviews where we sought narrative accounts of mathematics in people's daily lives. Narratives provide us with participants' accounts in their own words about their lives and mathematical activities. The total number of persons interviewed was 71 and included 35 children in 20 families. The families represented a spectrum of racial and economic diversity, with parents' educational levels ranging from some high school education through graduate school. All of the families included at least one child in middle school at the time of the interview, and we were interested in how students at this crucial juncture were experiencing mathematics at school and at home.

The Interviews

The MIAM stories were gathered as part of a semi-structured interview designed to generate conversation and elicit accounts about family members' uses of, and experiences with, mathematics. The interviews were meant to prompt discussions of the activities that families engaged in as part of family life, work, and school, allowing them to provide particularized versions of how they thought about and accomplished each life task (Linde, 1988). The interview was conducted with all available family members and two or more interviewers (including one camera operator).

The stories were collected early in the interviews, and were meant to be ice-breakers that allowed family members to talk about mathematics in their lives before our interview prompted them to talk about specific family contexts and activities. We asked for a story involving mathematics, good or bad, from any setting, including school, work, and home, that would take about a minute to tell. This task enabled each family member to relate to mathematics in his or her own way. MIAM stories covered all kinds of territory, from how people felt about mathematics in home schooling, to how people felt when they failed in school mathematics, to how they used mathematics while fixing something around the house.

To help interviewees understand the task, interviewers began by telling an example MIAM story that did not involve school mathematics. Each interviewer told a different story, about a recent experience. For example, one interviewer discussed helping some friends lay laminate flooring and the mathematical challenges involved in the project. Each interviewer's story undoubtedly influenced some of the participants' stories, perhaps by eliciting many nonschool mathematics stories (as intended), and likely in other subtle and idiosyncratic ways. Some family members followed on the interviewer stories through associations they made with them, but generally, the interviewer stories were not out of the common range for mathematics-involved descriptions we heard in other parts of the interviews. Likewise, because family members told stories sequentially, they influenced each other's stories. We take this interdependence as an interesting finding about how families tell stories (and perhaps as a clue to how family members are socialized into their understandings of mathematics).

Some stories were co-constructed by several family members, especially children's stories. Younger participants sometimes hesitated to offer a story and were prompted in a general way by the interviewer or for a specific story by an adult family member (e.g., "Remember the time we made the curtains?"). Also, although at times the interviewer participated in a family discussion of the mathematical content of a particular story, the stories themselves were always chosen by the families first. So, although the stories and their interpretations of mathematics were influenced by the interview setting, interviewers, and other family members, the choice of stories still reflects how participants represented situations in their lives in which mathematics played a role.

The interview went on to focus on different areas of family life where mathematics commonly appears (e.g., home improvement and repair, budgeting, shopping). We asked families to tell us about their experiences in these areas. Although in this part of the interview we did not specifically ask for mathematical stories, participants did focus mainly on mathematical aspects of their participation.

The analyses of the MIAM stories proceeded through several stages and data generation activities. Six team members completed interpretive analyses to identify characteristics of people's mathematics depictions. For example, early on we noted particularly salient differences between stories told about school mathematics and

Table 1.1 Number of stories per category

Category	Number of stories
Home	49
School	22
Total	*71*

those told about mathematics outside of school. That led us to create two groups of stories — what we call "school stories" and "home stories" — and search them for commonalities and differences based on emergent features, characteristics and themes. We take the categories of home and school stories separately, answering the questions "What is mathematics?" and "What does this story of mathematics say about me as an individual or us as a family?" by comparing and contrasting their characteristics. Families told between one and five stories, with a mean of 3.55 stories per family. Table 1.1 shows the number of stories we collected in each category.

The MIAM stories represent a powerful setting in which to analyze the family stories. Since we asked participants to tell a story about mathematics, their stories highlight what counts as mathematics for them (in the context of the interview), as well as how they used mathematics to tell us something about themselves and their families.

(Mathematical) Identity and Narrative

The mathematics stories we were told were not neutral, factual accounts of events people had experienced in the past. They were often emotional and evocative, whether they were told by children or parents, or about home or school. They were often tied to statements about the "kind of person" someone was, and they often related, both in the content of the stories and in the way they were told, to the ways people expressed and experienced their "togetherness" as a family.

These aspects of the MIAM stories pertain to their status as stories, albeit stories that were told in an interview situation. All stories are told in order to accomplish something – to present a view of the world, to entertain, to convince, to paint a picture of one's self or one's acquaintances (Schegloff, 2003). Some have argued that people construct their identities through narrative and that in telling stories they create and modify the identities available to themselves and others (Drake, Spillane, & Hufferd-Ackles, 2001, drawing from McAdams, 1993). We do not assume that the stories in our interviews reflect enduring self-portraits of this kind (although some of the stories may have done so); instead, we focus on the way the narratives allowed our participants to present themselves as certain "kinds of people" (Gee, 2000).

Several studies have investigated the stories that people tell about *school* mathematics, and considered the relation of these stories to the narrator's ongoing identity construction (e.g., Drake et al., 2001; Kaasila, 2007). These studies develop further the notion of the "mathematical identity" (or "mathematics identity"), a concept which has been variously defined. In some work, one's mathematical identity consists primarily of a set of beliefs about oneself and about (school) mathematics (Martin, 2000). In another example, a mathematical identity consists of a participative "mode of belonging" related to one's participation in a mathematical community of practice (typically, the mathematics classroom: Solomon, 2007, drawing on Wenger, 1998). Note that this second example considers what people *do*, in contrast to the first, which considers what people *think*. A third way defines identity to be the set of stories that we tell about ourselves (Drake et al. 2001; Sfard & Prusak, 2005) – forming a middle ground between thinking and doing, as stories are not stories until they are told to another.

Debates around appropriate definitions for identity continue in the scholarly community, and we make no attempt here to resolve them. In our analysis of MIAM stories, we do not need to rely on "what people believe" about mathematics or about themselves, nor do we argue that the stories in our data set reflect these inner beliefs. We do not have access to people's participatory identities because we did not observe people in action in a variety of settings. So, for our purposes, we consider storied identities, and how the stories that we were told provided a venue for our participants to construct such identities for themselves and for their families.

Stories can be used as resources for the identification and labeling of family members' personal characteristics (Gee, 2000; Holland, Lachiotte, Skinner, & Cain, 2001). In our data set, as we will see, family members self-identified ("I'm a responsible person," "I'm a numbers person"), labeled each other ("She's stingy," "She's easily frustrated") and co-identified in a myriad of ways ("We are that kind of people"). While it may not be surprising that narratives were used in this way, we found it intriguing that these stories of *mathematics* could be used as identity resources in such a wide variety of ways.

Another aspect of identity we wish to highlight here for our analysis is that because the narratives were co-constructed by multiple participants, including the family members and interviewers, so were the identities. That is, stories were used to create subject positions for one's self, for other family members present, and for story characters not present in the room (e.g., teachers, friends). These subject positions could be accepted and upheld, or challenged, modified, and altered as the story progressed.

With this chapter, we add to the growing body of literature considering mathematical stories and mathematics identities, through the consideration of these stories and identities *in out-of-school settings*. As described above, a number of studies have undertaken the description and analysis of school mathematics identities. For example, Drake et al. (2001) discussed three common story types (and thus, for them, identity types) for mathematics learners – *turning point*, *failing*, and *roller-coaster*. We consider it reasonable that if we broaden our examination of people's experiences to include experiences with mathematics outside of school, then we might find a broader set of possible stories.

The characteristics and themes that emerged around stories of mathematics at home and work differed from those at school. Mathematics at home was integrated with life activities, problem-solving, and people's values.[1] It was employed to accomplish goals that mattered to people. At home, mathematics was part of problem-solving and social activity, and, with the exception of homework, was rarely depicted like mathematics in school. As we will discuss, stories of school mathematics more often involved external evaluation and outcomes that were right or wrong. Because of these differences, we specifically separate and examine the characteristics of mathematics at home, and mathematics at school, and consider how families depicted their experiences of mathematics differently across these settings.

Home and Mathematics

"What Is Mathematics?" in the Family

One striking feature of people's home-centered mathematics stories was the diversity of mathematics applications they contained. About half of these stories focused on a protagonist competently resolving a problem that presented some difficulty or unexpected complexity. In these stories, mathematics was put to good use across a variety of important and valued activities, from measuring for home improvements, to budgeting, to figuring out best value while shopping, to deciding what college to attend. Some of these stories, such as one-time home improvement projects, involved substantial novelty. A second type of story accounted for about one third of the stories. These stories focused on routine mathematical tasks that family members faced at home. They did budgets over and over and claimed that they always did them in the same way. Like Lave's (1988) shoppers, they figured out the best value in items to buy in the supermarket. They approximated or always used the same proportions of ingredients when they worked with recipes. In other cases, mathematics was put to use playfully, in games and puzzles, where the problems needing solving were invented for fun. Across these examples, mathematics was embedded in solving problems that mattered to people, with the problems themselves driving the activity.

Diverse Kinds of Mathematics

People told us about many different kinds of mathematics. They created and maintained spreadsheets, and they used calculators and online tools. They rounded and estimated, worked with ratios and proportions, thought in two and three dimensions, and worked with patterns, geometry, algebra, multivariable analyses, and logic.

[1] We use "mathematics at home" as shorthand for mathematics that occurs outside of school. By this definition, "mathematics at home" occurs in quite a range of settings, including stores, neighborhood locations, the workplace, etc.

For example, one family described a kitchen cabinet remodel that involved balancing multiple constraints, including commercial constraints (a desired corner cabinet only came in one size) and usage considerations (the best placement of the dishwasher for efficient work flow in the kitchen) that in turn led to two-dimensional (2D) and three-dimensional (3D) geometric constraints (what to do with an awkward 6-in. gap, to make it usable), which then led to balancing financial constraints (the most attractive option for the gap, a spice rack, was also the most expensive, which had to be balanced against other expenses and a desire to avoid wasted space), all among conversions from metric to standard units, within the broader context of trying to create an attractive kitchen space. In this and other examples, people brought figuring and thinking together to solve problems alone and with others. Even so, they consistently privileged their memories of the situations over the mathematics.

Interestingly enough, people had little trouble identifying stories to tell about their mathematical experiences. Their descriptions revealed a great deal of mathematical thinking, processing, and communicating. Almost all of the stories described these processes in positive ways. Mathematics was part of being competent in their lives, and there was usually no single criterion for what counted as success. If they tried some mathematical strategy, and, if it did not result in an adequate solution, they did something else. In these stories, mistakes were not necessarily without cost, but many settings were forgiving enough to allow second and third tries. For example, recognizing that they might measure incorrectly while wallpapering, families could purchase some extra materials in case of mistakes. If they could not buy the clothes they wanted on sale, they could buy one less item to stay within budget.

Mathematics at Home Stories Are Social

Although some stories involved individuals, at least half involved multiple people in mathematical problem-posing and problem-solving. In one family, the father, Andre, was in charge of budgeting. The mother, Nia, wanted to retile the bathroom, and she knew she had to figure the costs before proposing the idea to her husband. Nia built an accurate 3D scale model of her bathroom using cardboard in order to figure out how many tiles they would need, and she got help from the person in the tiling shop to estimate the total cost. She was delighted that the person in the store was able to use her model to accurately estimate the cost. She recounts, "the measurements were all precise, and the number of tiles – you could tell how many would be per square foot. So that was, that was fun." Armed with the cost estimate, Nia and her husband could now discuss the renovation project. The story involved three people over the course of several days and settings, figuring and communicating around the model, the size of the tiles, and prices. Nia seemed absolutely delighted to share this coproduced and successfully solved problem with us.

Getting It Done Instead of Getting the Right Answer

Another feature of mathematics at home is that "right" and "wrong" are relative, not absolute. When Nikhil, a middle schooler, created a comparison chart to show his parents the differences between buying a hybrid or a conventional car, it mattered little if the purchase prices and gas consumption rates he gleaned from a news chapter were exactly right, as the structure was created to support a whole-family conversation about whether or not to purchase a hybrid. They could discuss whether all of the gas mileage information was correct and find more detailed information as needed. Similarly, another family told a story about covering a cylindrical birdcage with chicken wire so that a smaller bird could not escape. The project required measurement and geometrical reasoning for turning 2D wire into a cylinder shape, but fine precision was unnecessary, and measurements and calculations could be approximate. In the end, it was completed to satisfaction with approximation.

Mathematics as Part of Fun

Finally, we saw that mathematics could be part of family fun. Sometimes there was a multigenerational relationship around mathematics. In one family, the mother described how she and her father used to play mathematics games together when she was a child. Her father would make up silly mathematics story problems, which she would try to solve. Now that she had children, her father played mathematics games with his grandchildren over phone and over email. The granddaughter, a middle-school student, credited these games with shaping her relationship with mathematics, saying, "I think that encouraged me to like math."

Several families also mentioned using time in the car to do mathematics together, playing mathematics games or encouraging the children to calculate how long it should take to get to their destination given their current speed. The Echevarria family created a travel journal when they took car trips together, recording how many miles they drove each day. At the end of the trip they added up the total miles and they compared distances across trips. On a trip to Toronto, they used their previous trip to Las Vegas as a unit of measure. They told us they figured they had driven about the distance to Las Vegas every day of their trip. Both of these examples blur the boundaries between our "home" and our "school" stories, in that the mathematics problems were school-like in a dual sense: focused on either computations purely for the sake of computations or school mathematics applied in novel settings. Although we cannot be sure, we suspect that adult family members may have had pedagogical intentions as they embedded mathematics into the fabric of family life.

Other examples of family fun included board games and hobbies. In one family, the father and daughter worked together to program a computerized game spinner so that all three family members could play a game that usually required one person to stay out of the game to spin. A middle schooler in another family, Gaurav, enjoyed making complicated projects out of LEGO blocks and during the interview got into a discussion with his family about whether his hobby involved mathematics.

Another boy frequently checked the statistics of his favorite NBA star. A middle school girl described using mathematics in her sewing hobby at home, and discussed how learning dimensional analysis at school was like a "door opening" to help her convert between units of measurement. In all these examples, mathematical reasoning was part of activities people enjoyed doing.

In sum, mathematics in the home was used to help people with some problems that were routine and some that were unusual, some that were simple and some that were complex. Most of the stories portrayed successful problem-solving. Mathematics was integrated in social activity among family members and others across a variety of contexts. Mathematics was forgiving in the home context. It did not always result in absolutely correct answers, and people did not often speak of being evaluated by others or evaluating their mathematics performances. When evaluation was prominent, it was usually the task itself that was evaluated, not the specific mathematical techniques. People depicted mathematics as integrated into their activities, so that they were not always sure it was mathematics.

How Mathematics in the Family Relates to the Question of "Who Am I?"

When people told us quick stories of mathematics, they not only identified what they thought mathematics was, they also used the stories to tell us about who they were. As we discussed earlier, mathematics was not a neutral subject for people. Their MIAM stories were accompanied by emotion, statements about their values and ethics, and statements about the "kind of person" they were. As such, these stories sometimes invited participation by other family members, through prompts, elaborations, or corrections. The stories labeled traits, and people told us how they or others were "cheap," "stingy," "frustrated" by mathematics, or "a brain." Stories were occasions for being patted on the back by others, but they were also sources of bad memories and experienced conflicts. Our participants told us about mathematics in their lives and how it revealed characteristics of the family and its members.

Several themes relating to individual and family identity arose in the stories about home, including mathematics as part of developing character or personal responsibility, or fulfilling social goals and responsibilities. Mathematics was part of what families do together and integral to their shared experiences.

Being Personally Responsible

Several of our MIAM stories described mathematics in the context of helping to develop personal responsibility, particularly as related to budgeting and finance. For example, in the Echevarria family, the father described how he and his family used mathematics to determine whether they were making financially sound spending decisions, considering how much the family had to spend, as well as whether the item was a good value. "Is this too much? Is this appropriate? Do we need it? Basically, it's math." For the father, these decisions were an attempt to balance the family's needs with the desire to be financially responsible.

Related to this example, we also heard several stories involving parents giving their children money to spend. These practices seemed to be intended to teach children about staying within a budget and making responsible decisions. In one family, Hannah, who was in middle school, wanted to buy a dress for a father–daughter dance. The dress she liked was on sale, and she described how she calculated the percentage reduction to see whether it would fit within her budget.

Nia's elementary school daughter Giselle told us, "When I want to buy something I always have to think about how much money I still have to spend." Giselle's middle-school sister, Brianna jokingly described Giselle as "stingy," and said that Giselle often got "discombobulated" when she made a decision about spending her own money, especially when Giselle considered how much less she would have after the purchase. Her father Andre added, "[T]hey learn quickly, that, you know, if it's their money, then they get really stingy and very conservative." When Nia chimed in, "But when it's our [money]…," the family laughed. In these stories, children were learning to make value trade-offs when spending their own money. This theme was echoed in several of the stories. A father named Harold formalized the process. "It's interesting because they do, in theory they do a value–cost analysis in their head. They'll see something, and they'll say 'I want to buy this. How much does it cost?' So we find out. And [both children] say, 'you know, for four dollars, I don't want it. It's not that important to me.'"

Being Socially Responsible

Several of our MIAM stories blended talk of mathematics with talk of social or community responsibility. One mother, Swati, described how her financial discipline, combined with her shopping and budgeting prowess, allowed her to help her community. Swati had a weekly grocery budget, but typically ran under budget, and donated the excess money to charity or the church; something that she said made her proud. Her husband, Rupeni, described how Swati always knew which stores would have the best deals on which items, and she frequently bought in bulk. Swati also described her practice of buying necessities well in advance, so she would never need to rush out and pay full price. "I'm a good housewife. I do well at home. I know where to save money and how to save money." Swati credited her shopping expertise and discipline in not indulging in unnecessary expenses, such as pedicures or eating out, with allowing her to donate money to the community, which made her feel good.

Swati's husband Rupeni also talked about how mathematics helped him fulfill his social responsibility to his extended family. As a young man in Fiji, Rupeni's family owned a grocery store. Rupeni described how if they started with \$100 worth of goods one week, they would rollover their profits and buy \$200 worth of goods to sell the next week, and so on. In that way they built their store. But, when extended family members came to the house (which was a frequent occurrence), they needed to use supplies from the store to feed them, and they would not make any profit. They recuperated their costs by charging interest to customers who needed to buy goods on credit. In that way they were able to feed their extended family, and still balance the costs of the store.

In another family, Tania described how she hoped to use her mathematics experience to fill an important need for English language learner (ELL) students. Tania had recently transitioned from a job as a construction inspector to a teaching job. Tania hoped to teach Algebra 1, something she saw as the greatest need for ELL students. The principal assured her that she was a perfect candidate for the mathematics position, but at the last minute Tania was assigned to teach Spanish instead of mathematics, which surprised and disappointed her, as she felt teaching Algebra was an important way she could make a difference in her community.

We also heard stories from children describing how they used mathematics to contribute to their communities. For example, one middle school girl talked about how mathematics came into play when trying to create a quilt for needy children. One challenge involved subtracting a half-inch on every side to leave room for the seams. Across our stories, we were surprised at the number of examples of people using mathematics in service of the community.

Being a Family Together

Another theme that stood out to us in the MIAM stories was how much mathematics was a part of the family's shared experience. Families frequently described doing mathematics together (with over half of the stories involving multiple people jointly solving a problem), and in the telling of the stories, family members chimed in and embellished each other's accounts. It was evident that some of these stories had been told before and were enjoyed by all those present. In such cases, family members sometimes interjected what they saw to be general characteristics of the person telling the story and their relationship with mathematics. For example, in one story, the grandmother, Barbara, described getting a good deal on her cable bill. Her grandson jumped in and said that deal finding was something she was very good at, indicating that her deal finding skills were known throughout the family.

Sometimes, attitudes toward mathematics and problem-solving could be seen intergenerationally and at the family level. For example, a mother, Mahita, described how her family did not like to make major decisions based only on emotion. So they often tried to "translate a lot of things into numbers" to come to a more "objective decision." For example, when her children were young, she and her husband had to choose among three places where they wanted to move. They decided what criteria they cared about for quality of life (e.g., education, culture, weather, etc.) and gave each place a score for each criterion. In making their decision, they compared the scores across the three potential locations. In response to this story, the older daughter, Tara, described a similar numerical scoring process she went through when choosing which colleges to apply to. In the telling of the story, the rest of the family members chimed in about how the scoring system worked. This method that we now call "multi-attribute utility theory" was first described by Benjamin Franklin as an algebra to support decision-making.

Home and Mathematics Identity

Family members' participation in mathematical activities encouraged the development of identities that went beyond being good or bad at mathematics. For example, Swati described herself by saying "I'm a good housewife," in part because of her ability to stay under budget and give to charity. Similarly, because of her difficulty making purchasing decisions with her own money, Giselle's family jokingly called her "stingy" several times throughout the interview, such as when she was describing her Monopoly strategies. In other families, Gaurav, who enjoyed building with LEGO blocks, was known as being good at "figuring things out," and Barbara, who was able to negotiate a cheaper cable rate, was known as a bargain finder.

In addition to the development of different roles within the family, we saw attitudes toward mathematics being carried down from generation to generation. In one family, playing mathematics games with grandpa over email encouraged a positive attitude toward mathematics. In another, the daughter took a similar "objective" approach to making decisions by translating evaluative criteria for alternative decision choices into numbers, as her parents did years before. In several stories, parents encouraged the development of responsible attitudes toward financial decisions in their children by giving them control over spending their own money. Across the MIAM stories, participating in mathematical activities helped family members develop a sense of "Who am I?" that went beyond being someone who was good or bad at mathematics, to encompass issues of personal and social responsibility, as well as roles and characteristics, and attitudes toward mathematics.

School and Mathematics

"What Is Mathematics?" at School

Not surprisingly, many MIAM stories involved school (approximately one third of the stories). School stories told by adults especially were often specifically about experiences in mathematics classes. (All eight school stories told by adults were about experiences in mathematics class, versus 10 out of 14 by children.) As such, the question of "what is mathematics" was tied up, at least implicitly, in school-based definitions of mathematics and mathematical activity. It is easy enough to imagine what this "mathematics class factor" might mean for people's understandings of what mathematics is. Any one of a number of vices (authoritarian, formulaic, anxiety producing) or virtues (rigorous, elegant, powerful) of school mathematics might exert their influence on people's conceptions. We examined the stories for specific evidence of how school stories were distinct from home stories.

Generalizing About Experiences with School Mathematics

Eight of the school stories involved general talk about "getting it," or "not getting it," or about a great mathematics class or a terrible teacher, without addressing specific mathematical problems or topics. Four of these stories were negative, and four were positive. All eight cases either explicitly or implicitly involved a teacher or other authority figure evaluating the storyteller. For example, the grandmother in one of our families, Loretta, told a story about how she hated mathematics as a child. She told us of a time when, in high school, she received an "F" in mathematics, and when the report card arrived at home, she secretly changed the "F" to an "A." She said the bad grade was traumatic, as she was a good student and got good grades in her other classes, and in the interview told us that after receiving that grade she never took a mathematics class again.

Loretta's daughter Alisha followed with a story about struggling to help her son Marcus do his mathematics homework, in spite of her own dislike of mathematics. This story cycle continued when Loretta mentioned that Marcus was really good at mathematics. At the same time, Alisha suggested to Marcus that he used to like mathematics, but did not like it any more. He protested that he did like mathematics, and that he was good at it. To emphasize his point, he produced his school progress report and showed it to his grandmother. In each of these stories, mathematics was described in the most general of terms, with no differentiation among its varieties, and the emphasis was on whether one liked or disliked mathematics, as well as whether one was "good" or "bad" at mathematics.

A more positive, but still quite general, story was told by Brandon, a young participant who had attended a "Physics Day" at a local amusement park. He was proud that he and his classmates were the only sixth graders there, whereas the other participants were high school students. The implication was that Brandon and his classmates were doing more advanced mathematics than others their age. A similar experience concerned a more advanced mathematics course, as told by Harold. He recounted his calculus experience by retelling the fun he had as he suddenly realized all the things that he could do with calculus – calculating volumes of cubes, how much water goes into a shape, and so on. Harold likened his experience to "a light going off" when he began to realize many things about mathematics that he had not known before. He then explained that when he "coached" his own children in geometry, he wanted to make that light go off for them. His wife Harriet, in contrast, told us about when she took a semester-long accounting class in graduate school and really struggled with the mathematics, in part because she could not understand the instructor. Harriet described the content as "hard," even more so because of the instructor, but did not provide any more detail about the mathematical challenges she faced.

Evaluation by a teacher or other figure was less prominent in these three stories, although in all three of them, teaching and teachers played a central role. In the amusement park story, Brandon talked about asking questions of teachers to help him solve the calculation problems he had to do. The calculus story was introduced by Harold discussing how he liked to coach his children so that they would experience the fun of mathematical "lights going off." Harriet's accounting story was all about the perceived shortcomings of the instructor. In these general stories about school experience, teaching and evaluation were central.

Mathematics as Specific Problems, Teachers, and Grades

In addition to the more general stories just described, there was also a set of eight school mathematics stories in which people explained a specific problem that they were trying to solve, their methods of solution, and how their solution was evaluated by others, especially the teacher. One young person, Felix, described a problem that involved dividing a cake five ways so that there were equal amounts of frosting and cake. Felix devised a creative solution, saying that he would initially separate the frosting and the cake, divide each of those into five equal pieces, and then pair the cake and frosting back together. The teacher disagreed with Felix's method, saying that it violated common sense. Felix felt that the teacher's assessment of his solution was wrong, and that he had actually followed the constraints of the problem.

Victoria recounted a specific event with a strong negative effect on her perceptions of mathematics. She remembers being given a single-digit multiplication question by her teacher when she was quite young. Victoria said that her teacher had been "a fierce old lady," who "whacked" her on the knuckles when she gave an incorrect answer. Victoria continued her story by explaining that this experience stayed with her for a long time and affected how she felt about mathematics in general. Her daughter Madison, in turn, told a story about doing sets of division problems in school, and being proud of being the fastest in the class in completing problem sets. Madison's brother Jay, who was a little bit older, interjected to boast that he was even faster than his sister.

In a different family, one of the children described how his desire to "get to the next level" on timed multiplication tests was thwarted by his teacher, who said he was not ready, even though he had met the criterion they had previously agreed upon. Jay had also complained that while his class learned fractions by doing boring worksheets, another class learned fractions using graham crackers and icing. He thought it unfair that some children got to learn mathematics by using food while he languished in a world of worksheets. In these stories, mathematics was a source of either positive or negative feelings (more on this below), but in each case the mathematics was intimately tied up with issues of authority and evaluation.

Mathematics for Mathematics' Sake

One salient factor common to the majority of these stories is that mathematics learning was the primary focus. Mathematical activity was an end in itself. This finding is not surprising, since the stories took place in school mathematics classes. It provides a stark contrast with the home stories, where mathematics was a means to an end rather than the focus of the activity itself. While the home stories generally focused on mathematics in service of a particular goal, the school stories were about the mathematics experience itself. In particular, the explicit focus on mathematics as an end in itself provides a partial answer to the question, "what is mathematics?" – namely that it is something to be pursued in its own right.

This pattern was not without exception, as we collected several stories in which mathematics appeared in other school subjects as a means to nonmathematical ends. For example, Darren, a middle-school-aged participant, described a project he had done for one of his classes, in which he had decided to make a poster that was shaped like a pyramid. His story recounted the challenge of constructing this difficult shape. Although mathematics was at the foreground in the telling of the story, the actual purpose of the activity was just to make a creative poster; mathematics in this story was used as one tool to achieve these ends. We also heard stories from two families about time management and homework, where the young protagonists discussed having to budget their time so that they could get all of their homework done. In their stories, mathematics was used as a means to an end, but was not as an end in itself.

In summary, school stories painted a picture of mathematics as something requiring an ability of some sort (for it is possible to be bad at it), as something to like or dislike, as something that institutions and their agents (especially teachers) have special authority over, and as a potential source of pride or trauma. Mathematics was primarily portrayed as something to be studied as a free-standing entity. Stories of mathematics as a tool for accomplishing quantitative goals were rare, especially in comparison to their prevalence in the sample of home stories.

How Mathematics at School Relates to the Question of "Who Am I?"

One of the notable features of the school stories described above is that so many had a significant emotional component, ranging from like to dislike, from pride to shame. In addition, these stories were an occasion for interview participants to describe themselves in terms of their mathematics competence, reporting that they were "good," "bad," or even "terrible" at mathematics. While in some cases these self-identifications seemed relatively static and long term, our data set suggests that a view of mathematical identity as individual and enduring is too simplistic.

As discussed in the introduction, we use the term identity to refer not only to people's *beliefs* about themselves in reference to mathematics, but to the ways in which they are socially and situationally positioned with respect to mathematics.

School and Mathematics Identity

A number of our stories exemplify the shifting and socially constructed nature of one's mathematical identity. For example, in the Medrano family, it was taken for granted that the middle-school-aged boy, Ismael, was good at mathematics. Ismael told us this directly, and his mother and sister both made reference to Ismael's mathematical competence. Ismael sister's MIAM story showed a much more complex mixture of confidence and uncertainty. Leticia began by reporting that her teacher told her to be careful with her "steps" when she solved problems, because Leticia was not as good at mathematics as her brother Ismael. The rest of her story was about helping her classmate with mathematics, and Leticia appeared to be proud of her ability to help another person solve a mathematics problem on the geometry of parallelograms. When asked to show us how she solved it, she began to write out the problem, but when the camera moved closer to capture what she was doing, Leticia covered her work with her hand, perhaps out of shyness. After a brief reassurance, Leticia continued, and she and Ismael spent some time discussing the purpose of the problem, finally deciding that the goal was to determine the area and the perimeter. She closed by saying it was not hard to teach her classmate about how to solve the problem.

We do not wish to overinterpret such examples, as a few minutes of storytelling can only shed so much light on important theoretical issues concerning identity. Nonetheless, it is remarkable that so much – from unequivocal assertions of competence to softer mentions of mathematical accomplishment, from the pride of helping to the embarrassment of doing mathematics on camera – can be seen in such a brief snippet of storytelling. Mathematical identities need not be as straightforward as they may appear from stories of being "good" or "bad."

These stories also amply demonstrated the social and situational nature of mathematical identities. Earlier in this chapter, we discussed how Marcus was variously positioned by himself, his mother, and his grandmother in terms of his feelings about mathematics, demonstrating that even within a family, mathematical social identities could be controversial. Alisha's story contrasted her general dislike of mathematics, as a student, with her strong desire, as a mother, to learn and understand mathematics well enough to help her son succeed in school. The role mathematics played in her identity was not unitary or monolithic, but closely tied to the social roles that she (and that mathematics) took in everyday life.

These stories reveal how mathematical identities are social and situational. They are also historical. In another family, one man's MIAM story about a difficult budgeting job at work quickly transitioned into a reflection of his own mathematical history in school. He recalled a specific word problem from his high school mathematics class as an example of the curriculum, one which he felt did not adequately prepare him for college mathematics. His story, as brief as it was, was populated with people and institutions whose definitions and expectations of "what mathematics is" persisted over time as important elements of his mathematical identity. His memories are reminiscent of the literary theorist Mikhail Bakhtin's (1981) concept of "heteroglossia," in which the self is conceived to be literally peopled with memories of past interactions with significant others that continue to live on in present-day thinking, feeling, and interactions.

To summarize, the nature of the mathematics-related identities revealed through school stories was intimately related to the nature of the mathematics they described. Since mathematics in the school stories was strongly tied to teaching and learning situations in which evaluation was central, our participants reported mathematics-related identities also related to being a teacher, a learner, and a person evaluated by others. Across the stories, other people's opinions about one's mathematics competence – especially a teacher's opinion – had an effect on identity and on subsequent life choices about whether or not to pursue further mathematical study or careers.

Design Implications of the MIAM Stories

The mathematical stories we elicited from families varied in mathematical content, in terms of whether mathematics was a means or an end, and in implications for one's identity. One striking distinction between the stories of mathematics at home and the stories of mathematics at school was that the home stories were overwhelmingly stories of using mathematics *competently* to achieve desired ends. Whereas school stories were mixed, all the home stories were quite positive, even for family members who recounted difficulty with mathematics in their school experience.

People were able to draw on their school mathematics learning experiences to solve problems encountered in daily life, and their accounts of solving the problems of home, hobbies, and work were in fact more successful than their stories of solving the mathematics problems of school. Given this finding, we would argue that stories of mathematics at home provide a candidate model of mathematical success. From this model, we may draw some tentative design implications for more effective mathematics instruction in schools.

The stories of home mathematics described situations in which there were a wide range of allowable solution methods and resources, more so than in the stories of school mathematics in which featured teachers prescribed which solution paths were allowed. Problem-solving practices in the home were social, involving multiple people and tools as resources. Family members often had multiple opportunities to try to work out a solution, and if one problem-solving approach did not work out, they could try again in a different way. Consider the problem that one schoolchild reported, of having to divide a cake into five perfectly equal pieces – by the amount of cake and amount of icing. Not only is this problem unlikely to be encountered in the home, but if it were, multiple solution strategies would probably be allowed. If one used the child's strategy – to divide up the icing and the cake separately, and then recombine – one's answer would be perfectly acceptable. In a school context, the teacher placed restrictions on solution methods, so that the child's strategy was deemed incorrect.

With an emphasis on "getting it done," and not necessarily having a completely accurate answer, family members estimated or "eyeballed," but still applied mathematical reasoning to judge the validity of their results. In cases where careful measurements were required, people were up to the challenge and used a number of carefully constructed representations (a 3D model, a scale drawing) to ensure accuracy. Yet even in these cases, our family members described how they allowed for error in their calculations, by purchasing extra materials, by budgeting more than they needed, and so on. There was an adaptive flexibility to mathematics in the home.

Considering the nature of the problem-solving process in the home, we might consider designing mathematics classes differently so students would have access to more resources, and to more creative ways to solve problems. By allowing students multiple attempts to solve problems, we might alleviate the anxiety of being evaluated by others, and encourage more risk-taking and experimentation in their methods (Hatano & Inagaki, 1992). By allowing students to mathematize problems in multiple ways, we might find that different students in the classroom develop different kinds of mathematical skills, perhaps fostering some interesting discussions when comparing these multiple strategies (Lampert, 2001). These features are integral in reform classrooms that pursue mathematics learning by fostering mathematical inquiry and discourse (Yackel & Cobb, 1996), with collaborative activities (Boaler, 1998), or in model-building projects (Lesh & Doerr, 2003).

So far, our discussion of design implications has focused primarily on the resources and solution methods available for mathematical problems in school. A second provocative area of difference between home and school mathematics stories resides in the nature of the problems themselves, and how solving those problems reflects people's developing identities. In stories of mathematics in the home, we found examples of mathematics being used to support one's sense of personal and social responsibility. Family members used their stories to illustrate their sense of fiscal responsibility, caring for others, and desire for precise and thoughtful answers in the context of family values. In the school stories, people's identities were generally summed up as either being "good" and "fast" at mathematics, or being "bad" at it. Whereas in the school stories, a wrong answer might lead to a sharp slap on the wrist and public exposure as a dunce, in the home stories, a mistake often led to a reevaluation of the problem and a second attempt. These are forms of accountability of very different types.

If school mathematics problems were more like the home problems, then mathematics would be introduced as one tool (among many) to demonstrate one's care and responsibility for the world, as in curricula focused on investigating social justice issues through mathematics (Enyedy & Mukhopadhyay, 2007; Gutstein, 2006), or project-based learning environments (Greeno & Middle School Mathematics through Applications Project [MMAP], 1998; Stevens, 2000). Students might be less likely to leave behind a difficult problem, saying "It's not for me," but instead might work with renewed efforts to solve it.

Another area in which we might learn from mathematics problems encountered in the home is the integration of mathematics with fun hobbies and activities.

In low-risk settings like a family car ride, parents and children engage in playful problem-posing and problem-solving activities together. These stories differed remarkably from a prototypical "fun" school mathematics activity – a competition – that serves mainly to differentiate winners (smart students) from losers (dumb students). In family mathematics games, mathematics often served a valued end, as when one family created a computerized spinner so that everyone could play Twister together. When mathematics was the end goal of the game – like comparing distances on a long family road trip, or sending funny mathematics problems to one another via email – problem-solving was supported by multiple resources, multiple people, and everyone had a chance to be successful.

Far from being mathematically barren spaces, we have found that home environments abound in mathematical activities that almost all family members participate in. The MIAM stories suggest some key differences in the nature of mathematics in the home and in school, differences that rebound to influence people's socially constructed identities. School mathematics stories were often structured around mathematics as an end in itself, involving external evaluation. By contrast, stories of mathematical activities in the home showed how problem-solving was a social activity, involving multiple people coordinating activities over multiple contexts and with many chances for revision and success. These stories demonstrate how people structure their environments to maximize competent and successful problem-solving, and highlight the function of school mathematics in constructing success and failure that may not appear in other facets of daily life (Varenne & McDermott, 1998). As Lave (1988) and Saxe (1990) helped show us in the 1980s in their pioneering work on everyday mathematics, "understanding how successful mathematical activities work will ultimately contribute more to advancing effective learning practices than repeated diagnoses of failures" (Pea, 1990, p. 31).

Our data clearly show the importance of school mathematics as a source for mathematical competence and mathematical identity. In considering everyday life as a model of successful engagement with mathematics, we hope to reveal aspects of everyday problem-solving, which, despite their promise, are often overlooked. Schools might become better places for thinking and learning about mathematics if they shared some of the meanings, the values, the social nature, and the adaptive flexibility of mathematics in family life.

Acknowledgments The Family Math Project is part of the research portfolio on learning in informal and formal environments from the LIFE Center (http://life-slc.org/), supported by National Science Foundation Grant No. 0354453. Any opinions, findings, and conclusions expressed in the chapter are those of the authors and do not necessarily reflect the views of the NSF.

References

Bakhtin, M. M. (1981). Discourse in the novel (C. Emerson & M. Holquist, Trans.). In M. Holquist (Ed.), *The dialogic imagination: Four essays by M. M. Bakhtin* (pp. 259–422). Austin, TX: University of Texas Press.

Boaler, J. (1998). Open and closed mathematics: Student experiences and understandings. *Journal for Research in Mathematics Education, 29*(1), 41–62.

Brown, J. S., & Duguid, P. (1996). Stolen knowledge. In H. McLellan (Ed.), *Situated learning perspectives*. Englewood Cliffs, NJ: Educational Technology Publications. (Originally published 1992, Educational Technology)

Drake, C., Spillane, J. P., & Hufferd-Ackles, K. (2001). Storied identities: Teacher learning and subject-matter context. *Journal of Curriculum Studies, 33*(1), 1–23.

Enyedy, N., & Mukhopadhyay, S. (2007). They don't show nothing I didn't know: Emergent tensions between culturally relevant pedagogy and mathematics pedagogy. *The Journal of the Learning Sciences, 16*(2), 139–174.

Gee, J. P. (2000). Identity as an analytic lens for research in education. *Review of Research in Education, 25*, 99–125.

Goldman, S. (2005). A new angle on families: Connecting the mathematics in daily life with school mathematics. In Z. Bekerman, N. Burbules, & D. Silberman-Keller (Eds.), *Learning in places: The informal education reader*. Bern, Switzerland: Peter Lang Publishing Group.

Greeno, J. G., & Middle School Mathematics through Applications Project. (1998). The situativity of knowing, learning, and research. *The American Psychologist, 53*(1), 5–26.

Gutstein, E. (2006). *Reading and writing the world with mathematics: Toward a pedagogy for social justice*. New York: Routledge.

Hatano, G., & Inagaki, K. (1992). Desituating cognition through the construction of conceptual knowledge. In P. Light & G. Butterworth (Eds.), *Context and cognition: Ways of learning and knowing* (pp. 115–133). Hillsdale, NJ: Lawrence Erlbaum.

Holland, D., Lachiotte, W., Jr., Skinner, D., & Cain, C. (2001). *Identity and agency in cultural worlds*. Cambridge, MA: Harvard University Press.

Kaasila, R. (2007). Mathematical biography and key rhetoric. *Educational Studies in Mathematics, 66*, 373–384.

Lampert, M. (2001). *Teaching problems and the problems of teaching*. New Haven, CT: Yale University Press.

Lave, J. (1988). *Cognition in practice: Mind, mathematics, and culture in everyday life*. New York: Cambridge University Press.

Lave, J., & Wenger, E. (1991). *Situated learning: Legitimate peripheral participation*. Cambridge, UK: Cambridge University Press.

Lesh, R., & Doerr, H. M (Eds.), (2003). Beyond constructivism: Models and modeling perspectives on mathematical problem solving learning and teaching. Mahwah, NJ: Lawrence Erlbaum Associates.

Linde, C. (1988). The quantitative study of communicative success: Politeness and accidents in aviation discourse. *Language in Society, 17*(3), 375–399.

Martin, D. B. (2000). *Mathematics success and failure among African-American youth: The roles of sociohistorical context, community forces, school influence, and individual agency*. Mahwah, NJ: Lawrence Erlbaum.

McAdams, D. P. (1993). *The stories we live by: Personal myths and the making of self*. New York: Morrow.

Pea, R. D. (1990). Inspecting everyday mathematics: Reexamining culture-cognition relations. *Educational Researcher, 19*(4), 28–31.

Saxe, G. B. (1990). *Culture and cognitive development: Studies in mathematical understanding*. Hillsdale, NJ: Erlbaum Press.

Schegloff, E. A. (2003). "Narrative analysis" thirty years later. In C. B. Paulston & G. R. Tucker (Eds.), *Sociolinguistics: The essential readings* (pp. 105–113). Malden, MA: Blackwell Publishing.

Sfard, A., & Prusak, A. (2005). Telling identities: In search of an analytic tool for investigating learning as a culturally shaped activity. *Educational Researcher, 34*(4), 14–22.

Solomon, Y. (2007). Experience mathematics classes: Ability grouping, gender and the selective development of participative identities. *International Journal of Educational Research, 46*, 8–19.

Stevens, R. (2000). Who counts what as math? Emergent and assigned mathematics problems in a project-based classroom. In J. Boaler (Ed.), *Multiple perspectives on mathematics teaching and learning* (pp. 105–144). Westport, CT: Alex.

Varenne, H., & McDermott, R. (1998). *Successful failure: The school America builds*. Boulder, CO: Westview Press.

Wenger, E. (1998). *Communities of practice: Learning, meaning, and identity*. Cambridge, UK: Cambridge University Press.

Yackel, E., & Cobb, P. (1996). Sociomathematical norms, argumentation, and autonomy in mathematics. *Journal for Research in Mathematics Education, 27*(4), 458–477.

Chapter 3
What Counts as Science in Everyday and Family Interactions?

Maureen Callanan, Megan Luce, Lara Triona, Jennifer Rigney, Deborah Siegel, and Jennifer Jipson

Introduction

What counts as science? As evidenced by the focus of this entire volume, this is a complex and controversial question and the answer necessarily depends on the contexts and the stakeholders involved (Aikenhead, 2001; Stevens, 2000). Other chapters in this volume take the valuable step of examining in some depth how various stakeholders seem to count certain talk and action as science. Our goal in this chapter is somewhat distinct from the other chapters in this section, because our starting place is different. We ask how children first encounter and begin to think about science without even knowing about science as a domain (see also National Research Council [NRC], 2007, 2009). And we ask how, while engaging in a variety of everyday activities, parents may contribute to children's early thinking about science, sometimes without even meaning to do so (see also Ash, 2003; Crowley & Galco, 2001; Ellenbogen, 2002).

In our data, we investigate everyday talk and action within families where the participants are very rarely labeling what they are doing as science. An immediate problem we encounter is deciding whose definition of science we should use in our exploration of this talk and action. The solution we choose in this chapter is to substitute the general question of whether families are doing "science" with more specific

M. Callanan (✉) • M. Luce • J. Rigney • D. Siegel
Developmental Psychology, University of California, Santa Cruz, Santa Cruz, CA, USA
e-mail: callanan@ucsc.edu

L. Triona
Developmental Psychology, California State University, Fresno, Fresno, CA, USA

J. Jipson
Department of Psychology and Child Development, California Polytechnic State University, San Luis Obispo, CA, USA

B. Bevan et al. (eds.), *LOST Opportunities: Learning in Out-of-School Time*, Explorations of Educational Purpose 23, DOI 10.1007/978-94-007-4304-5_3,

questions about activities that may be considered, by at least some definitions, to be part of science. We investigate the extent to which we see these particular science-related activities occurring in family interactions in four different research settings, and describe what these activities look like for different families. In particular, rather than asking, "Are they engaging in science?" we ask more specific questions about four science-relevant activities: (1) Are they using and discussing scientific terms? (2) Are they generalizing beyond the specific event to broader classes of events either through analogy or generalizing language? (3) Are they discussing causal explanations for events? (4) Are they seeking evidence to test hypotheses or experimenting? In each case where we find examples of these sorts of science-relevant talk and/or actions, we further describe how these conversations and activities unfold. Later we integrate across settings, and consider what children may be learning from these everyday activities, as well as how these findings may inform the development of learning environments for children in museums or classrooms.

We begin the chapter by presenting examples from four different studies, each exploring family conversations in a different setting. The settings are: an aquarium tank at a marine science center, a car track exhibit in a children's museum, reading a picture book about a snowman, and a booth at a community health fair. In each of these settings we found at least one of our four types of science-relevant activities (discussing science terms, generalizing beyond the present, explaining, experimenting). We first discuss the examples from each setting, and then compare across settings in the second part of the chapter.

Examples of Science-Relevant Activities in Everyday Life

A growing body of research indicates that family conversation and activity are essential contexts for early cognitive development (Ash, 2003; Callanan & Valle, 2008; Ochs, Taylor, Rudolph, & Smith, 1992). Inspired by Vygotsky (1978) and contemporary sociocultural approaches (Rogoff, 2003; Wertsch, 1979), this research suggests that a complete picture of children's developing understanding of the natural world requires not just an account of how children think, but also of the social context of their early learning and thinking. In particular, there is evidence that early engagement with science topics occurs in some families in everyday conversation, for example, in the form of causal explanations and definitions for scientific terms (Callanan & Oakes, 1992; Jipson & Callanan, 2003). Following sociocultural theories, we argue that children's participation in these early experiences is a cultural practice that develops over time within families (Gutiérrez & Rogoff, 2003; Ochs et al., 1992). Our research focuses on how children and parents engage in scientific activity and discourse as they navigate through the course of everyday activities. Specifically, we investigate family interactions that arise spontaneously in the designed, somewhat structured, yet nonschool, science-learning environments of a marine science center and a hands-on children's museum. We next explore whether and how families infuse science-related activity into their engagement in two settings that are not overtly marked as related to science: a book-reading session and a visit to a booth at a community health fair. We find that the conversations and activities take different forms depending on the setting and on the habits of reasoning and talk that are familiar for different families. We now consider examples from these four settings to illustrate ways that families engage in discussions of scientific terms, generalization, explanation, and experimentation.

Conversations About Sea Creatures at an Aquarium

In one study we explored parent–child conversations at a marine science center (Rigney & Callanan, 2011). These conversations were recorded at a viewing tank, where visitors could see a wide variety of sea animals. Many of these conversations were very brief, about 1–2 min. The following examples contrast talk about a fish between a parent and her 1-year-old versus another parent and his 2-year-old.

Example #1: Using Pretense
[A 1-year-old boy and his mother are both looking into a fish tank.]

Mother: *Oh fishy. Say hi fishy. Is the fishy gonna move? They're saying 'where's my food?' Do you have any food for them?*

Example #2: Correcting Labels
[A 2-year-old girl and her father watch a fish swim along the glass.]

Girl: *Why is he moving his fur?*
Father: *His fur? Those are, those are gills.*
Girl: *Why is he moving his gills?*
Father: *That is how a fish is able to breathe under water.*

We first ask whether the parents and children in the interactions above are engaging in any of the science-related activities we have identified. In the conversation between the father and his 2-year-old daughter in Example #2, we find evidence of three of our target activities: using scientific terminology, generalizing beyond the current situation, and explaining. The father in Example #2 provides his daughter with information that fish have something called "gills" and that these gills enable a fish to "breathe" underwater. A scientific term is introduced, and an explanation for what it does is present. The child quickly appropriates the word "gills" into her next question, suggesting that some learning may have taken place.

Interestingly, instead of responding to his daughter by referring to the single fish they are attending to, the father replies "That is how *a fish* is able to breathe underwater." By using the generic form, this father is telling his child not just about this particular fish, but that fish in general use gills to "breathe." Cognitive developmental research informs us that by age 2 children understand such generic statements as descriptions that apply beyond the individual item being talked about (Cimpian & Markman, 2008; Gelman & Raman, 2003). Generic statements have been shown to be frequent in parent–child talk and are a concise way for parents to convey important information about categories or kinds rather than individuals (Gelman, Coley, Rosengren, Hartman, & Pappas, 1998).

It is important to point out that while Example #2 contains more of our science-related activities than does Example #1, the father is still not talking in a manner consistent with how a biologist would talk about fish. The father does not point out to the child that fish do not have fur (and provide information that only mammals have fur), and he uses the verb "breathe" without explaining how fish "breathing" is different from how humans breathe. Indeed, Gleason and Schauble (2000) might argue that the father is missing opportunities to provide such science information. However, the father's goal in this short interaction may be to emphasize that gills are an important feature of fish. Perhaps providing more information would detract from this immediate goal. Indeed, research on how scientists talk about the phenomena they study shows how highly contextualized it is (Dunbar, 1995; Ochs, Gonzales, & Jacoby, 1996).

Although arguably using fewer of the science-relevant activities we are seeking, the mother in Example #1 could be interpreted as providing more subtle biological information in a way that may be developmentally appropriate for a 1-year-old. We would argue that this mother, by using an analogy to humans, is actually highlighting some aspects of the fish that are important for understanding it as an animal. She guides the child's attention to self-generated movement ("is the fishy gonna move?") and provides her child with information entailing that fish are the kind of thing that engages in eating (as opposed to plants, which are living things that do not eat, for example). Previous research has shown that preschool-age children often do not understand that all animals grow, eat, and have babies (Carey, 1985). Self-generated movement and needing to eat are two important properties of animals, and this mother can be seen as providing information that these animal properties are also true of fish. Additionally, by introducing movement information in a question form ("is the fishy gonna move?"), this mother may be modeling how to predict subsequent events and guiding her child to think in terms of generating hypotheses. Finally, this parent embeds information about fish needing to eat in a statement attributing human-like characteristics to the fish ("he's saying 'where's my food?'"). Framing this as a human might say it may make it easier for this young child to understand and may perhaps guide the child to draw an analogy between this fish and himself. Further, comparing fish to humans may allow the child to make other important inferences about fish (Inagaki & Hatano, 2002).

Testing Predictions at a Car Track Museum Exhibit

We next consider family interactions at a children's museum exhibit. Children's museums are places where parents choose to take their children with a dual goal of learning and fun (Falk & Dierking, 1992). The examples in this section are from a study of a museum exhibit that was part of a larger exhibition focusing on various aspects of circles at Jose Children's Discovery Museum of San Jose (Triona & Callanan, 2008). At the car track exhibit, visitors could roll foot-long cars down a 15-ft ramp.

The cars varied in the shape of the front wheel (e.g., circle, square, hexagon, octagon, off-center circle) allowing families to test the effect that different wheel shapes had on how the car rolled. At the end of the ramp the track flattened out with numbered slats, which could be used to measure the distance that a particular car rolled. To encourage reflection about the differences between the cars, next to the exhibit was a table with paper and pencils and a wall chart that asked visitors to "Keep track of how the wheels roll: How bumpy? How far?" This exhibit was designed to encourage the visitors to make comparisons among the different wheel shapes and relate this experience to the surrounding exhibits about circles.

Children engaged with this exhibit in a variety of ways ranging from playful interactions, such as making car sounds while rolling cars on the ground (rather than the ramp), to systematic experimentation that compared the various cars. In some cases children explored the cars with an adult, while in other cases they explored the cars with other children or even on their own. In two examples, we compare how children engage in a science-related activity with adult guidance to children's engagement with only other children.

Example #3: Mother Coordinating an Experiment
[Two girls, 2-year-old and 5-year-old, are rolling cars down the ramp for about 2 min. They roll all five cars a total of seven times down the ramp, but on four of these trials the cars run into other cars or fall off the ramp. The two girls notice the chart and walk to the table to use the paper and pencils. This is when their mother approaches the exhibit and rolls the oval-wheeled car down the ramp.]

Mother: Aqui ira bebe, aqui (Look here baby, here.)

[The girls do not watch the car roll down the ramp; they continue marking on the paper]

M: A ver cual carro corre mas rapido? (Let's see which car runs faster.)

[As the mother releases the octagon-wheeled car, the 5-year-old girl turns around to watch it and walks to the ramp end where all cars are lined up after having been rolled.]

M: Cual, [inaudible]? (Which one ... ?)

[5-year-old points foot at the hexagon-wheeled car.]

Girl-5: Este. (This one.)
M: Mira (pause) este es el, oh este es el. (Look (pause) This is the one. This is the one.)

[As mother is looking at the cars; the 5-year-old turns around and walks to the table. The mother picks up the off-center-circle-wheeled car and walks to the top of the ramp as her 7-year-old boy walks to the exhibit, grabs the circle-wheeled car, and rolls it backward up the ramp.]

M: A ver (boy's name), vamos a hacer un experimento. (Let's see (boy's name), we are going to conduct an experiment)

[The boy moves next to his mother as she sets the off-center-wheeled car at the top of the ramp.]

M: Vamos a ver cual corre mas recio. (We are going to see which one runs faster.)

[Mother gestures for the boy to move the cars from the ramp end and bring them to her – which he does. She tells the girls to come to the end of the ramp so they can watch where the car ends up. The 2-year-old girl stays at the table, but the 5-year-old girl goes to the end of the ramp to wait. Both mother and 5-year-old turn to the 2-year-old, who is still at the table.]

M: Vamos a mirar cual carro llega mas primero. (Lets see which car gets there faster.)

G-5: [inaudible] no me taches mi eso, ok. (… don't touch my that one, ok.)

[Mother releases the off-center-circle-wheeled car down the ramp.]

M: Ese como se, ¿cual? ¿que carro es? (How did that one, which one? What car is it?)

[The boy runs to the end of the ramp to see where the car ended up.]

Boy-7: El verde, llego hasta, quedo aqui 39. (The green one, it went until, it was here 39.)

[The 2-year-old walks to the ramp bringing the paper and pencil and uses the middle of the ramp to do her "writing." The mother notices and calls her name and points her back to the table. Then she releases the square-wheeled car. The boy squeals excitedly and picks it up and brings it back to the top of the ramp.]

B-7: Si, asi, o asi. (Yes, like this, or like this.)

[Boy releases it with a little bit of a push and it goes further down the ramp.]

M: Number, what number? [in English]

B-7: Quedo, catorce. (It was on fourteen.)

M: Ok, fourteen … [in English]

[The 2-year-old goes back to the ramp to write as the mother releases the off-center-circle-wheeled car again. The mother then notices the girl will be hit by the car if she does not move quickly.]

M: (2-year-old girl's name)…

[The car bumps into the girl and she cries.]

M: Mira [pointing at car wheel]…Ven. (Look…Come here.)

[The mother picks her up out of the way of the ramp and over to the table while comforting her. The 5-year-old also returns to the table to keep her sister occupied.

After both the girls are occupied at the table the mother goes back to the experiment. The mother rolls a few cars while the boy reads out the numbers and then they switch places and the boy rolls the cars and the mother reads out the number where the car stopped. The mother goes to the girls to mark on the paper the distance the different cars rolled and the boy continues rolling the cars and looking at the distance each of the cars rolled. While the boy is still rolling cars the mother gestures that it is time to leave.]

M: Vamos a comer, vamos a comer. (*Let's go eat, let's go eat.*)

In this example, the children on their own gather evidence about the manner that the different wheeled cars roll down the ramp. When the mother approaches the exhibit, she appropriates the goal of comparing the different car wheel shapes and engages her children in the process. At first only her 5-year-old daughter seems only peripherally interested, switching between watching the cars her mother is rolling down the ramp and drawing on the paper. But after a short time both the 5-year-old girl and her 7-year-old brother are working together, gathering data. The mother models for her children how to roll the different cars one at a time and how to measure the distance using the numbers, an approach that the children did not follow when testing on their own. In some ways, this is an ideal example of the parent working with her children on informal inquiry. The mother poses the main question, but asks her daughter which car to roll. The mother breaks up the task by assigning different roles for each of the children. Thus, the mother encourages systematic testing of each car with an empty track and collecting evidence from multiple trials of each car. Additionally, the complexity of the situation is also evident – the youngest child does not participate in the experimenting role that she was initially "assigned" (i.e., to watch the cars) and then gets hurt when she is in the way of the cars' path. The mother is trying to juggle the interests of the children and multiple goals (such as experimenting, having fun, and keeping the children safe). We would argue that this family is engaging in a multiparty version of experimenting and testing hypotheses that is seamlessly overlapping with just having fun with the provided materials. The scene is somewhat chaotic, and yet the interconnectedness of the individual roles is impressive.

In the next example five children each take turns rolling different cars and discussing the distance each went, without directly interacting with any adults.

Example #4: Collaboratively Collecting Evidence
[A 7-year-old boy picks up the off-center-circle-wheeled car from the ramp.]

Boy-7: Hey, get a car. Everybody get a car.

[Four other children – a 7-year-old girl, 5-year-old boy, 4-year-old girl, and 3-year-old girl – go to the cars and all, except the 3-year-old girl, pick up a car.]

B-7: This is yours here. [Pointing at an unclaimed car while looking at the 3-year-old girl.]

[The 5-year-old boy places his car at the top of the ramp and lets go while the 3-year-old girl watches.]

Several children [to 3-year-old girl in Vietnamese]: This is yours over here. Don't touch it. [They refer to a car going down the ramp.]

B-7: Go check where you are. [Points to the end of the track while looking at 5-year-old boy.]

[Both 7-year-old and 5-year-old boys walk to the end of the track where the car has stopped.]

B-5: 32, er, 34

[Then the 7-year-old boy removes the car from the track and then the 7-year-old girl places the hexagon-wheeled car on the top of the track and speaks to the 3-year-old girl.]

Girl-7: This is yours here.

[The 3-year-old girl pushes the hexagon-wheeled car down the track and the children watch it roll to a stop at the end of the track.]

B-5: 40.

[The 7-year-old girl talks to the boys at the end of the track about the hexagon-wheeled car.]

G-7: Get her car out.

[The 7-year-old boy attempts to remove car with foot while talking to the 5-year-old boy.]

B-7: Get out.

[The 5-year-old boy pulls the car with hands instead. The 7-year-old girl rolls the square-wheeled car down the track, but it tips off the side of the ramp.]

G-7: Oopsie.
B-7: Zero.

[The 4-year-old girl laughs. The group rolls a few additional cars down the ramp and says the distance each rolled aloud.]

In Example #4, the children experimented collaboratively by rolling different cars. While the children seem to take ownership of individual cars ("Get hers out"), they also carefully observed all the cars as they rolled down the ramp. In addition, to be able to compare the different cars, they identified relevant evidence (e.g., reading aloud the distance the cars went) and ensured that the tests were valid (e.g., by removing cars from the ramp before rolling a new one). At times, the children encountered some frustration with the car track, such as having their car fall off the ramp on its way down, or encountering difficulty when trying to remove their car from the end of the ramp when other cars were rolling down. However, the children easily used humor to handle these frustrations and usually continued with their data collection. The total interaction was only about a minute long, but the children rolled five different cars down the ramp – testing all of the cars that were available. The children might think that they were just having fun playing with cars, but features of their experience indicate that they were engaging in the science-relevant activity of seeking and evaluating evidence to answer a question.

Conversations About Temperature and Melting in a Storybook-Reading Activity

One may expect to find parents and children engaging in science-relevant reasoning in settings such as science museums or aquaria, or even when reading books about science. In such settings or activities parents may assume that they should talk about science-related concepts or engage in scientific reasoning. However, in settings or activities where the focus on science is less explicit, or not present, parents may also spontaneously engage in science-related conversation. Storybook reading may be just one of these activities in which parents and children spontaneously engage in science-related conversation that is peripheral to the main point of the activity.

Next we present some examples from a study of parent–child storybook reading activity conducted in the families' homes (Luce & Callanan, 2010). Parents and children (3-, 4-, and 5-year-olds) from a city in Northern California read the wordless picture book, *The Snowman* (Briggs, 1978), which contains illustrations depicting a boy making a snowman that subsequently comes to life. The boy and the snowman have a night filled with adventures, which include exploring inside the boy's home and flying across foreign lands. When the boy wakes up in the morning the snowman has melted. Despite the nonscientific nature of this fantasy story, we did indeed find that parents and children spontaneously discussed the concepts of hot, cold, and melting. In the following examples, we can see parents and children using and trying to understand scientific terminology, constructing causal explanations, and seeking evidence to answer questions.

Example #5: Using Scientific Terms
[Mother and 3-year-old girl turn to the last page of the book, where the snowman has melted into a puddle of water.]

Mother: *I think he's going outside to see the snowman. [turns page] Awww. What happened to the snowman?*
Girl-3: *He fell down.*
M: *He fell down...He melted.*

As argued above, parents may introduce children to scientific terms and the ways that adults in their communities understand them. In this example, the mother is highlighting the correct use of scientific terminology by confirming the child's description of the event ("he fell down") and then extending it to the more conventional term ("he melted"). This kind of experience may engage children in trying to understand abstract concepts underlying scientific terms.

Example #6: Explaining Properties
[Mother and 3-year-old boy are discussing the page where the boy is making the snowman.]

Boy-3: *I just want to make a snowman.*
Mother: *You want to make one now.*
B-3: *My snowman will melt.*
M: *Most snowmen do melt.*
B-3: *But will mine melt?*
M: *After awhile, but it will, if it's cold out it will stay up for a long time. If it's cold outside you can keep adding more snow to it.*

In this example, the parent introduces the concept of "cold" as a causal factor in whether or not a snowman may melt. Although there is considerable debate in the field regarding how children learn the scientific concepts of heat and temperature (e.g., diSessa, 1993; Slotta & Chi, 2006; Wiser & Amin, 2001), young children's early linguistic experience with the concepts of hot, cold, temperature, and heat is a potentially important aspect of this process (Luce & Callanan, 2010). We do not know how the child interprets the concept of cold in this interaction. Is it a property of the snowman or a property of the world "outside," as his mother implies? Situations like these, where parents attempt to explain the causal mechanisms behind phenomena, may be relevant for children's later understanding of "cold" and its relation to heat energy and temperature. In addition to encountering scientific concepts, this child is seeking out a form of evidence (here in the form of "testimony" from a more knowledgeable person, see Harris & Koenig, 2006) to answer his question about whether his snowman is an exception to the generic statement made by his mother. In science, identifying the conditions under which phenomena occur is an important aspect in understanding the phenomena themselves. The parent and child in this example seem to be invested in understanding the nature of snow and snowmen, and the phenomenon of melting is important in doing so.

Example #7: Multiple Names for a Concept
[Father and 5-year-old girl are discussing the page where the snowman discovers the kitchen stove.]

Father: *Then they go over to the stove, and the snowman turns on the stove, and again it's very hot, and that could... melt him.*
Girl-5: *Burn him.*
F: *It could melt him.*
G-5: *Melt him.*
F: *Besides burning him.*

This example highlights the use of scientific terminology and seems to suggest that the concepts of *burn* and *melt* may be confusing to young children. The stove would burn the child, but how does one characterize how a snowman would be affected? The child is interpreting the experiences of the human-like snowman as similar to her own when she uses the word "burn." The father seems to be correcting the child when he introduces the term "melt"; however, interestingly, the father seems to indicate that both concepts are relevant to the snowman. The father probably knows that snow does not burn, but may want to reiterate that stoves can burn because it is important that the child learns that stoves can be dangerous. This example highlights that parents and children engage with science topics in the midst of achieving myriad other goals in everyday activities, which may lead to understanding science concepts in the context of how they apply to safety issues.

In these examples, parents and children are talking about concepts (temperature, heat, melting) that have been shown to be difficult to understand for students in elementary school and beyond (e.g., Wiser, 1995). For example, children's ideas about what happens to molecules during changes in states of matter (e.g., when ice melts) are developing toward a scientific view during the elementary and middle-school years (Driver, 1985). Some researchers argue that while children can correctly identify the scientific terms to indicate changes in state (e.g., melting, evaporation, condensation), they often hold superficial understandings of these ideas (Osborne & Cosgrove, 1983). Children's everyday experiences may help them learn these terms and gain a beginning understanding of the processes, which can be built upon in the science classroom.

Additionally, in the research on children's understanding of heat and temperature, it has been suggested that the ways we talk about hot and cold in everyday language leads to misconceptions about the nature of heat (Slotta & Chi, 2006). However, others argue that our everyday language may be a resource for understanding such abstract concepts (e.g., diSessa, 2000; Lautrey & Mazens, 2004). Thus, the tension between scientific understandings of heat and temperature and the everyday ways that we talk about hot and cold is an excellent example of a contentious debate in "what counts as science." Even though parents and children may not demonstrate the same conceptions as the scientific community, getting opportunities to engage in reasoning processes about such concepts is arguably important for learning to reason scientifically. After all, science is not static – at one time, views of the nature of heat and temperature that are now seen as "incorrect" were the most prominent scientific views (see Wiser & Carey, 1983, for a review of historical concepts including caloric theories and an undifferentiated concept integrating our now distinct concepts of *heat* and *temperature*). We argue that learning to participate in scientific reasoning processes (e.g., seeking evidence to answer questions) is just as important as learning the most current scientific conceptions.

Together these examples about hot, cold, and melting begin to illustrate that in the course of everyday parent–child activities, even those not explicitly about "science," parents and their children engage in scientific-relevant reasoning practices about physical phenomena.

Conversations About Sun Safety at a Community Health Fair

Finally, we consider family conversations about science during visits to a community event organized to promote healthy behaviors in children (Jipson et al., 2009).

This project focused specifically on family interactions at an informational booth and hands-on activity built to encourage sun-safe behaviors. The "Sun Savvy" booth consisted of signs and pamphlets that provided information about sun safety and skin cancer. In addition, the booth offered an activity in which families could make bracelets out of color-changing ultraviolet beads (The UV-sensitive beads contain a pigment that changes color when exposed to ultraviolet radiation). Sunscreen with a range of SPF factors was made available to enable families to experiment with whether and how applying sunscreen to the beads influences the way they react to the sun. Analysis of the ensuing conversations revealed that families engaged in multiple modes of scientific inquiry, including the introduction of terminology, systematic experimentation, and explaining everyday behavior in light of scientific evidence and inquiry.

Example #8: Introduction of Scientific Terminology
[Mother and 5-year-old boy are creating a bracelet out of UV-sensitive beads.]

Mother: *We need to put the ultraviolet ones [beads] on.*
B-5: *What's "ultraviolet"?*
M: *The ones that change color in the sun.*

In this exchange, the mother introduces the term "ultraviolet" as if her son were already familiar with its meaning. This is not the first time that we have observed parents treat children "as if" they already had understandings that they did not (e.g., Callanan, Jipson, & Soennichsen, 2002). In this case, by assuming that her child already knew what "ultraviolet" meant, the mother presented her child with a puzzle to be solved. He could either ignore this puzzle, or he could select from a variety of strategies in an effort to better understand his mother's instructions. The child, a 5-year-old boy, takes the initiative to ask for an explanation. In her response, the mother does not discuss electromagnetic radiation or wavelengths of light, and instead focuses on a more local and practical answer. This finding is consistent with other work that suggests that parents rarely articulate complex scientific principles (e.g., Callanan & Jipson, 2001, Crowley et al., 2001), but this does not mean that they are not potentially contributing to their children's science learning. The response of the mother in the current example focuses not on what ultraviolet radiation *is*, but rather what it *does*. Consequently, her response is critical to the continued scientific nature of the dyad's present task engagement. By specifying that the beads she wants to use have the ability to change color in sunlight, she makes the observation that sun exposure can cause an effect on distal objects. Although she does not make an explicit connection to potential effects on people, the idea that the "sun changes things" is a potentially central concept in this activity. Thus, this mother's casual introduction of a scientific term, paired with her young child's curiosity about unfamiliar words, worked together to create a seedbed for further scientific exploration.

Example #9: Directive Scientific Experimentation
[Bypassing the bracelet-making activity, a father and a 4-year-old boy jump right into an exploration of the properties of the UV-sensitive beads.]

Father: *Do you want to do an experiment (child's name)?*
Boy-4: *Yeah.*
F: *The experiment is put some sunscreen on these four beads but don't put it on these [other] beads, ok?*
F: *We're going to see what happens after we put sunscreen on them.*

[Father and boy talk about the process of applying sunscreen]

F: *Now go put this in the sun, ok?*
F: *The ones with sunscreen will be protected from the sun and the ones without sunscreen might get a sunburn.*

[Time passes with some conversation]

F: *See, these are the ones with sunscreen on them and they didn't get sunburned, but this one over here got a little sunburned.*
F: *So, we want, if you wear sunscreen you won't get sunburned.*
F: *Do you want to wear some?*
B-4: *Yeah.*
F: *There, this way you won't get sunburned....this will protect you from the sun.*

This father appears to define the activity as "traditional science" when he proposes an "experiment." He then suggests a method to investigate the influence of sunscreen on beads, hypothesizes a result, and describes the resulting evidence. Thus, he is modeling the process of scientific inquiry, although one might argue that he is only minimally engaging his attentive son in the conceptual aspects of the enterprise (see Gleason & Schauble, 2000); for example, leaving little room for the child to articulate his own observations of the effects. Interestingly, the father deviates from a traditional science stance and extends the term "sunburn," usually applied to human skin, to describe the beads that change color. This domain-blurring may be intentional on the part of the father. He may be using an analogy that he expects the child to understand, and/or may be attempting to preview his later extension of the bead experiment to health-related behaviors. Our prior work has revealed a similar tendency for parents to offer scientifically inaccurate information as they engage in spontaneous conversations with their children (e.g., Callanan & Jipson, 2001; Jipson & Gelman, 2007). Often, they seem to do so in a figurative or pretend way. At other times they seem to be using language that they believe will be more understandable to their children. An important question is whether and how children separate credible statements from those made in a more figurative way.

In addition to offering potentially valuable instruction in the scientific method, and potentially confusing information about the type of change observed, the father pursues a health socialization agenda by directly relating their evidence to the adoption of preventative health measures (i.e., wearing sunscreen). The child agrees to wear sunscreen and the father reinforces the behavior by predicting a positive health result. The manner in which the father weaves together multiple goals demonstrates the importance of considering the relationship between those who collaborate in scientific activities. Would a teacher or older peer have acted similarly? As a parent, this father seizes the opportunity to infuse his "science lesson" with a preventative health lesson, thereby making a potential difference in the everyday life of his child.

Example #10: Mismatched Goals and the Importance of Timing
[A mother introduces a variety of science-related concepts to her seemingly uninterested 5-year-old son as they make a bracelet out of the UV-sensitive beads.]

Mother: Why do you think the sun might not always be safe?

[no response]

M: See how they're changing color? [pointing to the beads]

[several minutes of talk about making bracelets]

M: So, guys, why do you think the sun may not always be safe?
Boy-5: I don't know
M: What does it give us?
B-5: Light
M: It gives us light.
B-5: The whole world light (continuing to make bracelet)
M: Mm Hmm
M: Do you know what UV is?
B-5: Hey, this one's changing colors!
M: Do you know what UV is (child's name)? You know what it stands for?
M: Ultraviolet
M: And the sun is super helpful because it gives us light but it also –
B-5: Yep-the whole world light.
M: Mm Hmm, but it can also burn us, right?
M: When we're sunburned?

[Child ignores and focuses on bracelet making]

M: What can the sun not be so healthy?

[ignored]

M: Why do you think we put sunscreen on when we go to the beach?

[ignored]

M: So, guys, why do you think we put sunscreen on?
B-5: [inaudible]
M: To protect you from getting what?
M: Sunburned
M: Why does it hurt when you get sunburned?
B-5: Cuz the sun.
M: Yeah the sun has ultraviolet rays in it that can burn our skin and make us sick.

[ignored, family leaves without applying sunscreen]

This example illustrates a situation in which participants do not seem to share a common task objective. The mother in this interaction is determined to engage her child in discussion about the relationship between the bracelet-making activity and personal health. She first points out the central observation of this activity, that the beads change color. She then attempts to guide her child toward a finish line she seems to have clearly defined. The child, however, is focused throughout the event on constructing bracelets and resists his mother's attempts to engage him in discourse about science and health. Then, as if in a badly choreographed dance routine, the child excitedly makes the discovery that the beads change color. Instead of building on the excitement of this discovery, the mother ignores his observation and continues to question him about scientific terminology. Predictably, the child tunes her out again; yet the mother persists until the end of the interaction when she answers her own questions with a passably scientific explanation. The mismatched agendas apparent in this session undermine both participants' efforts to engage in meaningful scientific inquiry. The mother's noncontingent persistence, paired with the child's withdrawal, leaves both participants isolated as they attempt to make sense of the activity. The exchange is likely not an optimal example of collaborative scientific reasoning, yet it likely represents the reality of many of children's scientific explorations both in and out of school. In this case, the mother may have been responding to being videotaped as part of a study, or to the perceived expectation that they should engage in experimental activity in this booth. Finding ways to extend children's scientific inquiry, without intruding on their goals and process, is critical to promoting engagement and enthusiasm on the part of both learner and teacher.

The interactions we observed at the community health fair demonstrate the varied ways that conversations between parents and children about science unfold. Although the differences between the conversations presented here may seem greater than their similarities, it is noteworthy that examination of the larger data set revealed that scientific inquiry was a clear goal for the families visiting the Sun Savvy booth, with 100% engaging in at least one informal "experiment" with the materials provided. And this is not because science was the only thing to demand families' attention; 67% of the families also made connections to sun-safe behaviors. Thus, even when multiple possible topics of conversation compete for "family air time," families in this study embraced science-related talk and activity at least at some level.

Conclusions

In the diverse examples we have considered across four settings, we found evidence of family conversation and activity that touched upon each of the science-relevant activities we sought: use of scientific vocabulary, generalization beyond the current situation, causal explanation, and experimentation. At the same time, it is clear that the conversations and activities represented here are quite different from both school science and professional science. Most importantly, perhaps, these examples appear within the complexity of everyday activities, with their myriad intersecting goals. This is quite different from science content that is carefully constructed within a curriculum, with clear learning goals for children, and it is also quite different from the ongoing practice of science by scientists.

Despite these differences, there are some ways that these activities might arguably relate to science learning and even "count" as science. When parents introduced technical vocabulary such as "gills" in the context of meaningful activity, we would argue that children are in a better position to learn the meaning of those terms than when introduced out of context. When parents move seamlessly from a particular event, like a snowman melting, to a more general rule (snowmen usually melt), children are likely to be building up a knowledge base of cultural beliefs upon which they can build when they study science later. In suggesting that such experiences are science-relevant, we certainly do not wish to argue that this type of everyday science learning should be a replacement for school science learning. On the contrary, everyday science is encountered in a spontaneous way and it would be very difficult, if not impossible, for children to reach deep levels of understanding in a science discipline without a more structured approach. However, we would argue that these sorts of everyday science-relevant experiences will interact with and support classroom science learning, and that it is crucial that we take these into account in order to make more productive links between home settings, museum settings, and school settings for thinking about science.

Admittedly, parents may not always provide children with helpful guidance, as suggested in the "mismatched goals" example from the Sun Savvy study. Further, some might argue that the examples provided indicate that parents are "missing opportunities" to provide more thorough science content. On the contrary, we would argue that in most of these conversations parents are responding appropriately to their children's developmental level. The parent of the 1-year-old who talks about what the fishy might say is perhaps communicating subtle information about biology in a way that is not only subtly informative, but fine-tuned to the child's interest and ability.

A question that is beyond the data presented here is whether the children in these examples *learned* anything about science by participating in these activities and conversations. Assessing learning in complex and varied everyday activities and contexts is just as complex as the activities and contexts themselves. Roth and Calabrese Barton (2004) make an intriguing argument that "scientific literacy" is not a property of individuals. Children do not "have" or "not have" scientific literacy and thus it is not something to teach and test, per se. Scientific literacy may rather emerge during collective activity where knowledge is distributed among people and thus scientific literacy is a property of such collective activity. Each person engaged in the activity likely contributes in some way to the scientific literacy that occurs. In the second example from the Car Track study, the conversations that occurred and the experimentation that took place seemed to emerge as each person contributed to the conversation. Without each person's contributions, the activity and conversation may have been much different. To attribute any one piece of scientific knowledge or reasoning process to any one person is difficult and takes meaning away from the collaborative interaction. This raises challenges for sociocultural approaches to develop more appropriate ways to "assess" science learning or scientific thinking ability.

Integrating across the four studies represented here, we emerge with three claims about children's engagement in everyday science-related experiences. Our first claim is that virtually all children begin to participate in exploring science-relevant ideas early in life (NRC, 2007, 2009). In particular, they learn the conventional ways that science-related terms are understood by adults in their cultural communities, and sometimes these novel terms lead to more open discussion about science-related concepts. They are also exposed to reasoning that is relevant for science, such as explaining causal relations. In many families, children may be less likely to hear about science for its own sake and more likely to hear about science concepts that are meaningfully connected to parents' goals, such as safety for their children. Regardless of the family attitudes regarding science, however, our claim is that most children participate in conversations that at least occasionally include both the topics of science (e.g., nature, mechanics), and the practices of science (e.g., questioning, predicting, explaining).

While we argue that science-relevant experiences are ubiquitous, our second claim is that everyday experiences with science are embedded in a messy world, and usually not a central focus of activity, as they might be in school classrooms. One of the key differences between everyday settings and other more formal science settings is that the primary goal of these everyday activities may not often be to learn science. Science-related topics can come up spontaneously when doing everyday household activities or when sharing interest with someone you care about (e.g., a child may want to be involved when a parent is gardening). In contrast to school settings, however, we contend that everyday conversations may be more likely to be fine-tuned to the child's individual experiences, long-standing interests, or even interests "in the moment" (Crowley & Jacobs, 2002; Ellenbogen, 2002; Leibham, Alexander, Johnson, Neitzel, & Reis-Henrie, 2005; Palmquist & Crowley, 2007). For example, many parents go to great lengths to provide experiences to young children that connect with their individual interests, such as going to visit construction sites, or horse farms, or purchasing books and videos on their child's favorite topics (Leibham et al., 2005). In everyday settings children's scientific investigations are often motivated by their interests and the focus is often on specific topics such as dinosaurs, rocks, or worms (Johnson, Scott, & Mervis, 2004; Palmquist & Crowley, 2007). Learning science might be one goal that parents have in such activities, but it can compete with many other goals (such as having fun or completing a task). And while some parents in some settings may have pedagogical intent (Palmquist & Crowley, 2007), in other cases families are instead "having fun," "exploring a museum," or "just reading," and in some cases parents and children may not even recognize that science or science learning is taking place.

Finally, our third claim is that different children have very different experiences with science-relevant ideas. For example, parents with science background may lead children toward different conceptualizations of a domain than would parents with less formal science education (Tarlowski, 2006). Parents with science background may also focus more on evidence whereas parents with humanities background may focus more on ethical issues related to science (Valle, 2005). Further, although some children may engage in talk about science only peripherally, other children may conduct experiments at home or use other school-like science concepts during everyday activities. Regardless of the focus on school science, however, we argue that virtually all children have experiences that can serve as a foundation for learning school science. Tenenbaum and Callanan (2008) found that families with varying formal education background were nonetheless equally likely to engage in explanatory conversations when in a familiar (home) setting rather than an unfamiliar setting. As a result, we propose that it is important to look at everyday science settings when aiming to address issues of equity in science opportunities (Aikenhead, 2001; Warren et al. 2001). These contexts may help us to understand how children from different backgrounds and personalities may approach science in different ways.

Making meaningful connections between the family settings we have identified and school science settings is, of course, a huge challenge. Gutiérrez, Baquedano-Lopez, and Tejeda (1999) have discussed after-school programs and other "hybrid" settings, which are neither classrooms nor homes, where students might encounter science concepts in ways that bridge more directly to more familiar everyday concepts. We would include museum visits and other family activities among these hybrid settings, and we suggest that they may be especially important for children whose family activities have less obvious overlap with classroom science. As we, as a field, struggle with the question of what counts as science, it is crucial to consider the kinds of examples discussed here. Regardless of how children begin to define science for themselves, they may need guidance to recognize the rich background they themselves bring to the science classroom by virtue of their participation in conversations and activities from their everyday lives.

Acknowledgments This research was funded by the Center for Informal Learning and Schools (ESI-0119787), as well as research grants from NSF (ESI-0229872), NICHHO (26228), NIH (T32HD046423) and an AERA-IES Postdoctoral Research Fellowship to L. Triona. We are grateful to children's Discovery Museum of San Jose and Seymoul Marine Discovery Center for their Collaboration.

References

Aikenhead, G. S. (2001). Students' ease in crossing cultural borders into school science. *Science Education, 85*(2), 180–188.

Ash, D. (2003). Dialogic inquiry in life science conversations of family groups in a museum. *Journal of Research in Science Teaching, 40*, 138–162.

Briggs, R. (1978). *The snowman*. New York: Random House.

Callanan, M., & Jipson, J. L. (2001). Explanatory conversations and young children's developing scientific literacy. In K. Crowley, C. D. Schuam, & T. Okada (Eds.), *Designing for science: Implications from everyday, classroom, and professional settings* (pp. 21–49). Mahwah, NJ: Lawrence Erlbaum Associates.

Callanan, M., Jipson, J., & Soennichsen, M. (2002). Maps, globes, and videos: Parent-child conversations about representational objects. In S. Paris (Ed.), *Perspectives on object-centered learning in museums* (pp. 261–283). Mahwah, NJ: Lawrence Erlbaum Associates.

Callanan, M. A., & Oakes, L. M. (1992). Preschoolers' questions and parents' explanations: Causal thinking in everyday activity. *Cognitive Development, 7*, 213–233.

Callanan, M. A., & Valle, A. (2008). Co-constructing conceptual domains through family conversations and activities. In B. Ross (Ed.), *The psychology of learning and motivation* (Advances in Research and Theory, Vol. 49, pp. 147–165). San Diego, CA: Elsevier Academic Press.

Carey, S. (1985). *Conceptual change in childhood.* Cambridge, MA: MIT Press/Bradford Books.

Cimpian, A., & Markman, E. M. (2008). Preschool children's use of cues to generic meaning. *Cognition, 107*, 19–53.

Crowley, K., Callanan, M., Jipson, J., Galco, J., Topping, K., & Shrager, J. (2001). Shared scientific thinking in everyday parent-child activity. *Science Education, 85*, 712–732.

Crowley, K., & Galco, J. (2001). Everyday activity and the development of scientific thinking. In K. Crowley, C. Schunn, & T. Okada (Eds.), *Designing for science: Implications for everyday, classroom, and professional settings* (pp. 393–413). Mahwah, NJ: Lawrence Erlbaum Associates.

Crowley, K., & Jacobs, M. (2002). Building islands of expertise in everyday family activity. In G. Leinhardt, K. Crowley, & K. Knutson (Eds.), *Learning conversations in museums* (pp. 333–356). Mahwah, NJ: Lawrence Erlbaum Associates.

diSessa, A. (1993). Ontologies in pieces: Response to Chi and Slotta. *Cognition and Instruction, 10*, 261–280.

diSessa, A. A. (2000). *Changing minds: Computers, learning, and literacy*. Cambridge, MA: MIT Press.

Driver, R. (1985). Beyond appearances: The conservation of matter under physical and chemical transformations. In R. Driver, E. Guesne, & A. Tiberghien (Eds.), *Children's ideas in science* (pp. 145–169). Philadelphia: Open University Press.

Dunbar, K. (1995). How scientists really reason: Scientific reasoning in real-world laboratories. In R. J. Sternberg & J. E. Davidson (Eds.), *The nature of insight* (pp. 365–395). Cambridge, MA: MIT Press.

Ellenbogen, K. M. (2002). Museums in family life: An ethnographic case study. In G. Leinhardt, K. Crowley, & K. Knutson (Eds.), *Learning conversations in museums* (pp. 81–101). Mahwah, NJ: Lawrence Erlbaum Associates.

Falk, J. H., & Dierking, L. D. (1992). *The museum experience*. Washington, DC: Whalesback Books.

Gelman, S. A., Coley, J. D., Rosengren, K. S., Hartman, E., & Pappas, A. (1998). Beyond labeling: The role of maternal input in the acquisition of richly structured categories. *Monographs of the Society for Research in Child Development, 63*(1), Serial No. 253.

Gelman, S., & Raman, L. (2003). Preschool children use linguistic form class and pragmatic cues to interpret generics. *Child Development, 74*(1), 308–325.

Gleason, M. E., & Schauble, L. (2000). Parents' assistance of their children's scientific reasoning. *Cognition and Instruction, 17*, 343–378.

Gutiérrez, K., Baquedano-Lopez, P., & Tejeda, C. (1999). Rethinking diversity: Hybridity and hybrid language practices in the third space. *Mind, Culture & Activity: An International Journal, 6*(4), 286–303.

Gutiérrez, K. D., & Rogoff, B. (2003). Cultural ways of learning: Individual traits or repertoires of practice. *Educational Researcher, 32*, 19–25.

Harris, P. L., & Koenig, M. A. (2006). Trust in testimony: How children learn about science and religion. *Child Development, 77*, 505–524.

Inagaki, K., & Hatano, G. (2002). *Young children's naïve thinking about the biological world.* New York: Psychology Press.

Jipson, J., & Callanan, M. (2003). Mother-child conversation and children's understanding of biological and non-biological changes in size. *Child Development, 74*, 629–644.

Jipson, J. L., & Gelman, S. A. (2007). Robots and rodents: Children's inferences about living and nonliving kinds. *Child Development, 78*, 1675–1688.

Jipson, J.L., Clark, L., Etcheverry, R., Jones, H., Lebs, J., Reid G., Slavik, L., & Ziemba, R. (2009). Parent-child considerations of science and health. Poster presented at the biennial meeting of the Society for Research on Child Development. Denver, CO.

Johnson, K. E., Scott, P., & Mervis, C. B. (2004). What are theories for? Concept use throughout the continuum of dinosaur expertise. *Journal of Experimental Child Psychology, 87*, 171–200.

Lautrey, J., & Mazens, K. (2004). Is children's naïve knowledge consistent? A comparison of the concepts of sound and heat. *Learning and Instruction, 14*, 399–423.
Leibham, M., Alexander, J., Johnson, K., Neitzel, C., & Reis-Henrie, F. (2005). Parenting behaviors associated with the maintenance of preschoolers' interests: A prospective longitudinal study. *Journal of Applied Developmental Psychology, 26*, 397–414.
Luce M. R., & Callanan, M. A. (2010). *"Heated" discussions: Family conversations about heat and temperature*. Manuscript under review [unpublished paper].
National Research Council. (2007). In R. A. Duschl, H. A. Schweingruber, & A. W. Shouse (Eds.), *Taking science to school: Learning and teaching science in grades K-8* (Committee on Science Learning, K-8). Washington, DC: National Academies Press.
National Research Council. (2009). In P. Bell, B. Lewenstein, A. W. Shouse, & M. A. Feder (Eds.), *Learning science in informal environments: People, places, and pursuits* (Committee on Learning Science in Informal Environments). Washington, DC: National Academies Press.
Ochs, E., Gonzales, P., & Jacoby, S. (1996). "When I come down I'm in a domain state": Talk, gesture, and graphic representation in the interpretive activity of physicists. In E. Ochs, E. Schegloff, & S. Thompson (Eds.), *Interaction and grammar* (pp. 328–369). Cambridge, UK: Cambridge University Press.
Ochs, E., Taylor, C., Rudolph, D., & Smith, R. (1992). Storytelling as a theory-building activity. *Discourse Processes, 15*, 37–72.
Osborne, R. J., & Cosgrove, M. M. (1983). Children's conceptions of the changes of state of water. *Journal of Research in Science Teaching, 20*, 825–838.
Palmquist, S., & Crowley, K. (2007). From teachers to testers: How parents talk to novice and expert children in a natural history museum. *Science Education, 91*, 783–804.
Rigney, J., & Callanan, M. (2011). Patterns in parent-child conversations about animals at a marine science center. *Cognitive Development, 26*, 155–171.
Rogoff, B. (2003). *The cultural nature of human development*. New York: Oxford University Press.
Roth, W.-M., & Calabrese Barton, A. (2004). *Rethinking scientific literacy*. New York: Routledge.
Slotta, J. D., & Chi, M. T. H. (2006). Helping students understand challenging topics in science through ontology training. *Cognition and Instruction, 24*, 261–289.
Stevens, R. (2000). Who counts what as math? Emergent and assigned mathematics problems in a project-based classroom. In J. Boaler (Ed.), *Multiple perspectives on mathematics teaching and learning*. Westport, CT: Ablex.
Tarlowski, A. (2006). If it's an animal it has axons: Experience and culture in preschool children's reasoning about animates. *Cognitive Development, 21*, 249–265.
Tenenbaum, H., & Callanan, M. (2008). Parents' science talk to their children in Mexican-descent families residing in the United States. *International Journal of Behavioral Development, 32*, 1–12.
Triona, L. M., & Callanan, M. A. (2008). Science play: Comparing children alone, with peers, and with adults. In B. C. Love, K. McRae, & V. M. Sloutsky (Eds.), *Proceedings of the 30th annual conference of the cognitive science society* (p. 1378). Austin, TX: Cognitive Science Society.
Valle, A. (2005). *"How do you know?" Communicating ideas about science and scientific reasoning in parent-child conversations.* Unpublished doctoral dissertation.
Vygotsky, L. S. (1978). *Mind in society: The development of higher psychological processes.* Cambridge, MA: Harvard University Press.
Warren, B., Ballenger, C., Ogonowski, M., Rosebery, A., & Hudicourt-Barnes, J. (2001). Rethinking diversity in learning science: The logic of everyday sense-making. *Journal of Research in Science Teaching, 38*, 529–552.
Wertsch, J. V. (1979). From social interaction to higher psychological processes. A clarification and application of Vygotsky's theory. *Human Development, 22*, 1–22.
Wiser, M. (1995). Use of history of science to understand and remedy students' misconceptions about heat and temperature. In D. N. Perkins, J. L. Schwartz, M. M. West, & M. S. Wiske (Eds.), *Software goes to school: Teaching for understanding with new technologies* (pp. 23–38). New York: Oxford University Press.
Wiser, M., & Amin, T. (2001). Is heat hot? Inducing conceptual change by integrating everyday and scientific perspectives on thermal phenomena. *Learning and Instruction, 11*, 331–355.
Wiser, M., & Carey, S. (1983). When heat and temperature were one. In D. Gentner & A. L. Stevens (Eds.), *Mental models* (pp. 267–297). Hillsdale, NJ: Lawrence Erlbaum Associates.

Chapter 4
What Counts as Mathematics When "We All Use Math Every Day"? A Look at *NUMB3RS*

Indigo Esmonde

When students learn mathematics in school, they learn more than just the procedures, algorithms, and concepts. They also learn about what counts as mathematics, who practices mathematics, how they do so, and for what ends. But school is not the only source for students' developing notions of what mathematics is and what it is for. Students also learn about mathematics from the societal curriculum: "that massive, ongoing, informal curriculum of families, peer groups, neighborhoods, churches, organizations, institutions, mass media, and other socializing forces" (Cortés, 2004, p. 211). One pervasive source of education about mathematics is the media, including movies, television, newspaper and magazine articles, and YouTube. While popular media are often considered to reflect typical views of what mathematics is, they also influence popular opinion and convey information about what counts as mathematics, and what kinds of people engage in mathematics.

The question of "what counts as mathematics" in the media is important precisely because of the complex relationship of the media to popular opinion. Representations of mathematics in the media both inform and are informed by social views of mathematics, and can, therefore, influence students, parents, educators, and policy-makers as they consider the goals and the principles of mathematics education. In this chapter, I consider how mathematics, and people who use mathematics, are represented in popular television and film.

I begin by considering mathematics in popular media fairly broadly, and then consider one television crime drama, *NUMB3RS*, because of its popularity in North America and its mathematical focus. I present an analysis of how this series portrays mathematics and people who do mathematics. In addition, I offer results from a pilot interview study to understand how viewers of the show might interpret "what counts as mathematics."

I. Esmonde (✉)
Ontario Institute for Studies in Education, University of Toronto, Toronto, ON, Canada
e-mail: indigo.esmonde@utoronto.ca

B. Bevan et al. (eds.), *LOST Opportunities: Learning in Out-of-School Time*, Explorations of Educational Purpose 23, DOI 10.1007/978-94-007-4304-5_4,

Mathematics in Movies and Television

One way to consider what counts as mathematics in the media is to examine the full range of representations of mathematics, of which there are surprisingly many in popular television shows, books, and movies. A quick web search with keywords "mathematics" and "movies" reveals several web pages devoted to movies and television shows that involve mathematics. These sites provide data about how their authors and readers decide "what counts as mathematics." Two such sites were examined. One site provides "a guide to major motion pictures with scenes of real mathematics" (Reinhold, 2007). Reinhold lists 31 movies and 2 television series (one of them *NUMB3RS*) along with a description of the mathematics contained in each. A second site, housed on the Harvard Mathematics Department website, offers a collection of movie clips "in which Mathematics appears" (Knill, 2007).

There is considerable overlap in the movies cited on these two websites, implying that perhaps the authors define mathematics similarly. Both lists include the movies *Proof*, *A Beautiful Mind*, *Good Will Hunting*, and *Pi*, and the television show *NUMB3RS*. Almost all of the examples they include focus on events related to academic or school mathematics. Examples feature people lecturing, writing mathematical symbols and equations on blackboards or in notebooks, and doing mental computations like finding cube roots or adding large sums. Some clips involve very advanced mathematics like the "Snake Lemma" in homological algebra, whereas others involve simple mathematics like addition and multiplication of whole numbers. Specifically, of Reinhold's 33 examples, 23 take place in school or university settings. Several more examples seem relatively trivial: one involves samurais who "plan," one involves soldiers who "count," and one was included because a mathematics book is shown briefly, but not mentioned in the film's dialogue. In the remaining examples, ordinary people are portrayed as struggling to understand the mathematics they encounter in and outside of school (e.g., a middle-school student who needs help with his homework, police officers forced to solve a math problem in order to stop a bomb from exploding).

These two websites and their lists of films and television shows were constructed by mathematical experts themselves, and as such may not match up with ordinary people's views of mathematics. There may be significant differences in the ways different people view representations of mathematics in the media. To understand further the portrayal of mathematics in popular media, this chapter will focus on a single television series, *NUMB3RS*.

NUMB3RS is a crime drama that aired on prime-time television for 5 years (2005–2010), and was, at the time of writing, the only current scripted network television show with a consistent and clear focus on mathematics. In the remainder of this chapter, I consider how mathematics is represented in *NUMB3RS* by taking a close look at the television series itself, and by reporting on results from a pilot interview study, in which viewers identified and discussed mathematical moments in the series.

NUMB3RS

NUMB3RS features a fictional FBI agent, Don, whose brother Charlie is a professor of mathematics at nearby "Cal Sci University." Charlie is often called in to consult on the FBI's cases, and his mathematical analyses invariably play a key part in solving the case. Mathematics itself forms a central part of each storyline in the series, and the series offers ample opportunities to explore the portrayal of mathematics and mathematicians. Although popular, and currently being re-presented on cable television, the show has been controversial. The calculator manufacturer *Texas Instruments* was an enthusiastic supporter of the show and created a set of classroom-based activities that built on themes from the series. Some mathematicians have argued that the series stereotypes mathematical activity and professions, and have expressed doubt that the series is appropriate for schoolchildren to watch because of onscreen violence and sexuality (Greenwald, 2006).

The Mathematics of *NUMB3RS*

The series' tagline is "We all use math every day," suggesting that the series may take a broad view of mathematics. Instead of a theme song, each episode is introduced with a list of ways "we" use math: "to predict weather, to tell time, [and] to handle money." The voiceover reminds us "math is more than formulas and equations: it's logic, it's rationality, it's using your mind to solve the biggest mysteries we know." These examples resonate with the goals of math reformers, and of learning sciences researchers who have studied the ways in which ordinary people use mathematics (e.g., Brenner, 1998; Guberman, 2004; Lave, 1988; Masingila, 1994; Nunes, Schlieman, & Carraher, 1993; Saxe, 1991; Scribner, 1984). It seems reasonable to expect, then, that the show would demonstrate not only Charlie's advanced mathematics, but the more mundane ways ordinary people use mathematics. This chapter discusses in greater detail not only what types of mathematics are presented on the show, but whose activities are presented as being mathematical. I begin with a relatively narrow approach to understanding this question: a study of the use of the words "mathematics," "math," "mathematical," and "mathematically."

Methods: Word Counts

To conduct a study of the use of the word "mathematics," I viewed 24 episodes of the series *NUMB3RS* – the entire second season of the series. Each episode was watched at least twice. When words with the root "math" (e.g., mathematics, math, mathematical, mathematically) were uttered in an episode or displayed visually on the screen, I noted the episode, time, and a brief transcript indicating how the word was used. I excluded the word "mathematician" because it was used solely to identify individuals whose profession involved mathematics, and would perhaps skew the analysis. I then examined these brief transcripts to understand "what counts as mathematics" in the series: what sorts of activities were described as mathematical, and who was identified as doing mathematics.

I recognize that in a television series devoted to mathematics, these counts underrepresent the use of mathematics in the show. Other words, phrases, and symbols are also used to signal mathematical activity. Counting the word "mathematics" and other related words provides the least inferential method to understand how the scriptwriters and producers of the series chose to represent mathematical activity.

I found 63 clips in which math, mathematics, mathematical, mathematically, etc., were used – an average of approximately 2.7 instances per episode. Some episodes did not use the word at all, whereas one episode used the word 11 times.

We All Use Math Every Day: Who Is "We"?

The word "mathematics" (and variants of the word) was used almost exclusively to refer to the activities, ideas, and products of professional mathematicians and physicists (since one of Charlie's colleagues, also a prominent character on the series, is an astrophysics professor at Cal Sci). Of the 63 clips in which mathematics was used, 29 used the terms to refer to Charlie's work (almost half!), while an additional 23 refer to the work of other professional mathematicians, scientists, and serious students of mathematics. The remaining instances mostly use "mathematical" as an adjective to describe objects like formulas, problems, and technology. It is unclear, then, from these moments how "we" who are not mathematicians are to be understood to use math every day.

In fact, when other characters on the series discuss mathematical methods, Charlie would often "pull rank" and demonstrate how his (mathematical) analytic methods were superior to their (by implication, nonmathematical or less rigorous) methods. The following examples demonstrate how Charlie would position his analyses as superior.

In the first vignette, Charlie is about to describe a pattern that he discovered in the dates of a series of home invasion robberies. Colby, an FBI agent, watches and comments.

Vignette 1. Charlie and Colby both describe searching for a timeline pattern (Episode: Convergence, *begins approximately 9:25)*

Charlie is standing in a conference room at the FBI, taping pieces of paper to the walls and circling some dates in blue, some in red. Three FBI agents come in to speak with him about his results. When they ask him "What's up," he replies "Calendars, wonderful analog mathematical tools." After explaining the history of calendars, he says that he ran a "Fourier analysis" on the chronology of dates. Colby, one of the FBI agents, says that they had already tried to find a "timeline pattern" but were unsuccessful. Charlie goes on to explain further the results of his analysis, which leads the FBI team to discover a set of crimes that they hadn't known were related.

In this example, Colby and Charlie are portrayed as attempting to solve the same problem: finding a pattern in the timeline of seven robberies. Colby confesses he could not find any pattern. Charlie explains that he used a "Fourier analysis," thus upping the mathematical ante and implying that the agent's methods were not sophisticated enough. Charlie is successful, Colby unsuccessful, at applying mathematical methods to analyze criminal activity.

In the next excerpt, Charlie and Megan, the FBI behavioral profiler, are sitting at a computer, and Charlie is showing Megan some of his work.

Vignette 2. Charlie discusses social network theory and flock analysis
(Episode: In Plain Sight, *begins approximately 1:12)*

While Don and a team of field agents close in around a suspected meth lab, the scene changes to show Charlie and Megan seated at a computer, monitoring the team's progress. Charlie says that it's "kind of exciting ... modeling illegal activity using social network math." Megan smiles, leans back, and says, "Link analysis. We use that for the mob." Charlie widens his eyes and explains that Megan's link analysis "provides fairly basic connections." He leans towards the computer and explains that he uses "flock theory" to show "how the network moves, how it changes as a whole." He goes on to describe more about the application of flock analysis while on screen, a series of graphs, equations, and images of flocks of birds flash by in quick succession.

In this excerpt, Megan demonstrates her familiarity with social network theory – specifically, link analysis. Charlie describes her use of social network theory as "fairly basic" and contrasts it with his own, presumably more advanced use of "flock analysis" to describe connections between nodes in a network. Megan, the FBI's behavioral profiler, frequently uses her statistical and psychological knowledge to solve cases. Despite her status as an educated, successful professional, she is portrayed as having a "fairly basic" mathematical knowledge base, as compared with Charlie's more complex mathematics, even when discussing methods that fall directly within her area of expertise and not his.

In this next excerpt, Charlie's brother Don suggests a mathematical method that Charlie could use to help them find the students responsible for a series of school shootings.

Vignette 3. Charlie tells Don he was 'thinking like a mathematician'
(Episode: Dark Matter, *begins approximately 5:10)*

The FBI team is investigating a school shooting in which they don't know who all the killers were. They realize that the school identification cards have chips on them to take attendance and show where each child is in the school. Don suggests that they could use this system to "tell us where each kid is, then we should match it with witness statements and match the shooter's route. And by we, I mean Charlie." The scene quickly changes to a conversation between Don and Charlie, standing near math-covered chalkboards. Charlie proudly says, "I've got you thinking like a mathematician," but then goes on to criticize Don's method. He says "there's a natural flaw in witness statements, and statistics show that memory is often unreliable." Charlie's astrophysicist friend Larry suggests, "Neptune," a seeming non-sequitur that both Charlie and Amita (computer scientist/mathematician) immediately understand as a new suggested mathematical strategy. Even when they explain, Don has trouble following their explanation.

Don proposes a method for analyzing movements, but immediately defers to Charlie to carry out the plan, implying that he himself would not be competent enough. Charlie admits Don was "thinking like a mathematician," but points out a flaw in Don's mathematical methods. The flaw – "statistics show that memory is often unreliable" – has nothing to do with mathematics. Charlie refers to the use of quantitative analysis, but in the context of psychology, an area in which Charlie is not an expert.

In the final excerpt, Megan and Charlie are again discussing a case. Megan uses a colloquial mathematical term, and Charlie immediately corrects her, poking fun at her words.

Vignette 4. Megan uses 'exponential' colloquially
(Episode: Judgment Call, *begins approximately 2:45)*

The camera pans over a long table covered in file folders, as Megan explains that they have pulled all the files of suspects for a recent murder of a judge's wife. The FBI's hypothesis is that the judge was the real target. Megan comments that there is a very long list of possibilities, saying, "there's also family members, friends, co-conspirators, the possibilities are exponential." Charlie looks up and grimaces slightly, saying, "um, exponential would mean that the growth rate is proportional to its size, so, the mathematically correct term would be, 'more'." Everyone laughs.

Megan uses the mathematical term "exponential" as it is commonly used in every day settings. Charlie chastises and corrects her (and by extension, any audience member who might use the term as Megan does) and is presented as the arbiter of what is mathematically correct.

In each of the excerpts above, characters present mathematical ideas to Charlie, who quickly dismisses their work and proposes his own ideas. These excerpts exemplify the series' tendency to show Charlie and his colleagues as invariably smarter, quicker, and more sophisticated thinkers than the FBI agents. The FBI agents defer to Charlie's expertise, even when he does not provide full explanations of why his ideas are correct and theirs are not. Because of his status as professional mathematician, the other characters defer to him and accept his mathematical claims, even about non-mathematical topics (e.g., reliability of witness reports). This pattern of interaction strongly implies that mathematical expertise, or in fact even practical competence, is owned only by designated mathematicians. We may all do math every day, but the show seems to be saying that only some of us (i.e., mathematicians) do it right.

We All Use Math Every Day: What Is "Math"?

During most episodes, Charlie discusses some mathematical concepts, findings, or methods of analysis with other characters on the show. In this section, I analyze the characteristics of these mathematical discussions to consider what counts as mathematics. Vignette 5 provides an example of one such discussion. Charlie describes an analysis of romantic dating conducted by two mathematicians from the UK. Charlie's brother and father both seem to doubt that mathematics could be of any use in the realm of romance and dating.

Vignette 5. Using math to analyze courtship
(Episode: The O.G., *begins approximately 1:00)*

The scene opens in Charlie's home, where Charlie and his brother Don are watching their father Alan get ready to go out on a date. He asks their advice on what gift to get the woman for her upcoming birthday. Charlie smiles and says, "You know, two mathematicians from University College in London actually recently addressed this very same question." Alan and Don seem doubtful that two "math geeks who don't date much" might have any useful advice, but Alan encourages Charlie to explain more. Charlie begins by asking his father, "Well, assuming that the goal is the female's (pause) receptiveness..." to which his father nods yes. Charlie continues by saying that "the best chance of success" can be had by buying "an extravagant gift, one that's costly to the man but of no real financial value to the woman." He suggests flowers or dinner at an expensive restaurant. He finishes by leaning back in his chair with a smirk, and asks, "See? You guys still doubt the power of math?" Don smirks back and says, "Yeah, well when it comes to female receptiveness, yes."

Vignette 5 illustrates some characteristics of mathematics as it is often portrayed in *NUMB3RS*. First of all, this is an example of *mathematical modeling*, the use of mathematics to model some real-world phenomenon or problem. The key to creating a good model is to enumerate the underlying assumptions. Depending on the assumptions, radically different models might be created for the same phenomenon. In this case, the phenomenon to be modeled is "courtship" (framed entirely as a man's romantic or sexual pursuit of a woman). Other episodes contain references to a range of phenomena to be modeled mathematically: the dispersion of gas in a subway car, a seating map for a wedding, card-counting algorithms for blackjack, the hierarchical behavior of social groups, how neurons work together to create human consciousness, and how strands of spaghetti break when you bend them, among other topics.

In the mathematical model of courtship as reported by Charlie, the mathematicians have clearly made a host of assumptions, including what a date is and should be, how one shows affection, who does the gift-giving and who does the gift-receiving, what one's goals for dating are, and the like (assumptions that are heteronormative, sexist, and Eurocentric), not to mention assumptions that all men/gift-givers and all women/gift-receivers have the same interpretation of and appreciation for the particular gifts that might be chosen. Only one of these assumptions is explicitly stated: that the man's goal is the "female's (pause) receptiveness."

Mathematical models always have assumptions. The problem here is that Charlie describes only one of the assumptions, ignoring the other possible assumptions that could be made, and that were undoubtedly made by the mathematicians responsible for this model. Where *NUMB3RS* could have taught us a lesson about the values and assumptions that are embedded in every mathematical model, the series instead reifies these values and assumptions, and upholds an image of mathematical activity as objective arbiter. In the mathematics of romance as represented here, "values are built into an ostensibly value-free mathematical framework, which thus provides 'scientific' justification for the decision desired" (Martin, 1997, p. 161). The use of mathematics allows Charlie's assertions to be seen as objective fact, when in reality they are based on a host of untested and perhaps inappropriate assumptions.

Again and again, the writers of *NUMB3RS* present Charlie blithely making assumptions and treating his mathematical models as mirrors to reality. In another example, Charlie presented Don with a mathematical analysis pointing to a high probability that a particular suspect committed the crime. When Don seemed reluctant to accept this analysis as proof of guilt, Charlie asked his brother, "What's the difference between my math and a partial fingerprint?" (*Judgment Call*, 26:51). His brother is swayed by this argument and admits that there should be no clear distinction between mathematical reasoning and other, more traditional forms of evidence. In fact, on several occasions Charlie's mathematical chains of reasoning, built on tenuous assumptions that are not seriously investigated, trump the more typical forensic evidence. In *NUMB3RS*, the traditional crime drama trope of the competent professional is extended to include mathematicians and, surprisingly, the mathematician is presented as even more competent than the FBI agents, forensic scientists, coroners, and other professional crime-fighters.

Vignette 5 also exemplifies Charlie's tendency to "black box" the mathematics he uses. Several times in each episode, he is usually called on to explain a mathematical technique or problem solution. He does this in one of two ways. In some cases, as in Vignette 5, Charlie does not report on the reasoning underlying the finding. He just reports the result of the mathematical analysis, and the other characters must take it on faith that these results are accurate. The second way that Charlie typically explains his mathematical techniques is through the use of metaphor (as in Vignette 2 when he describes flock theory by comparing social networks of people to flocks of birds). These metaphoric explanations provide the reader with a general sense of the mathematical strategy – a big picture view. But the details – the underlying assumptions, the step-by-step mathematical reasoning – are hidden from view of the characters and of the show's viewers. By hiding the details of mathematical reasoning behind the metaphors, the script positions viewers as incapable of understanding these details. We, and the FBI agents, have to trust Charlie's authority on these matters.

To summarize this discussion of the use of the term "mathematics," "mathematical," and so on, *NUMB3RS* portrays mathematical activity quite narrowly, as a complex activity based on using formulas, algorithms, and other mathematical tools to solve very difficult problems. In addition, mathematical activity is most closely associated with the work of professional mathematicians and scientists, and is rarely associated with the activities of the FBI agents or other characters in the show. The scriptwriters and producers of the series seem to belie their tagline's claim that we all use math every day.

Still, as I have already argued, this method definitely undercounts the presence of mathematical activity in the series, since the word "mathematics" is not always used even when the action on screen is clearly mathematical. And, this method has presented just one interpretation of the underlying messages about mathematics. For these reasons, in the remainder of this chapter I present a brief analysis of a pilot interview study in which participants watched and commented on an episode of *NUMB3RS*. The process of "reading" popular media is informed not only by the content and structure of the media itself, but by the experiences and insights that the viewer brings; therefore, we must look to viewers to understand the variations in possible readings that they may take (Cortés, 2004).

Methods: Audience Response Study

In an individual interview, each participant viewed "Convergence," an episode of *NUMB3RS*, to discuss the mathematics that they saw. During the interview, which occurred while viewing the episode on a DVD, the participant was asked to press pause on the DVD player when he or she saw or heard someone "doing or using mathematics." Once the tape was paused, the participant was asked how he or she had identified that someone was doing or using mathematics. Play was resumed and participants were asked to stop again when they judged that the mathematical part was finished. When they did so, they were asked to describe what had happened mathematically in the clip, and to reiterate how they had identified that someone was doing or using mathematics.

Each participant watched the entire episode during the interview. To make sure participants would not forget the interview task, if a participant did not identify any mathematical moments for a span of 5 min, the interviewer would remind them of the task. Participants were assured that the reminder was a part of the interview process and not an implication they had missed any mathematical moments. Once the episode had concluded, participants were asked to reflect on what they had seen, and to describe what mathematics is, and what it isn't.

This particular episode was selected because several mathematical topics were discussed, and multiple characters were engaged in mathematical activity. This was important because if only the professional mathematicians seemed to be engaged in mathematics in the series, then the interview would not provide insight into how our participants viewed non-mathematicians' activity, and whether they would identify non-mathematicians as doing mathematics. The episode selected also included several ambiguous scenes, in which I thought there might be some variability in terms of whether participants identified mathematical activity or not.

Convergence follows the FBI team as they attempt to solve a string of home invasions. Mathematically, Charlie contributes a "data-mining algorithm" and also presents an analysis of the trajectory of a bullet to help the FBI agents to locate a bullet that had been shot into the air and that might provide useful forensics. In addition to the plot elements involving the crime, a subplot shows Charlie debating another mathematician, who presents a lecture to disprove one of Charlie's theorems.

For this pilot study, our participants were three Canadian adults. Their ages ranged from 19 to 20 and all were white, middle-class students at Canadian universities. Two of the participants were majoring in science-related fields, while a third was majoring in business.

Who Does Mathematics in NUMB3RS?

Overwhelmingly, participants identified Charlie and his academic colleagues as doers and users of mathematics. This is unsurprising since, of course, the premise of the series privileges Charlie's mathematical activity as his contribution to the FBI investigations. Of the 16 distinct types of mathematical activity that interview participants identified in the show (listed in Appendix), 14 were attributed to Charlie and his colleagues (an astrophysicist, a computer scientist, and a mathematician) and only two were attributed to non-mathematicians. The only non-mathematicians to contribute to the mathematics of this episode were David and Colby, two of the FBI agents. Their mathematical contributions came when they used the results of a program designed by Charlie to determine the landing spot of a bullet fired into the air. Even their mathematical work was directed by Charlie rather than being their own independent contribution.

Given the nature and plot of the series, it is not surprising that professional mathematicians were depicted as engaging in mathematics more frequently than other people. Some excerpts from the show depicted non-mathematician characters engaging in activities that I as an analyst considered mathematical – but the interview participants did not identify them as such. Analytically, it is helpful to describe both what interview participants described as mathematical and what they did not (Stevens, 2000). Below, I describe two examples of activity I consider mathematical, but that our participants did not discuss.

The first example was already discussed in Vignette 1. Charlie had taped some calendar pages to a wall, and circled dates in red and blue marker. Colby, an FBI agent, mentions that he had been looking for a "timeline pattern" but that "we didn't come up with anything." Charlie goes on to explain that he used a "Fourier analysis" to obtain his results, and describes the pattern he found. In our interviews, Charlie's work, and not Colby's, was described as mathematical.

There are aspects of the television production that might partially explain why participants did not identify Colby's work as mathematical. For one, the script and camera work do not emphasize Colby's work here, nor does he go into any detail about what he did to look for a pattern. Participants may have felt there was not enough information to decide whether Colby's work was mathematical. Since

Charlie uttered the words "Fourier analysis," a phrase that "sounds mathematical," participants may have decided that was enough. In addition, when describing what mathematically was happening in that clip, participants may have glossed over Colby's work – as one brief utterance that was not followed up – and focused on the main event of the clip, which was Charlie's explanation of his data-mining analysis and the timeline pattern he found.

In the second example, the team had located a bullet they believed to have been fired by the home-invaders during a carjacking. They entered characteristics of the bullet into a database to match it to bullets used in other crimes. Two FBI agents were shown watching a screen as a number of images of different bullets flash by, accompanied by a series of numbers. A large image of two bullets came up on the screen when the computer found a match.

Although it is not clear how the bullets were "matched," it is probable that the computer program conducted statistical analyses to find a bullet with a high probability of having been fired by the same gun. Once again, the script did not discuss the method by which the bullet was matched, providing little information for viewers to determine if they were watching a mathematical process. No participant identified this moment as a point where someone was "doing or using mathematics."

These two examples highlight the fact that the script itself does not explore the mathematical activity of non-mathematician characters on the series, and that our participants, despite unanimously declaring in their separate interviews that mathematics is "everywhere" and used in many areas of life, were not likely to identify routine police work with mathematics.

How Do Viewers Decide Something Counts as Mathematics?

Participants mostly agreed on which clips included people doing or using mathematics.[1] Each participant identified between 20 and 25 excerpts from the show as being related to mathematics. Most of the variation in numbers occurred when one participant might consider a stretch of interaction to contain two distinct topics and therefore two clips, whereas another might consider them to be just one.

When participants were asked how they had identified particular clips as mathematical, they often cited key words used in the script. These words included "math" and "mathematical," and a host of others. The words participants mentioned are listed in the table below. Participants explained that some of these words and phrases were familiar to them from prior study of mathematics, and some "sounded like math."

[1] One participant, Marcus, seemed to interpret the charge slightly more broadly than the others, and towards the beginning of the interview, pointed out several instances in which people used number words (such as saying there had been *seven* home invasions so far). However, as the interview progressed he began to mention more substantial mathematical interactions.

Mathematical words and phrases	
Math	Mathematics
Mathematical	Parallel
Algorithm	Asymptotics of random matrices
Convergence	Patterns and sequences
Fourier analysis	Emergence theory
Matrices	Quantum
Frequency	Analysis (e.g., simplistic analysis)
Velocities	Angle
Probabilities	One-dimensional lattice
Trilateration	Nodes (word from physics)
Complex database	Asymptote
Variable	Data mining
Number (or any number, e.g., one, two, three, etc.)	Commonalities (in the context of the data-mining algorithm)
Equation	

In addition, participants explained that when they saw mathematical symbols, diagrams, and equations on the screen, these images helped them decide that the action on-screen was mathematical. These images sometimes showed up in the action of the scene itself – e.g., when Charlie or a colleague was writing equations on a chalkboard – or they sometimes showed up in the background of a series of images that were providing a backdrop to a mathematical explanation. For example, participants mentioned that they recognized symbols like variables (especially x, y), cos (meaning cosine), sin (meaning sine), Σ (the Greek letter *sigma*, used for sums), and an equation $1+1=2$.

Thirdly, participants pointed out that a character's use of mathematical tools helped them to decide that some action was mathematical. For example, all three participants called attention to the graphing calculator used by two FBI agents. They reported using the calculator as evidence that the agents were doing mathematics. Other mathematical tools included computers and calendars.

Finally, comments made by participants indicated that in ambiguous circumstances, they were likely to identify some moment of interaction as mathematical purely because one or more mathematicians were involved in the discussion.

> It just sounded like math, like I recognized some of the words as being mathematical but mostly I didn't understand what was going on. And it's a conversation between two mathematicians. There's always a subtext of math there (laughs). (Julie)
>
> When he was talking about "convergence," I assumed that's something to do with math or physics maybe… since I know he's a mathematician, and if you tell him that his work is wrong, I assume it has something to do with math. (Olga)
>
> If they're mathematicians and they're talking about things I don't know, I assume they're talking about math. (Olga)

In these quotes, our participants described their propensity to believe that mathematician characters were talking about mathematics if the participants did not understand the dialogue on the series. This issue did not arise in our discussions of the FBI agent characters on the series, although there may have been moments when some of the technical language used by the FBI agents was confusing for viewers.

Participants probably drew on their own past experiences in school mathematics to help them decide whether someone was doing or using mathematics, as well as contextual clues from the episode itself. Mathematics instruction probably played a role in introducing participants to certain key words (e.g., equation, formula, algorithm), symbols (*cos, sin, Σ*) and to certain kinds of problems as mathematical. For example, one participant mentioned that Charlie's attempt to locate where a bullet had fallen was similar to problems she had solved in physics classes in the past.

Learning in Out-of-School Time: Mathematics and Media

This chapter has focused broadly on representations of mathematics in movies and television, and more specifically on the television series *NUMB3RS*, to demonstrate the relatively narrow conceptions of mathematics represented in popular media. The television series *NUMB3RS*, with its tagline "we all use math every day," might seem to present a broad view of what mathematics is, and who is mathematically competent. Because of the many examples of mathematical analysis in the show, this and many other film and television representations may indeed help people see that mathematics can be used to analyze virtually any phenomenon. However, the series may also reinforce some more problematic ideas: that mathematicians are precocious geniuses and that their work is far beyond the understanding of ordinary people; that mathematicians are "geeks"; and that mathematics provides value-free, objective tools to analyze complex and messy human problems.

This series is part of a cultural milieu, a societal curriculum (Cortés, 2004) that portrays mathematical activity in particular ways. Viewers at home learn something about what counts as mathematics as they watch, and children who engage in *NUMB3RS*-related activities in school learn more than the mathematics. While this is just one cultural representation of mathematics, it is embedded in and akin to the many representations highlighted in the Reinhold (2007) and Knill (2007) websites. These cultural representations undoubtedly influence how people think about and engage in (or avoid) mathematics. They interact with people's school experiences, as in our interviews in which participants drew on their own experience doing mathematics in school to identify mathematical moments in the show. And yet precisely what these cultural representations mean to their audiences is unclear.

This chapter, therefore, represents a small contribution to a larger project of understanding how people make sense of media portrayals about mathematics. In today's globalized and technological world, numeracy is deemed to be critically important for all people, yet unquestioned experts like Charlie contribute to many governmental and corporate decisions that affect ordinary people. Do shows like *NUMB3RS* influence school-age students to learn more, or less mathematics, or to engage with it in particular ways? Do they influence people to maintain narrow views about what counts and who can do mathematics? Or, is the slogan "we all use mathematics every day" the lasting message that viewers keep? Given the examples and interviews presented here, it is unlikely that the show contributes to broadened, more equitable notions of mathematics and mathematicians. Further research into people's views of mathematics can shed much needed light on this issue.

Author Note

This work was supported by the National Science Foundation under Grant No. SBE-0354453 to the Learning in Informal and Formal Environments Science of Learning Center, as well as by the Connaught Fund at the University of Toronto. Any opinions, findings, and conclusions or recommendations expressed in this material are those of the author and do not necessarily reflect the position, policy, or endorsement of the National Science Foundation or the Connaught Fund.

The study design and interview was developed in collaboration with Reed Stevens, who also provided valuable feedback on earlier versions of the chapter. I would also like to thank Rubén Gaztambide-Fernández, Lance McCready, Joseph Flessa, and Ray McDermott for their critical feedback.

Correspondence concerning this article should be addressed to Indigo Esmonde, Ontario Institute for Studies in Education, University of Toronto, 252 Bloor St. W., Toronto, Ontario, Canada, M5S 1V6. Electronic mail may be sent to indigo.esmonde@utoronto.ca.

Appendix

Table 4.1 Mathematical activities that interview participants identified in the episode

Character	Mathematical Activity

Charlie	• Construct a data-mining algorithm
	  ◦ Find a pattern in dates of crimes (done by data-mining algorithm)
	  ◦ Discuss commonalities between the crimes (commonalities located by data-mining algorithm)
	• Write mathematical symbols and equations on a chalkboard
	• Discuss ants and "emergence theory"
	• Create a program to graphically display probabilities for finding a bullet discharged into the air
	• Discuss general mathematical topics
	• Construct an equation that fixes the flaw in his work
	• Explain how cellphones are located using GPS chips
	• Create a new theory – "math of the brain"

Amita (astrophysicist and computer scientist) and Larry (astrophysicist)	• Discuss an upcoming lecture by Charlie's rival • Discuss the flaw in Charlie's work
Penfield (Charlie's rival)	• Discuss the flaw he found in Charlie's work (the "Eppes Convergence") • Discuss general mathematical topics • Discuss "deep current sets" and resolve a problem with Charlie's data-mining algorithm
David and Colby (FBI agents)	• Implement Charlie's program for locating bullet, by inputting muzzle velocity variable • Discuss relationship between muzzle velocity, gun type, and bullet type

References

Brenner, M. E. (1998). Meaning and money. *Educational Studies in Mathematics, 36*(2), 123–155.

Cortés, C. E. (2004). Knowledge construction and popular culture. In J. A. Banks & C. A. M. Banks (Eds.), *Handbook of research on multicultural education* (2nd ed., pp. 211–227). San Francisco, CA: Jossey-Bass.

Greenwald, S. J. (2006). Complex *NUMB3RS*. *Notices of the American Mathematical Society, 53*(7), 741.

Guberman, S. R. (2004). A comparative study of children's out-of-school activities and arithmetical achievements. *Journal for Research in Mathematics Education, 35*(2), 117–150.

Knill, O. (2007). *Mathematics in movies*. Retrieved January 16, 2008, from http://www.math.harvard.edu/~knill/mathmovies/

Lave, J. (1988). *Cognition in practice*. Cambridge, MA: The University Press.

Martin, B. (1997). Mathematics and social interests. In A. B. Powell & M. Frankenstein (Eds.), *Ethnomathematics: Challenging eurocentrism in mathematics education*. Albany, NY: SUNY Press.

Masingila, J. O. (1994). Mathematics practice in carpet laying. *Anthropology & Education Quarterly, 25*(4), 430–462.

Nunes, T., Schlieman, A. D., & Carraher, D. W. (1993). *Street mathematics and school mathematics*. Cambridge, UK: Cambridge University Press.

Reinhold, A. G. (2007). *The math in the movies page: A guide to major motion pictures with scenes of real mathematics*. Retrieved January 17, 2008, from http://world.std.com/~reinhold/mathmovies.html

Saxe, G. B. (1991). *Culture and cognitive development: Studies in mathematical understanding*. Hillsdale, NJ: Lawrence Erlbaum Associates.

Scribner, S. (1984). Studying working intelligence. In B. Rogoff & J. Lave (Eds.), *Everyday cognition: Its development in social context* (pp. 9–40). Cambridge, MA: Harvard University Press.

Stevens, R. (2000). Who counts what as math? Emergent and assigned mathematics problems in a project-based classroom. In J. Boaler (Ed.), *Multiple perspectives on mathematics teaching and learning* (pp. 105–144). Westport, CT: Alex.

Chapter 5
What Counts Too Much and Too Little as Math

Reed Stevens

Stupid Gerry

Gerry was a Vietnam veteran who had bounced around VA hospitals for a few years before he found his way to the school for adults where I was working. My job was essentially a kind of educational triage. My role at the school was to get students into educational and training programs and to help them establish themselves in these programs, since most had been out of school for a long time.

Gerry's goal was to work in computer-aided drafting. I needed to prepare him to pass the entrance exam for the school where he hoped to receive this CAD training. The entrance exam involved mostly mathematics problems, involving measurement, geometry, and arithmetic with decimals and fractions. To me, CAD work seemed like a pretty good direction for Gerry. He already could draft beautifully by hand. Measurement was no problem at all for him. Nor was intuitive geometry of common shapes. We worked on the addition and subtraction of decimals, and we both came to have enough confidence that he could handle these sorts of problems.

The challenge was fractions. His intuitive feel for fractions was pretty good, seemingly grounded in his understanding of measurement. When he could use a ruler and was dealing with problems with fractions of like denominator, he could solve the problems, albeit a bit slowly. But when he could not use the ruler, and when he faced problems of unlike denominator, he struggled and could almost never solve the problems. I worked with him many afternoons on these problems, trying everything I could think of to help him understand and "get" the standard algorithm. Weeks of practice and encouragement did not lead him to autonomous performance

R. Stevens (✉)
School of Education and Social Policy, Northwestern University, Evanston, IL, USA
e-mail: reed-stevens@northwestern.edu

B. Bevan et al. (eds.), *LOST Opportunities: Learning in Out-of-School Time*, Explorations of Educational Purpose 23, DOI 10.1007/978-94-007-4304-5_5,

of the kind he'd need to pass the test. So I tried different strategies. I used visual materials to help him see the fractions. Those visual strategies helped him estimate the sums and differences but not get the exact answer for which he needed an algorithmic approach.

One day, Gerry's frustration boiled over. He was trying to work a problem when he suddenly slammed his hands down on the table with a thunderous bang. He knocked his chair backward and fled the room. I was shaken and so were the other students who were working in the open room. I left the room for a while, and when I returned, Gerry was there again. He was working on a piece of paper but not the one with the fraction problems. I approached him and asked how he was doing, and he pushed the paper over toward me. On it was a grotesque caricature of himself with a caption in beautiful lettering that read, "Stupid Gerry can't do fractions."

That afternoon, Gerry and I did not do any more fraction problems. We just talked. That afternoon, Gerry told me about going to Catholic school and having his hands beaten with rulers by nuns when he repeatedly failed to accomplish an assignment. He also told me about being exposed to Agent Orange in Vietnam, the source of the disfigured skin on his face. And he told me that whenever he tried to solve the fraction problems, both of these memories rushed to his mind. I left work that day stricken.

When I returned to work after the weekend, I had resolved to do everything I could to help get Gerry admitted to the CAD training program at the local school. I also decided to try a different approach. I went down to the school and met with the curriculum director, a perhaps impetuous move for a 22 year old. I told him about the situation. The school was small enough that he knew about Gerry and his failure on a pretest that had set us on this course. I told the curriculum director that I wanted to learn a bit more about how CAD work made understanding these sorts of fraction problems necessary. I framed it for the curriculum director in terms of motivation, saying that I thought that if I could show Gerry how these were necessary to being a successful CAD draftsperson, it might have an effect on his performance on these problems.

So I sat with the curriculum director, and he demonstrated the CAD program. As he showed me the interface and took me through some of the standard tasks, I had this encroaching feeling that *these fraction problems were not to be found within these CAD tasks*. In fact, I was able to see ways that the CAD system had been set up to *avoid* these problems, as in the case of addition of fractional lengths. The computer allowed you to drag out each segment to a particular fraction and then "add" another by dragging from the end of the prior segment. As you added segments in this way, the CAD system did the addition automatically and represented the sum in the scale that was set for that particular drawing. In the context of my discussion with the curriculum director, I did see how estimating sums and differences of fractions with unlike denominators would be useful, primarily to avoid errors in the dragging and stretching operations that produced the sums. But Gerry could estimate; he just couldn't calculate the fractions precisely with an algorithm. Algorithms are, of course, what computational tools are made to do.

The curriculum director of that school was not an unreasonable person. When I remarked about the fact that it seemed that Gerry and others could probably get along fine without being able to solve those sorts of fractions problems, he agreed… mildly. I suggested that maybe the test should be determining if he could estimate fractional sums and differences because that skill seemed directly useful for this work. As we talked, the curriculum director produced two rationales for keeping things as they were. His rationales were, in essence, "Well, we can't change the test" and "Well, it is just important to understand the basic facts."

What Counts as Math Times Two

These events took place more than 20 years ago. Since then, I've thought about Gerry's story many times and retold it more than a few. I have often highlighted the charged associations of corporal punishment and Agent Orange that elementary mathematical practices brought up for Gerry. Those associations are surely the most dramatic features of this story, but I see them now as a by-product of two more conceptually central issues. Both are issues I've investigated over my career, and both were on the scene in these events with Gerry. These issues are the focus of this chapter and highlight two distinct but also related senses of what counts as math.

The first sense of what counts was suggested by those remarks from the curriculum director. One of his rationales, "Well, we can't change the test," asserted that the inflexible object in this scenario was the test, and Gerry and others like him needed to bend to it or potentially be broken by it. His second rationale about the importance of understanding "the basic facts" suggests a moral position operating in this claim. In a very real sense what that curriculum director was saying was that while he was willing to grant that Gerry might not need to solve those sorts of fraction problems in the way the test demanded, he still *should* be able to solve them. Years later in the context of some studies of mathematical practices among engineers, I heard something that would remind me a lot of that comment from the curriculum director. We interviewed a civil engineer whom we watched design a roadway system. My colleague Rogers Hall and I looked for and could not find the recognizable symbolic manipulations and algorithms of integral calculus or other higher-level mathematics. In a thoughtful conversation with the engineer, he acknowledged that he did not do that work—his computer did, and it was largely invisible to him—but that it was still important to "understand the raw calculus." This perspective, or at least its assertion, was for him something of a moral precondition of being an engineer, just as it was seemingly a moral precondition for the curriculum director that students admitted to his school could solve fraction problems. So one sense of "what counts as mathematics" is as a moral measure of a person or an activity, aside from whatever practical function it might have in the activities of particular people.

The second sense of what counts is the practical sense, where "counting" implies useful and consequential to the outcome of peoples' everyday activities. Here, I do not mean "everyday" in contrast with the disciplinary, elite, or expert practice; these are everyday activities as well. This practical sense of what counts comes across in the scene with the curriculum director and me trying to find particular types of fraction problems inside the work tools and problems of computer-aided drafting. We really couldn't.

Together, the curriculum director and I had identified a gap between that CAD work and those fraction problems. I came to see that gap as a more general phenomenon—that gaps between math that counted for getting life's work done and math that counted for sorting people into and out of educational contexts were common. And this led me back to graduate school. Working with Rogers Hall, we undertook a project that sought to go beyond identifying the existence of a gap between school mathematics and the mathematics that does *work* for people. What we were seeking to understand were the "forms and functions" (Saxe, 1991) in which mathematical ideas and tools were used in and out-of-school life—to understand, as our NSF-funded project was called, "math at work."

Math and "Our Language"

In the field studies Rogers Hall and I conducted research on how mathematical tools and ideas were used in high-status professions (Hall & Stevens, 1995; Hall, Stevens, & Torralba, 2003; Stevens, 1999; Stevens & Hall, 1998) and compared this to mathematical work in middle school project-based classrooms in which students fictively worked together in teams as these kinds of professionals. These professions included architecture, engineering, and field biology. This work was built upon prior studies of "everyday math." What was different about our studies was the focus on high-status professions, which enabled us to engage with arguments circulating in national policy debates about what "we" "needed" to survive and thrive economically. The argument was (and remains) that unless US students learned more advanced mathematics, thus preparing them to participate in scientific and technical professions (of the kind we were now studying), our nation's economic and political well-being would be threatened. This viewpoint is still circulating. A recent version of the same perceived threat can be found in the report "The Gathering Storm" (Committee on Prospering in the Global Economy of the 21st Century, 2007) and in other policy reports.

None of these arguments for more advanced mathematics among K-12 students was informed by direct empirical studies of how mathematical tools and ideas were used in scientific and technical professional work practices. At best, the reports that made these claims were written by committees on which sat practicing scientists, mathematicians, and engineers who drew upon their own experiences to support these arguments. Against this background, we decided to go and see for ourselves, conducting extended ethnographic fieldwork on the professional mathematical practices of engineers, architects, and population biologists.

I led the field study of architectural work at one Bay Area firm. The architects at this firm worked closely during my fieldwork with all sorts of engineers, and I followed their activities as well. On a chilly February morning in 1996, I ended my first day of fieldwork climbing a ladder through a small hatch onto the roof of a school in San Francisco. I was on the roof with Jackie, one of the principal architects at the firm I was studying. She needed to take a look at a drainage system. It had rained earlier that day and puddles of water had settled in pitted regions all over the 40-year-old roof. After a few minutes, Jackie realized the roof had no drainage system and that adding one would be a major part of the renovation her firm had been hired to do. As we walked on the roof, Jackie looked in specific places for specific things, which at the time were mysterious to me, and as she did, I asked her questions about the work she was doing.

What I began to learn that very first day on that roof were some basic facts about how 'mathematics'[1] functions in the everyday practice of architecture. On the roof, Jackie was discovering that the only viable drainage system for this wide flat roof would be one that slightly pitched the roof so that water could drain down through an existing hole in the center of the building. Using this hole for drainage meant that they wouldn't have to add expensive gutters and drainpipes to get the water to the ground and away from the building. As Jackie related this scheme to me, I recognized a math problem involving the familiar concept of slope. As a former math teacher, I had conditioned myself to identify "real-world" math problems, which were always more rare than I expected. From my field notes blurred by the rain that had begun to fall, it appears I asked Jackie what the "right slope" was for this type of problem. Then she told me that this was not the way they approached the problem. Talking in unfamiliar terms, she explained a "performance standard" and how the "contractor will specify slope to drain from specific roof points" (*Field note, 2/7/96, 4:30 pm*). My field notes here had a series of question marks bunched together, my shorthand for some confusion. The confusion I came to identify retrospectively was that the familiar term "slope" was there, and in a sense, it could be connected to the meaning I knew from mathematics of "rise over run," of a quantified measure of steepness. But in another sense, it had a different meaning because as I came to learn through my fieldwork, the way it was calculated, the way it was embedded in standardized devices, how it was seen in the built environment, and the representational forms in which it was expressed *were all different.*

This issue of same but different was made explicit for me the first time I attended a big meeting about the architecture firm's major project during my fieldwork, the renovation and seismic upgrade of two historically significant public libraries. I was introduced to the team that included the architects from the firm, engineers, cost estimators, and two historical preservation architects. At the end of the meeting, I was in conversation with one of the historical preservation architects, who, in reference to my expressed interest in the mathematical aspects of their work, said, "Every two minutes we were saying something math-related, but because we were

[1] I have placed the single quotation marks around 'mathematics' once in this chapter to highlight the issue of whether and to whom these practices were rightfully counted as mathematics. For readability, I will drop the quotation marks in the rest of the chapter, but that question "Is it and to whom?" should linger in the reader's mind throughout (Lynch, 1991; McDermott & Weber, 1998; Stevens, 2000).

talking in our language you wouldn't have known that" (*Field note, 4/4/96, 11:51:36 a.m.*). This quote nicely captures one of the trickiest and persistent issues in fieldwork studies of this kind: finding math or other subject matter topics "in the wild" means trying to find it in language, embodied practices, and material forms that are sure to be different than in schools or stereotypical imagery.

Architect A and Architect B

Mathematical practices in architecture are different from those of school mathematics, but one might assume some internal consistency, within architecture, as to what counts as math. This assumption was dashed during my field study of the architecture after I met two architects. Let's call them Architect A and Architect B. Architect A at some point told me that, "I use math all the time." Architect B at some point told me, "I don't do math. That's why I hire an engineer." This presented quite an analytic puzzle hearing such diametrically opposed perspectives on the presence of math in the working lives of architects. The seriousness of this analytic problem is perhaps enhanced if I reveal that these seemingly conflicting utterances were made by one and the same architect. Architect A was Architect B.

These different utterances set up a choice of analytic strategy. As an analyst, I had the option of disregarding my informant as inconsistent. Alternatively, I could treat both utterances as sensible in relation to whatever referent they had for the architect at the time he made those statements. Choosing this second ethnographic perspective on my informant set for me the analytic and empirical tasks of discovering in which situations the first comment was accurate and in which situations the latter was accurate. This leads me to describe, in the next section, how mathematical work is both distributed (to other people and tools) and expressed in mundane but diverse forms in daily work, what I called *form diversity*.

Math Is Distributed and Expressed in Diverse Forms in Everyday Life

In this section, I offer examples of the ways that mathematical work is *distributed* and diversely *expressed*—often in forms unfamiliar in school mathematics—in architectural practice. These terms roughly capture two key differences that I will highlight about school math as it is typically enacted in mathematical work outside of school. First, school math is largely organized to be accomplished by individuals. In school, individuals' performances are compared and assessed. In contrast, mathematical activities outside of school are often *distributed* among people and other tools. Second, school math performances are *required* to be expressed in narrow range of forms. Out-of-school mathematical work is expressed in a much wider diversity of forms. Just as important is the fact that forms of expression in school math are not well aligned with those outside of school. In school mathematics, the narrow range of common forms includes two-column proofs in geometry, written symbolic algorithms, coordinate graphs, and (perhaps most prominently) equations. To get credit in school math most of the time, a specific problem must be solved using a unique, prespecified form of solution.

Distribution of mathematical activity is a feature of architectural practice. Mathematics is distributed within a firm, with principal architects being more practiced in some forms of mathematical activity, while associate- and entry-level architects are more practiced in other forms. Even more prominent evidence of distribution of mathematical work is evident among the architects and other professionals they hire to help them move projects forward, as exemplified in architect A's statement, "That is why I hire an engineer." In every project I observed, engineers worked with architects as consultants, subcontractors, or project team members.

Engineers are the most frequent collaborators in architectural projects to whom mathematical work is distributed, and there are many reasons that architects hire engineers and distribute mathematical work to them. The most obvious is that divisions of mathematical labor reflect divisions of knowledge and specialization. That was surely the case across architects and engineers. As one of the principal architects named Charles told me in an interview about their major project during my field study, he probably *could* do structural calculations and in fact he had learned to do them in architecture school, but it would take him a long time and his judgment would be less assured than that of a consulting structural engineer. A second explanation for distributing mathematical work to engineers is to *distribute liability*, which means that if something goes wrong like a structural support system of a building in an earthquake, the engineers will share the legal liability with the architects. Distributing work in professional practice, in general, means distributing accountability. It is also the case that the state and local "code" for architectural work require that for certain design features, engineers must be consulted and "sign off" on those design features; distribution is required by law.

Distributions of mathematical work among architects and engineers were most clearly visible when the architects proposed a "scheme" for a building, and the engineers took design drawings away to their offices to determine if the proposed scheme would "calc out." These calculations were literally distributed right out of the architects' offices. This example points to a general feature of distributions of mathematical work—that of trust. The architects had to trust the calculations that the engineers did; they did not have to do the mathematical work or have the mathematical know-how themselves. This suggests that trust and know-how can be seen in a kind of inverse relationship; if someone is prepared to fully trust the mathematical labor of another person to whom you have distributed it, you will need to know little or nothing about the actual mathematical work. To the degree that you are not willing to trust a person's mathematical work, then your own specific mathematical know-how becomes decidedly relevant. This is really a general property of the social distribution of knowledge and not unique to mathematical activity; it is true when we go into a restaurant and eat food prepared behind closed doors, when we leave our children with a babysitter, or when we have our cars fixed. Ingrained beliefs may keep us from applying this perspective to mathematical know-how.

What can be said of distributions of mathematical labor to other people can also be said of distributions of mathematical labor to tools and machines like computers, with some important differences. Earlier, I told a story from my fieldwork experience in which a civil engineer endorsed the values of "raw calculus" in the same conversation in which he acknowledged that his computer handled all the calculus-like mathematical labor when he designed roadways, since these computations are built into CAD tools. Computational tools "black box" (Latour, 1987) many calculations and shift the amount of algorithmic calculation required of users; this shifting is by design. As calculators and spreadsheets redistribute calculation, so too do CAD systems. CAD systems have an additional property that suits the visual style of architects; they shift the mathematical labor to a visual form of activity; users manipulate geometric shapes and objects directly. The CAD system calculates changes and then displays relevant quantities that result from the manipulations. Users must know how to read quantities, must know what they mean, and must relate them to other quantities, but they rarely if ever do the calculations themselves. They trust their computers to do this work.

This example brings me to the second term that is helpful for characterizing mathematical practices in out-of-school settings—*form diversity*. I am using this term to describe the observably diverse forms that people use to accomplish mathematical work in out-of-school settings. These forms don't typically resemble those of school math problems. The purpose of this term is to expand what we count as math, to look beyond recognizable symbolic forms to others in which mathematical work is accomplished. All of us recognize the correct execution of pencil and paper algorithms as mathematical. So too do we recognize feats of mental calculation. But also mathematical, we should insist are: gestures of the hands that represent comparative quantities; a carpenter's precise use of a bundle of sticks stretched and then marked across pairs of sticks to make analog measurements; and speech that formulates a quantitative relationship or works out an estimated value for a calculation, even if that speech occurs in informal, nonspecialized language.[2]

An example of how architects' mathematical practices are expressed in forms that don't look like school math comes from their uses of CAD tools. In making project drawings with CAD, architects have a menu-based palette of Euclidean shapes (rectangles, circles, linear segments, etc.), which they transform through coordination of hand and eye, via a mouse. For example, an architect might choose a rectangle from a menu and then manually resize it to fit some desired dimensions. This is achieved by dragging the shape (e.g., a rectangle) by a certain point (e.g., a corner) and visually monitoring its change in dimensions in a dialog box. It is important to note that the process of creating objects from a geometric vocabulary and transforming them (e.g., rotate, reflect, shear, etc.) is neither a verbally explicit process nor one that is based on selections from a mathematical vocabulary; the systems use their own

[2] See Stevens and Hall (1998) for an earlier description of form diversity in mathematical work, showing how productive mathematical work is "embodied" in and coordinated across different bodily modalities and media.

vocabularies, and the work itself becomes the tacit hand and eye work of the user. Mary, one of the firm's architects, remarked to me during a field observation session and follow-up interview that "it's funny to think about all these things in CAD that have become second nature" (Field note, 9/8/96: 2:36 p.m.).

Another common situation in which I observed mathematical practices in unconventional (from the perspective of school mathematics) forms among the architects was in design and design revision. This is when architects recognize themselves as "us[ing] math all the time." These design practices typically took place over the surface of standardized architectural representations called plans, sections, and elevations. These representations were produced through an extended temporal chain of actions that began with "on-site" measurements of "existing." From these measurements, a scaled CAD drawing was made and then these drawings, when printed, were the surfaces over which architects would layer new possible designs. This layering practice involved mathematical practices and materials that elsewhere I have called a "package" (Stevens, 2000). The package consists of the underlying base drawings (e.g., plans, sections, and elevations), rolls of trace paper, a six-edged scale ruler, and a pencil. Together, the coordinated use of the package allowed architects to pose design possibilities and explore their quantitative entailments like, "Will this fit in the space?" "Is this to (building) code?" and "Would this work in our budget?"

An example of this mathematical practice comes from one of the projects I followed, a design of a winery and tasting room for a relatively small two-story warehouse space. The architects first generated ideas, in conversation with the client, of what elements would constitute the "program" (e.g., tasting room, wine production area, office spaces for salespeople, and bathrooms), and then they began sketching on the trace paper over the top of the scaled base drawing. The sketch below captures a moment in this process, with the gray displaying the handmade marks on trace layered over a CAD-generated plan, which can be seen in darker lines. Architects envision a possible "space" by drawing it on the trace paper over a base drawing (Fig. 5.1).

If a particular design idea seems plausible, architects will then typically measure the possible space on the drawing using the scale ruler. This measurement involves laying the correct edge along the drawing. The correct edge is the one that corresponds to the scale of the base drawing. With the scale ruler, the architect makes linear measurements, reading directly from the edge of the scale ruler. (The scale ruler is another example of a tool to which mathematical labor is distributed). Often, the relevant calculation is made in linear feet but just as often it is square footage. In the figure pictured, the architects made an "area count" (i.e., a measurement of square footage) on a region of the proposed structure where they considered locating office spaces for the winery's salespeople. They first calculated the overall sum of 640 square feet by adding up the subregions, took the estimated number of people in the space (i.e., 20 people), and divided to infer that each salesperson would have about 32 square feet of office space. To envision this as some version of lived space, the architects then drew a rectangle, using the trace and scale ruler, with roughly this area to see how it might work. This involved another calculation, usually mental, to decompose a number like 32 into two factors and like 8 and 4 or an approximation

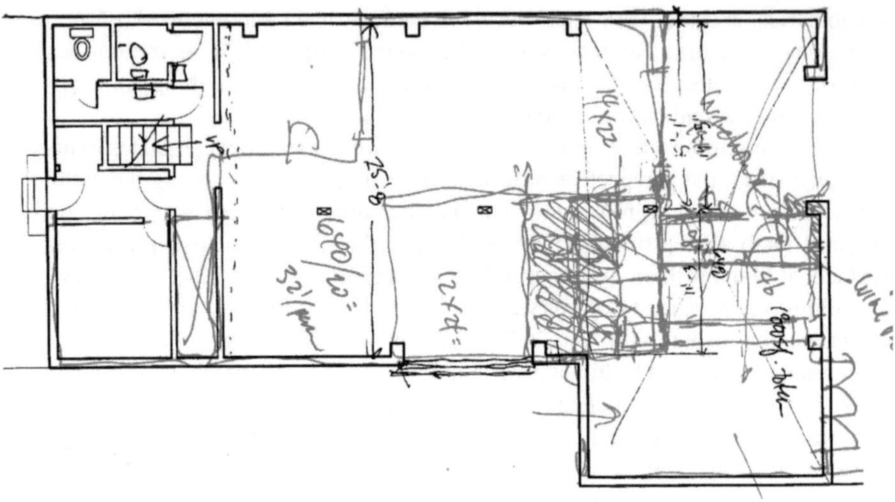

Fig. 5.1 Trace drawing (*in gray*) over base drawing (*in black*)

of square factors like 5.5 and 5.5. This allows the architect to get a sense about whether they are allocating sufficient space to a particular use. In this instance, the architect Jackie concluded that, "32 square feet per person might be enough." In my observations of the work that generated these drawings, I watched the architects move fluidly back and forth between sketches on trace, measurements, calculations, and design inference. *The distinctive quality was the fluidity.* Mathematics that could be identified as that which architects "do all the time" included measurement, calculations of area, factoring of a number into two exact or approximate factors (not as common), and frequent calculations with basic arithmetic operations of addition, subtraction, multiplication, and division. These mathematical practices were ubiquitous. But in nearly a year of a fieldwork at the architecture firm, I never saw any of the architects write down an equation and then manipulate it. I rarely saw calculations, made by the architects themselves, more complex than these I have just described. The heart of the architects' geometric practice was a deep familiarity with basic Euclidean shapes—because these are buildable—and an ability to visually and physically transform them in the extended process of architectural design.

Mathematics In-Not-As the Activity

Architects are just one of the social groups whose mathematical practices are best understood *in-not-as* the activity. Studying these social groups outside of school is a different research enterprise than studying mathematics *as* the activity—to study, for example, whether a group of "students" understand and can demonstrate successful performances on a set of already written down, curricularly specified mathematics problems in school (e.g., multidigit arithmetic, factoring, or solving differential equations). The in-not-as problematic has been at the center of a generation of "everyday mathematics" studies. These include, among others, studies of candy sellers, carpet layers, engineers, basketball players, families handling money, nurses, milk-processing plant workers, grocery shoppers, and *Weight Watchers* participants (see, e.g., de la Rocha, 1986; Esmonde et al., this volume; Hall & Stevens, 1995; Hoyles, Noss, & Pozzi, 2001; Lave, 1988; Masingila, 1994; Nasir, 2000; Nunes, Schliemann, & Carraher, 1993; Scribner, 1984; Stevens, 1999; Stevens & Hall, 1998; Stevens et al., 2006).

The collective accomplishment of these everyday math studies is to make visible forms and social functions of mathematical activity that differ from what we know from schools. These everyday math studies show quite clearly that the forms and functions of 'mathematical' activity—or whatever we call what people do with number, calculation, geometricized shapes, predictive projection, and quantitative inference—don't often directly resemble those of school math. Mathematical practices are embodied in expressive forms and bodily modalities, distributed to other people and technologies, and are embedded in the language of the locals. These differences should now be clear enough.

What remains a work in progress is how to make analytic or practical use of these differences. Can we rethink the relationship between out-of-school practices and those of school math? Can we possibly influence the reorganization of school mathematics? These issues will be the focus of the remainder of this chapter.

Dumb Ted

A reasonable instructional approach to closing the identifiable gaps between traditional school mathematics and the many ways that mathematical tools and ideas are used outside of school is sometimes called "project-based math." The idea is to enlist students in projects, often in teams, in which they take on goals—like designing something—that are not directly about performance on mathematics problems and tests but require students to learn and use mathematical tools and ideas to be successful on projects. The idea is that they learn "the math" because it helps them get something else done that is important to them and their peers. This is certainly how and why we learn many, if not most, things in everyday life. Yet, in the case I am about to present, readers will see the duality of math appearing again—on one hand as a practical resource to solve problems and on the other as ritual measure of a person's worth and intelligence.

Early in my career as an educator, I met stupid Gerry, and years later, while collecting data for my dissertation, I met dumb Ted. I met Ted during the other half of my dissertation research; the half I've discussed so far is the ethnographic fieldwork in the architecture firm. This other half of the study involved parallel ethnographic fieldwork in a project-based math classroom. The mathematical practices I report on were architectural design projects in a middle school classroom. In this classroom project, which lasted about 6 weeks, student teams of four participated in a simulated world of architects "hired" to design a research station for scientists who winter over in Antarctica. Again, the animating educational idea was that as student teams designed and tried to address the constraints and requirements for the research station set out in curricular materials, the students would pose and solve mathematical problems in order to achieve their design goals.

I followed one of the student teams in great depth, video recording all their activities during the project, interviewing them informally along the way, and collecting all their written work. In this group was a boy named Ted. His teammates were Marsha, Cathy, and Henry. As described elsewhere (Stevens, 2000), Ted was a fully vested participant in the project, taking the colead with Marsha on the design. He and Marsha had a familiar 7th-grade relationship. It was animated by frequent ritual insults that were nonetheless received and returned with a playfulness that was rarely shaming or taken as truly unkind. Marsha was a master of the form. When a student came around handing out blank worksheets for some assigned work, Marsha turned to the student as he was handing a paper to Ted and said, as one might hear at the zoo, "Please don't feed the animals." Despite this sort of banter, Ted and Marsha were very effective codesigners. They debated ideas and made changes on the basis of these debates. Later, when I organized to have professional architects come to the classroom to do a final "crit" of two classrooms' designs, Ted and Marsha's design was selected as the best of about 16 projects. The architects did not know that this was a team I was studying, so this judgment on the team's work was both expert and unbiased. All this is to say that Ted's participation in the project was full and his ability to collaborate on the design was evident.

Ted used mathematical ideas, embodied in forms other than those recognizable in school math, to debate and improve the design. As described elsewhere, one quantitatively based verbal argument and classroom-measuring demonstration of Ted's was particularly important to the ultimate design that Ted's team produced. In a prior chapter (Stevens, 2000), I called the mathematical problems that I observed in the design activities *emergent problems* and contrasted those with *assigned problems*. Assigned problems are those that come from the teacher or curriculum and are typically represented in a particular form, have explicit instructions about the form that solutions should take, and have, if not a single right answer, a very narrow range of possible right answers. Unlike emergent problems, they don't admit form diversity. In the fieldwork at the architecture firm, *all* I had seen were emergent math problems. Here in the classroom, I saw official, assigned math problems coexisting in parallel with the emergent math problems. The reason that assigned problems were here in this classroom—despite the fact that a different kind of curricular experiment was under way—was that like many research-based school interventions, we were able to displace some of typical school math activities but not all of them. Why? Because the teacher needed by law and institutional expectation to give her students a grade, which meant needing scores from tests or other assessable performances. No assessment system adequately extracted grades from the emergent problems and design activities.

One day, the emergent and the assigned crashed into each other and left Ted bumped and bruised in the process. That day, Ted was working on the team's design with alacrity but was told that he and teammate Henry needed to be working on assigned mathematical work, while the girls, Marsha and Cathy, were expected to be doing design-related work. The official piece of work involved a worksheet that asked groups to make posters that showed the relationship between area and perimeter among differently dimensioned rectangles. There was clear potential use for the knowledge this activity might provide in their architectural design projects, because the explorations on the poster could lead them to see that certain rectangles provided more square footage while minimizing material costs because of a smaller perimeter, with the optimal rectangle being a square. This group had done their poster incorrectly the first time, working as a whole group. And now, they and other groups were being told by the teacher that they needed to correct their poster to turn it in for a grade at the end of the class period in question. Ironically, this assignment that the teacher was using to assign a grade was explicitly identified as "not a worksheet" in the curriculum guide, but because the teacher needed to assign grades to students she took opportunities to use the most school math-like aspects of the project-based curriculum to assign them grades. Equally ironic, at the time of the reported events, the knowledge that the activity could provide for the design projects was no longer really useful, because by this point, all of the groups had already committed to the basic footprint of their structure (i.e., shape of the outer perimeter of the structure) and were building within it, so really the only use for the poster was for the teacher to record an assessable performance.

Ted took a look at the assignment, made a quick judgment that he could not do it, and returned to his design work, which at the moment was going well. During the class period, he occasionally picked up the math worksheet that provided instructions for the poster, announced his inability to do it (to Henry who was supposed to be doing it with him), and then returned to the design. When the end of the period arrived and the girls realized that Ted had not completed (or even started) the poster and that they could receive a bad grade because of it, they were very upset and let Ted know it. Ted said he wanted to do it but didn't know how, that he needed help. Marsha said to him, "See. That's your problem, you don't know what to do and you can't help us. It's not our fault you're dumb." Ted's audibly nonironic response was, "Yeah." Ted's reputation as someone who was unable to do what counted as "math" did not recover. He failed the pre-algebra test at the end of the year and found himself headed for "regular" math in eighth grade. What he had accomplished using math in service of design was effaced by a system that could not capture these skills and capacities, even though under the leading criterion of practical use and allowing for form diversity, they should have counted as math.

In conclusion, I offer a question about Ted's case. On what evidential basis are we to say that the math skills that Ted used effectively to coproduce the best design are any less critical than the skills of typical school math problem solving? Ted was *sorted out* for not performing well enough on typical school math problems, and there was no mechanism for *sorting him in* on the basis of his effective performance on these emergent problems. As was the case with "stupid" Gerry, do we really want "dumb" Ted's social futures overly determined by counting his performance on mathematic tasks of this kind against him.

Where Do We Go from Here?

In this chapter, I have drawn on stories from my teaching and research experience to show at least two faces of math: on one hand, the practical, useful, productive aspects of mathematical work and, on the other hand, the symbolic violence (Bourdieu, 1991) that can be carried out with mathematics as the weapon, when it serves as a ritual test of a person's moral worth, social future, or intelligence. I chose purposefully to write in a more conversational style to display that I feel these concerns personally, as a former math teacher and as a father of school-aged children and as a mathematics learning researcher. For more than 20 years, I have struggled, first as a teacher and then as a researcher, with the relationship between school mathematics and the practical contexts everywhere else in which "mathematical" tools and ideas are learned and used. I've struggled with what counts as math and what does not, with what counts too much as math and what counts too little. And the struggles are amplified each time I meet a Gerry or a Ted.

In this final section, I want to try to move forward to unravel some of the contentious and seemingly entrenched oppositional positions in this complex space. I also want to try to identify the very real limits of my own understanding and perhaps those of the broader field, which in turn may serve to help set an agenda for future work. I have organized this final section in the form of brief dialogues. These dialogues reflect conversations I have had over the years with others about these issues.

Voice #1: Yes it's too bad what happened to Gerry and Ted, but some people just aren't good at some things and for these guys it is math. It's tough but it's reality.

Reply: The point of these stories will have been missed if readers have taken away the view that Gerry or Ted was actually bad at something that demonstrably mattered, or mattered more than what rational analysis and direct observation suggest should matter as much or more in these situations. In each case, there were two available senses of what counted as math, and in both cases, what both Gerry and Ted *could do* and what contributed to productive activity went largely unrecognized because of what they *could not do*—which had no clear demonstrable use. This led to both being labeled and attached to unfortunate social identifications (i.e., "stupid" and "dumb").

Voice #2: School math may not be immediately useful but it is "obvious" that people need it in everyday life and in a lot of occupations. They are learning it now and they will use it later.

Reply: It is surprising how often I have heard the word "obvious" in statements like this. But it is not obvious at all that forms and functions of school math are those that people use later. Or if they do, it is not obvious which ones and in what times and places, with what frequency and with what consequentiality. And there is substantial evidence that there are many ubiquitous school mathematical practices that are not critically useful in this sense. These include school mathematical practices that are very central to the allocation of social opportunities through schooling, like algebra and calculus. This is why the everyday mathematics studies are so important; they allow us to ask and answer questions about alignments between forms and functions of mathematical activity in different contexts because we still don't know enough about which parts of school math are used in out-of-school life and which parts are academic parlor games and sorting mechanisms.

Voice #3: Okay, school math may not resemble math outside of school, but you're not considering the important mechanism, what psychologists call "transfer." What people are learning in school math are the concepts and those are what they use.

Reply: There are many versions of transfer, but the one that is most relevant for this chapter's concerns holds that if people learn mathematics correctly, they "store" it in the form of schematic mental structures. These structures are later activated and allow people to apply and instantiate these structures in various forms in different contexts. This view of transfer is relevant here because if it operated in this way, it would make less worrisome my findings, and those of many everyday math studies, that *observable* mathematical activity outside of school infrequently resembles mathematical activity in school. While debates about transfer are fierce, current evidences offer paltry support for the traditional mental abstraction view, especially in the specific case of the transfer of school math (e.g., Lave, 1988). And it is hardly a parsimonious hypothesis to suggest people spend a decade practicing the specific forms of school math (e.g., from grades two to twelve), and even though they can't routinely be observed to use most of these specific forms in everyday life, we still assume these forms are there operating in their minds, invisibly beneath the surface.

I don't wish to fully dismiss the mental/conceptual view of transfer. I believe it is probably part of the story. But I also believe that there is an inadequate evidential basis for transfer as *the* operative mechanism relating school math to out-of-school activity. It is certainly insufficiently supported as a rationale for retaining school math's central role in the allocation of social opportunities within the educational system.

Voice #4: Okay okay, all of this may be true, but I think you're just missing the point of school math. School math is not supposed to be practical. School mathematics is an encounter with a culturally important discipline, an intellectual practice with an amazing history. It's a form of intellectual activity that we should value its own right.

Reply: While I agree with this perspective in large part—as I do about filmmaking, music, cooking, painting, history, poetry, and all forms of creative human expression—this perspective provides little justification for the mandatory aspect of school math and math's disproportionate role in allocating social opportunities through schooling. If we collectively took this math-as-valued-intellectual-practice view, then we'd treat mathematics as important. But we also would probably dislodge it from its prominent role in allocating social opportunities and judging people's overall intelligence. We might also start to ask some interesting comparative questions about what "should" be in the school curriculum; is math any more important in school than learning to practice and appreciate cooking or visual communication? It may also be said that if valuing math as an intellectual practice were the goal of school math—to make it something that young people appreciate for the intellectual pleasure and challenge it provides—then school math is almost certainly an utter failure for most young people. How many young people form this sort of intellectual, creative, appreciative connection to mathematics based on their experiences in school?

Voice #5: Okay, okay. You are right about all this; school math is a "cultural arbitrary" (to use Bourdieu's term) that is used for the allocation of social opportunities. But few of us can change such things; that is simply how social power works. So it's our responsibility as educators to help those who lack resources to achieve within the system, regardless of its arbitrariness. Only then are we contributing to social justice.

Reply: I am sympathetic to this perspective. And I think a commitment to this project—taking the educational systems standards, values, and gates as more or less given and helping people navigate them—is both worthwhile and necessary. But this perspective should never be the enemy of a parallel enterprise that seeks to critically examine and change those standards, to cast skeptical eyes on the cultural arbitrary. To give up on this project is to slip into a hall of hopeless mirrors.

Coda

I look at the mathematics education my elementary school-aged children are receiving and see very little difference from my own school math education, more than three decades ago, even in the very highly regarded public school my children attend. I see the social sorting already beginning for both my children, probably to one's benefit and to the other's detriment. I see one of my children starting to decide that she/he is bad at math, based mostly on a struggle with the ubiquitous arithmetic facts and the way they are taught and tested. In a brief window of schooled experience, timed tests have become a litmus test of her/his mathematical intelligence and in part her/his moral worth as a student. It's not her/his teacher's fault; it's in the water.

A recent editorial by the president of the National Council of Teachers of Mathematics (NCTM) lamented what he sees as an "epidemic" of people in the United States who say that they are "bad at math" (Shaughnessy, 2010).[3] He argues that as right-thinking adults, US mathematics teachers can no longer allow people to say this—as if saying it is the source of the beliefs and feelings. In my view, the *institution* of school math—and by school math I mean the entire entrenched infrastructure that includes the textbooks, the worksheets, the teaching practices, the

[3] "Rather than sympathizing with people who publicly—and proudly—make this pronouncement, it's time for us to take them to task…Enough, I say!…saying 'I was never good at math' is unacceptable…spread the message to delete this offending statement from any social discourse" (Shaughnessy, 2010). I suspect my sympathies for Gerry and Ted, who were anything but proud of their "pronouncements," would not endear me to the NCTM president.

curricula, the tests, the binary beliefs about students who "get it" and students who don't, and the system of allocating social opportunities—needs to take responsibility for its central role in having produced this epidemic, if it is indeed one.

How to move forward? I have two modest proposals. First, I think mathematics education needs to get its purposes clear and possibly reenvision its role in the education of young people. A collection of informative historical accounts of mathematics education in North America (Stanic & Kilpatrick, 2003) suggests that most tensions this chapter identifies are not new. Abstractionist and form narrow versions of mathematics education in the cloak of mental training, college preparation, disciplinary authenticity, and national security have been around for more than a 100 years. And, surprisingly, a practical and vocational mathematics education seemingly had its historical day as well, in the first half of the twentieth century up to the Second World War (Kliebard & Franklin, 2003). World War II provided an opportunity for a shift in rhetoric. That rhetoric, which runs through Sputnik, *A Nation at Risk*, and *Rising Above the Gathering Storm*, links the literal survival of the United States to the purported essential role of particular versions of mathematics education.[4] I believe strongly that this rhetoric needs to be critically examined at our current historical moment and that mathematics education researchers, both those who focus on school math and those who work in out-of-school contexts, should be at that table.

My second modest proposal is that the mathematics learning and cognition research community needs to significantly rebalance its portfolio. The field needs a much more complete and broad ecological accounting of mathematical practices in out-of-school settings. There is a small but substantive corpus of such studies from the last few decades; important additions to this corpus are represented in this volume. But the size of this research corpus pales in comparison to research on the teaching and learning of any of the standard school math topics. For example, based on some estimates from research databases, there are probably at most 200 published studies about mathematical practices in all of the settings in the world that are not school but there at least 9,000 studies on the teaching and learning of fractions alone. (Little good these studies did Gerry.) In my view, only with such a broad, strategic sampling of mathematical practices in the world will we have a way to really answer questions of how, when, and why school math counts. Or to argue why school math might need to count less or be reorganized to count differently.

It falls to the mathematics learning and cognition research community to continue to build a conceptual vocabulary that does not take school mathematics as the exclusive reference frame for understanding mathematical work across society and that can follow mathematical practices in and across time and place, including school. In this chapter, I have advanced the concepts of *distribution*, *form diversity*,

[4] "As the clouds of war gathered over Europe and Asia during the 1930s, mathematics educators in the United States engaged in their own struggle. They sought to restore the place of their subject in the curriculum. Theirs was not a struggle of life and death or for the future of the world, although it would soon come to be portrayed as having dire implications, especially for the future of the United States" (Garrett & Davis, 2003: 494).

mathematics-in-not-as, and different senses of *what counts as math* to contribute to this conceptual vocabulary.

As far as expectations for substantive change in school mathematics, it might be prudent not to look to the existing educational infrastructure or the standard math education community, at least as represented by the current leadership of the NCTM. I worry that these institutions would be largely indifferent to (which is not to say its people would be uncaring about) the issues raised in this chapter. School math seems remarkably inertial in rolling along, pursuing minor variations in how to teach the more or less the same old topics in the same old sequence and chasing test scores as proxy measures of human capacity and meaningful learning.

What to do? Sometimes, change comes from without. I wonder if, in the context of this volume, it will fall to the out-of-school educators to really step up to plate on these issues because they have the liberty to be free of much of the apparatus of school-based testing and the sorting of young people. School teachers rarely have this liberty. So it may be the out-of-school educators who discover and create new ways to arrange activities and learning environments that make it possible for math to count as one or more of the many things that it rightfully should—such as a resource for *making ideas and things* that have meaning, use, and value outside of school or, alternatively, as a form of challenging, satisfying, intellectual, creative work with a history and a future. If the out-of-school educators manage to show that these things are possible, then maybe the school-based educational infrastructure might start to take notice and come around. What else to do? It seems also that while science has come recently to be re-understood, at some level of consensus, as something that is learned in and out of school and to have grappled with the fact that the relationships between science in and out of school are bidirectional and complex (Bell, Lewenstein, Shouse, & Feder, 2009), a similar consensus reconceptualization of math lags behind. That lag might be something worth fixing.

References

Bell, P., Lewenstein, B., Shouse, A. W., & Feder, M. A. (Eds.). National Research Council. (2009). *Learning science in informal environments: People, places, and pursuits*. Committee on Learning Science in Informal Environments, Board on Science Education, National Research Council. Washington, DC: The National Academies Press.

Bourdieu, P. (1991). *Language and symbolic power*. Cambridge, MA: Harvard University Press.

Committee on Prospering in the Global Economy of the 21st Century: An Agenda for American Science and Technology. (2007). *Rising above the gathering storm: Energizing and employing America for a brighter economic future*. National Academy of Sciences, National Academy of Engineering, Institute of Medicine. Washington, DC: National Academies Press.

de la Rocha, O. L. (1986). *Problems of sense and problems of scale: An ethnographic study of arithmetic in everyday life*. Doctoral dissertation, University of California, Irvine. *Dissertation Abstracts International*, *47*, 4198A.

Garrett, A. W., & Davis, O. L. (2003). A Time of uncertainty and change: School mathematics from WWII until the New Math. In G. M. A. Stanic & J. Kilpatrick (Eds.), *A history of school mathematics* (Vol. 1, pp. 399–440). Reston, VA: National Council of Teachers of Mathematics.

Hall, R., & Stevens, R. (1995). Making space: A comparison of mathematical work at school and in professional design practice. In S. L. Star (Ed.), *Cultures of computing*. London: Basil Blackwell.
Hall, R., Stevens, R., & Torralba, A. (2003). Disrupting representational infrastructure in conversations across disciplines. *Mind, Culture, and Activity, 9*(3), 179–210.
Hoyles, C., Noss, R., & Pozzi, S. (2001). Mathematising in practice. In C. Morgan & G. Woodhouse (Eds.), *Rethinking the mathematics curriculum* (pp. 48–62). London: Falmer Press.
Kliebard, H. M., & Franklin, B. M. (2003). The Ascendence of practical and vocational mathematics, 1893–1945: Academic mathematics under siege. In G. M. A. Stanic & J. Kilpatrick (Eds.), *A history of school mathematics* (Vol. 1, pp. 399–440). Reston, VA: National Council of Teachers of Mathematics.
Latour, B. (1987). *Science in action*. Cambridge, MA: Harvard University Press.
Lave, J. (1988). *Cognition in practice: Mind, mathematics, and culture in everyday life*. New York: Cambridge University Press.
Lynch, M. (1991). Method: Ordinary and scientific measurement as ethnomethodological phenomena. In *Ethnomethodology and the human sciences: A foundational reconstruction* (pp. 77–108). Cambridge, UK: Cambridge University Press.
Masingila, J. O. (1994). Mathematics practice in carpet laying. *Anthropology and Education Quarterly, 25*(4), 430–462.
McDermott, R., & Weber, V. (1998). When is math or science? In J. G. Greeno & S. V. Goldman (Eds.), *Thinking practices in mathematics and science learning* (pp. 321–339). Mahwah, NJ: Lawrence Erlbaum Associates.
Nasir, N. (2000). Points ain't everything: Emergent goals and average and percent understandings in the play of basketball among African-American students. *Anthropology and Education Quarterly, 31*(3), 283–305.
Nunes, T., Schliemann, A., & Carraher, D. (1993). *Street mathematics and school mathematics*. Cambridge, UK: Cambridge University Press.
Saxe, G. B. (1991). *Culture and cognitive development: Studies in mathematical understanding*. Hillsdale, NJ: Lawrence Erlbaum Associates, Inc.
Scribner, S. (1984). Studying working intelligence. In B. Rogoff & J. Lave (Eds.), *Everyday cognition: Its development in social context* (pp. 9–40). Cambridge, MA: Harvard University Press.
Shaughnessy, J. M. (2010, October). It's time to stamp out the phrase "I'm Bad at Math". *NCTM Summing Up*, published online at http://www.nctm.org/about/content.aspx?id=26737
Stanic, G., & Kilpatrick, J. (2003). *A history of school mathematics*. Reston, VA: National Council of Teachers of Mathematics.
Stevens, R. (1999). Disciplined perception: Comparing the development of embodied mathematical practices in school and at work. Unpublished PhD dissertation, Graduate School of Education: Cognition and Development, University of California, Berkeley.
Stevens, R. (2000). Who counts what as math? Emergent and assigned mathematics problems in a project-based classroom. In J. Boaler (Ed.), *Multiple perspectives on mathematics teaching and learning* (pp. 105–144). Westport, CT: Alex.
Stevens, R., & Hall, R. (1998). Disciplined perception: Learning to see in technoscience. In M. Lampert & M. L. Blunk (Eds.), *Talking mathematics in school: Studies of teaching and learning* (pp. 107–149). Cambridge, UK: Cambridge University Press.
Stevens, R., Mertl, V., Levias, S., McCarthy, L., Goldman, S., Martin, L., et al. (2006). At home with mathematics: Meanings and uses among families. In S. Barab, K. Hay, & D. Hickey (Eds.), *Proceedings of the seventh International Conference of the Learning Sciences (ICLS)*. Mahwah, NJ: Lawrence Erlbaum Associates.

Chapter 6
When Is Mathematics, and Who Says So?: A Commentary on Part I

Ray McDermott

Ralph Waldo Emerson stated the emotional center that the papers in this section are trying to recover:

> *We lie in the lap of an immense intelligence,*
> *which makes us receivers of its truth*
> *and organs of its activities.*
>
> (1841/2012, p. 173; line format is mine)

Conventions for applying the term mathematics (or science)[1] to certain activities have been more institutional than functional, more political than pedagogical. By these conventions, if activities display the formal operations of school mathematics, then they can be called mathematics; if they are evaluated positively by school personnel or testing agencies, that is, if they produce right answers with the mechanical algorithms of school mathematics, then they can be honored as mathematics. Although there is much workable commonsense in these beliefs and procedures, there is also a downside: the conventions lead to a loss of many students whose everyday mathematical reasoning goes unrecognized, unappreciated, and unused. Treated as an absence, its presence goes untapped by future development, and local, on-call, situation-specific mathematical reasoning can go untapped as a source of everyone's visible and institutionalized identity. Emerson's "immense intelligence" gets pushed aside and rendered unproductive and alienated.

[1] Three papers in this section report on mathematics and one on science. The issues are both different enough to invite a separate discussion, for which there is no space, and similar enough (formal, specialized vocabulary, procedure-focused, calculations-dependent) for the discussion of one to stand for them both. So I discuss the mathematics papers and only occasionally include a "(and science)" reminder that what is being said might apply to both topics.

R. McDermott (✉)
Stanford University, Stanford, CA, USA
e-mail: rpmcd@stanford.edu

B. Bevan et al. (eds.), *LOST Opportunities: Learning in Out-of-School Time*,
Explorations of Educational Purpose 23, DOI 10.1007/978-94-007-4304-5_6,

The papers in this section investigate two levels of operation in the conventions for defining mathematics: first, in the ways people use mathematical reasoning to solve problems in their own lives and, second, in the cultural resources available for nonmathematicians talking *about* mathematics in ordinary life and on television shows. The findings deliver good news and bad news. The good news is that people do much more mathematical reasoning—with and without the algorithms of school mathematics—than has been appreciated until recently (most recently, just by authors in this section; see Goldman & Booker, 2009; Martin, Goldman, & Jiménez, 2009; Pea & Martin, 2010; Stevens, 2010). The bad news, related to the good news by inversion and perversion, is that to the degree that schools have been the primary source of setting up and interpreting mathematics, the mathematics talk co-occurs all too often with unnecessary degradation, put-down, and not a few crises of confidence. To the extent that schools dominate the view, when it comes to mathematics (or science), we lie in the lap of an immense terror.

The papers by Stevens, Esmonde, and Esmonde et al. seek different conventions for applying the term mathematics. They seek conventions that appreciate the mathematics done in passing by perfectly ordinary people, handling the ordinary, distributed, and embodied demands of buying, selling, measuring, planning, fixing, showing off, at home, at work, and even, solving crimes on television.[2] The papers offer delightful stories tied to theories of situated learning. In a dreamed-of land, far, far away from a No Child of the Well-to-Do Left Behind educational policy and top-down Tea Party class warfare, they would have immediate practical influence.

In his paper, Reed Stevens delivers four new portraits of people doing mathematics. Two of them are called Architect A, who did not do mathematics at work ("that is why I hire an engineer"), and Architect B, who did do math at work ("all the time"). In a surprise move, it turns out Architect A and Architect B is the same person; they share a single body, but inhabit quite distinct work trajectories in mathematics. Architect A has to worry about getting the mathematics just right, but does not have to time to do it. Liability looms and so it is easier to hire an engineer. Architect B has to be "all the time" around mathematics, or at least around the people who do the mathematics. Stevens delivers some surprising news:

> *In nearly a year of a fieldwork at the architecture firm, I never saw any of the architects write down an equation and then manipulate it. I rarely saw calculations more complex than the basic arithmetic operations. The extent of geometric practice was a deep familiarity with shapes, often Euclidean ones, and an ability to visually and physically transform them in design practice.*

This news is surprising even to those of us who have done similar studies and found the exact same thing. It is counterintuitive to think of mathematics as something done along the way, as a subset to other activities. Even when we know better, it is difficult to see mathematics in activities that do not look like a mathematics test

[2] On ordinary as a term of compliment, see classic American thought from Emerson and Dewey to Stanley Cavell, the Marxism of Antonio Gramsci and Raymond Williams, and most versions of cultural anthropology.

from school. These facts allow Stevens to wonder just how many kinds of mathematics there might be. Just when is math and who gets to say so?

By the same logic of two architects in one body, the other two mathematics biographies given by Stevens might just as well be four. "Stupid Gerry" was terrible at the fractions examination he had to pass to become a professional draftsman, but he was smart and excellent at drafting, even at the parts that seemed to require a green-thumb knowledge of fractions. His work was excellent as long as he did not see fraction problems written down on paper. "Dumb Ted" could be found looking incompetent wherever teachers or other students were pushing mathematics, but he could work diligently and productively on mathematics projects. His work was excellent as long as he did not have to solve mathematics problems on a test. Both Gerrys and both Teds, like both Architects, enable Stevens to ask great questions about whether we can locate "the distribution of mathematical practices in and across the activities of the everyday world, in places of work, play, and domestic life. What specific ways is math, or whatever we call it, a resource for people's activities?" What are the conditions for a mathematics problem to be exactly the kind of problem we think it to be, when simply looking at it on different occasions would show that "the way it was calculated, the way it was embedded in standardized devices, and the forms in which it would be expressed were very different?"

Each biography in the Stevens paper can be divided into two parts: one struggling to solve a problem in life and using mathematics to get the job done, the other struggling to solve a school mathematics problem and making things worse by avoiding mathematics. The next two papers report how people—all kinds of people, including television scriptwriters—talk about mathematics in relation to problem-solving and sometimes about school mathematics as a problem in its own right. In a nice fit, the main division Stevens found in mathematics practice—important to solving a problem versus important to performing a school mathematics task—emerges again in the other two papers on people's talk about mathematics practice.

The paper by Indigo Esmonde et al. uses a cute eliciting device for surveying how people talk about mathematics in their lives. They ask people for short stories, or Math-In-A-Minute (or alternately, Math-I-AM) stories, about their experiences with mathematics. They are careful to not make too much of the stories as personal identities. They are dealing less with "enduring self-portraits" than a generalized terrain of communicative resources for talking about mathematics. The stories were not studied in association with the lives of the people who told them, and there are no claims to their truth or ecological validity. The only claim is that the stories reveal some patterns that might circumscribe the experiences available to people when they use mathematics, whether in school or elsewhere. Even with this limited probe into people's lives, the results are stunning. People's lives are filled with mathematics:

> *They created and maintained spreadsheets, and they used calculators and online tools. They rounded and estimated, worked with ratios and proportions, thought in two- and three-dimensions, and worked with patterns, geometry, algebra, multivariable analyses, and logic.*

Would it not be nice if teachers in school could know and appreciate how much mathematics was available to their students at home? School, unfortunately, is not just about making the best of things. School is where everyone has to do better than everyone else, where no one cares what you know, only how much more of it you know than others—as measured on arbitrary examinations with little connection to the real world of problem-solving beyond the school walls. The Math-I-AM stories reveal the contrast. Stories about school revealed the currency of ideas about ability, authority, competitive pride, and mathematics as a set of procedures disconnected from purpose other than school achievement:

> *...school stories painted a picture of mathematics as something requiring an ability of some sort (for it is possible to be bad at it), as something to like or dislike, as something that institutions and their agents (especially teachers) have special authority over and as a potential source of pride or trauma. Math was primarily portrayed as something to be studied as a free-standing entity.*

Stories of mathematics at home were based on quite different assumptions:

> *Math was integrated in social activity among family members and others across a variety of contexts. ... When evaluation was prominent, it was usually the task itself that was evaluated, not the specific mathematical techniques.*

Evaluation seems to be the heart of the matter. It is not that school has strict methods of assessment and people at home do not. It is rather that school has a rather singular view of what counts as right and people at home have a view more tied to the outcome of the larger task at hand. Mistakes come in many forms and they are received in various ways, many of them productive. An equivalent case has been made for jazz, where there is little talk of mistakes. Jazz masters hit wrong keys and play the wrong notes often, and sometimes on purpose, but everyone's responsibility is to make the best of it. Thelonious Monk complained one night that he played all the wrong mistakes! The only real mistake would be to stop playing (Klemp et al., 2008). So it is for mathematics and home and at work. Mistakes have to be moved along. The job at hand is still to be done, and people have to keep going. With school mathematics, without an ongoing and immediate tie to purpose, the miracle would be that anyone keeps moving at all. In fact, most children vote with their feet, and the great majority of them are out of what counts as year-to-year progress in mathematics by the tenth grade.

Esmonde's solo paper finds a different source of popular views of mathematics: the popular "crime-fighting with mathematics" television show, *NUMB3RS* (now RIP). Mostly I have enjoyed the show, and find the arrogant omniscience of the mathematicians less marked than on the average university campus. What I find more preposterous is the assumption that the complex mathematical models always have an application and are a rich data bank for testing hypotheses. In reality, it would take years of data production to run the applications that lead to catching the bad guys. Suddenly, the arrogance becomes more obvious. The mathematicians not only believe in their own wisdom, they do so against the odds, against any real test of verifiability. Esmonde finds this, accordingly, annoying and false. The mathematicians are sainted in the television scripts with power far beyond the possible:

> *[Their] chains of reasoning, built on tenuous assumptions that are not seriously investigated, trump the more typical forensic evidence. In* NUMB3RS, *the traditional crime drama trope*

of the competent professional is extended to include mathematicians, and surprisingly, the mathematician is presented as even more competent than the FBI agents, forensic scientists, coroners, and other professional crime-fighters.[3]

The show was perhaps most outrageous in directing mathematics to a study of courtship. Good perhaps for comic relief, I assume, but better for revealing attitudes toward mathematics. Esmonde is wise in pinpointing what the scriptwriters assume.

In the mathematical model of courtship ... the mathematicians have clearly made a host of assumptions, including what a date is and should be, how one shows affection, who does the gift-giving and who does the gift-receiving, what one's goals for dating are, and the like (assumptions that are heteronormative, sexist, and Eurocentric) ...

Instead of a mathematical study of courtship, we need a cultural analysis of mathematics. Take that Larry Summers. What a mess. Mathematics can be such a helpful friend. How annoying that it has been turned into a tool of arbitrary social hierarchy.

Taken together, the papers encourage a confrontation with three powerful assumptions that complicate American discourse on mathematics (or science) education:

- Mathematics is difficult to acquire.
- Although difficult, offering mathematics to as many as possible is worth the effort, because mathematical skills can aid the development of objective and rational minds.
- Mathematics should be taught in classrooms to those who need it.

First we can cover what seems right enough about these assumptions. Then we can notice what they leave out and even distort.

First, of course, mathematics seems difficult. This is why so few can do mathematics despite it being so useful. The success and failure rates of children learning mathematics in school correlates with success and failure in life. The data are everywhere. If mathematics were easy, everyone would be on the way to the top. Even at a basic level, a display of mathematics skills is a reliable sorting tool for separating those who can learn best from those who cannot.

Second, of course, mathematics seems cognitively good for the individual mind and institutionally enhancing for the economies in which mathematics-savvy individuals operate. Mathematics could be a name for civilization itself. The Chinese and the Babylonians likely had a Pythagorean Theorem long before Pythagoras, the astronomical calculations of the ancient Maya and Inca are still being unpacked and medieval seafarers could calculate their way through the "time and tide" of the North Atlantic (Frake, 1985). From mathematics to military might, monumental architecture, and expanding markets, some might argue, it is almost a direct line.

Third, of course, get those who need to learn mathematics into classrooms. No evidence is necessary. Where else could they learn material so technical and arbitrarily precise?

[3] For the same omniscience attributed by television scientists, see Kruse (2010).

If each assumption, when addressed, is supported by research, so too the concepts and methods dictated by the assumptions, when unaddressed, support the research (Lave & McDermott, 2002). When the topic is mathematics (or science) education, commonsense makes a nasty circle. We can undress the situation if we ask, as the papers in this section have, about the high ritual place of mathematics displayed in the time and tide of American social structure. We can validate the assumptions, but to that extent they can lead to false places. However true in general, the assumptions may be least true—and false in consequence—in societies in which people believe them too completely:

- The more people believe mathematics is difficult to acquire, the more they find reasons to celebrate those who learn mathematics better than others and, correspondingly, those who do better in social mobility.
- The more people believe mathematics is cognitively and culturally transformative, the more they can degrade those without such powers.
- The more people believe mathematics is best learned in classrooms, the more they can ignore other sources of mathematics.

Although the three assumptions feel true—mathematics can be difficult (particularly in competitive school situations), mathematics can help transform a social information processing system, and mathematics can be taught in classrooms—an insistence on their truth can interfere with building on them reasonably (for the same argument on literacy, see McDermott & Varenne, 1995). Suddenly, *NUMB3RS* is less occasionally annoying than it is ideologically dangerous.

This discussion opened with a quote from Ralph Waldo Emerson and can close with John Dewey's use of the same quote 85 years later.

> *We lie, as Emerson said,*
> *in the lap of an immense intelligence.*
> *But that intelligence is dormant*
> *and its communications are broken,*
> *inarticulate and faint*
> *until it possesses the local community as its medium.*
>
> (1927, p. 219; line format is mine)

Move ahead, another 85 years, to our current situation. We lie in the lap of communities broken and rendered inarticulate by divisions hard to overcome: divisions of race, gender, social class, and, the increasingly important measure on the block, test scores. With a school system that hands out success and failure accolades as much by the distribution of financial resources as by the distribution of school performances on arcane vocabulary and mathematics tests—oh, how they lie—it is time to rediscover the immense intelligence of ordinary mathematics (and science).

References

Dewey, J. (1927). *The public and its problems*. New York: Henry Holt.

Emerson, R.W. (1841/2012). Self-reliance. In D. Mikics (Ed.), *The Annotated Emerson,* (pp. 160–185). Cambridge, MA: Harvard University Press.

Frake, C. O. (1985). Time and tide in the cognitive maps of medieval seafarers. *Man, 18*, 254–270.

Goldman, S., & Booker, A. (2009). Making mathematics a definition of the situation. *Anthropology and Education Quarterly, 40*, 369–387.

Klemp, N., McDermott, R., Raley, J., Thibeault, M., Powell, K., & Levitin, D. (2008). Plans, takes, and *mis-takes. Critical Social Studies, 10*, 4–21.

Kruse, C. (2010). Producing absolute truth: CSI science as wishful thinking. *American Anthropologist, 112*, 79–91.

Lave, J., & McDermott, R. (2002). Estranged ~~labor~~ learning. *Outlines, 4*, 19–38.

Martin, L., Goldman, S., & Jiménez, O. (2009). The tanda: A practice at the intersection of mathematics, culture, and financial goals. *Mind, Culture, and Activity, 16*, 338–352.

McDermott, R., & Varenne, H. (1995). Culture *as* disability. *Anthropology and Education Quarterly, 26*, 324–348.

Pea, R., & Martin, L. (2010). Values that occasion and guide mathematics in the family. In W. Penuel & K. O'Connor (Eds.), *Learning research as a human science* (National Society for the Study of Education Yearbook, Vol. 109, pp 34–52, No. 1).

Stevens, R. (2010). Learning as a member's phenomenon. In W. R. Penuel & K. O'Connor (Eds.), *Learning research as a human science* (National Society for the Study of Education Yearbook, Vol. 109, pp. 82–97, No. 1).

Part II
Understanding How and Why People Learn Across Settings as an Educational Equity Strategy

Part II
Understanding How and Why People Learn Across Settings as an Educational Equity Strategy

Chapter 7
Introduction: Understanding How and Why People Learn Across Settings as an Educational Equity Strategy

Philip Bell

People circulate across a broad range of learning environments as a routine matter of daily living. In the LIFE center, we have referred to this "learning across settings" phenomena as the "life-wide" dimension of learning—as people circulate from moments of elective family life, compulsory schooling, participation in online communities, or other patterned routines of daily practice (see Banks et al., 2007 for a detailed description). Many of the variations that exist in educational pathways and the associated educational inequalities relate to differences in these life-wide patterns of learning and associated consequences. For example, access to academic language in and out of school is a strong predictor of academic achievement (e.g., Suárez-Orozco, Suárez-Orozco, and Todorova 2008). Situational interests of children are often cultivated, recognized, and supported within the context of family life (e.g., Crowley & Jacobs, 2002). Moments of formal instruction can pique children's interest and lead to subsequent learning outside of school (e.g., Reeve & Bell, 2009). In order to develop a theoretical understanding of the life-course outcomes of particular individuals, we need to better understand how they move and learn across a varied set of cultural niches with variable practices, materials, and evaluation systems that are used to gauge human behavior (Lee, 2008).

The scientific study of learning across settings supports a crucial educational equity agenda. Educational environments writ large—with the broad variety of formal and informal environments—vary dramatically in their ability to surface, adapt, and leverage the educational assets that learners bring to that environment as a result of their life-wide learning behaviors over developmental time—which can have profound consequences for learners. For complex reasons, learners may elect to not share aspects of their identities, language, and expertise in a given learning environment. For these reasons among others, learning is largely not coordinated

P. Bell (✉)
University of Washington, Seattle, WA, USA
e-mail:pbell@u.washington.edu

B. Bevan et al. (eds.), *LOST Opportunities: Learning in Out-of-School Time*, Explorations of Educational Purpose 23, DOI 10.1007/978-94-007-4304-5_7,

for individuals across environments—and the most vulnerable students in society are all too often marginalized. And yet, a long history of empirical studies has documented the educational assets that all learners bring to learning experiences and the educational possibilities associated with helping youth develop increased degrees of freedom to navigate their life as they so choose.

Our research literatures remain balkanized and, thus, underrepresent the complexity of life-wide learning processes, phenomena, and outcomes. Learning studies of schooling have tended to focus on relationships between narrow achievement measures and the details of instruction, teacher background, or educational policy choices. Learning studies of informal education have tended to focus on the educational impacts of specific interventions—specific exhibits or exhibit elements, organized programs, institutional practices, or produced media and communication strategies. Learning studies of family life—of which there are significantly fewer—have tended to focus on the qualities of discourse practices and strategies to support learning.

Efforts have been made across teams of researchers to work against this balkanization over the past decade by initiating research from a learning ecology perspective. Specifically, a range of efforts are underway to evolve our theoretical and methodological approach to understanding how, why, and where people engage in learning that is consequential to them. The four papers represented in this section highlight different methodological strategies for pursuing the learning across settings research agenda.

In the chapter by Barron, Wise, and Martin, the research strategy is to navigate into the current project work of students in an after-school setting and then conduct detailed "technobiography" interviews that allow for a historical reconstruction of the related life experiences of the learner. The theoretical frame used to interpret the work and the personal histories documents how the implicated expertise has developed from a learning ecology perspective. Special attention is given to the settings in which the expertise developed and the range of social and material supports that aided in its development. The approach lends itself to understanding longer timescale phenomena associated with the stabilization of interests, the refinement of disciplinary practice, the organization of social networks to support the work, and, ultimately, the formation of an expert identity. Such accounts help orient us to the corresponding educational supports that could be developed to more directly support youth in the development of desired forms of expertise.

In the chapter by Bell, Bricker, Reeve, Zimmerman, and Tzou, the research strategy is to more directly document learning among the same core set of youth across a broad variety of social settings over significant timescales. The unique data corpus was pieced together through extensive relationship building, the videorecording of activity during fieldwork across dozens of sites, the self-documentation of everyday life by participants, and coordinated design-based research in the school setting in partnership with teachers. On the one hand, the effort allowed for the documentation of the developing expertise of youth in relation to the cultural histories of the immigrant families and the social and material resources that were arranged to support learning. The study also allowed for an investigation of the barriers affecting how people might learn across settings and allowed for the testing of instructional techniques to promote bridging in life-wide learning (e.g., the leveraging of students' everyday repertoires of practice during formal instruction in the classroom). The study helps open up a new theoretical terrain in relation to a deep empirical accounting of learning dynamics and related learning strategies.

In the chapter by Mehus, Stevens, and Grigholm, the research strategy is to conduct the direct videodocumentation of learning of the same learners across home and preschool settings. The study uniquely documents how early learning is socially arranged and accomplished across family life and preschool environments. We lack detailed accounts of the life-wide learning conditions of early childhood. The study develops detailed accounts of different social arrangements for early learning—the instrumental role of parent-guided learning interactions at home and the peer-regulated group activities that are afforded in the preschool environment. Recent research syntheses highlight that very young children can think in sophisticated, abstract ways (Duschl, Schweingruber, & Shouse 2007), but we need further documentation of how domain-specific knowledge and sense-making practices start to form in early childhood. In this sense, this study represents an important line of research to help us interpret how moments of early childhood potentially relate to long-term domain learning and identification.

In the chapter by Razfar, the research strategy is to leverage a theoretical perspective for the detailed analysis of talk and action of students in an after-school setting that allows for the interpretation of social meaning and math learning from the perspective of the multiple discourses that students are familiar with and/or are learning. A life-wide orientation to learning is accomplished by simultaneously attending to the primary discourses that students have learned early in life—which they are encouraged to leverage in their sense-making in the after-school instruction—and the secondary discourses that they learn later in settings like school and after-school (i.e., associated with the academic understanding of mathematics and other domains). By interpreting students' activity from this layered perspective as they engage in a counting game over time, the analysis is able to document students' growing understanding of probability within the context of their problem-solving, relative to their linguistic repertoires. The chapter is emblematic of a tradition of discourse analytic work that operates from the premise that student activity should be interpreted relative to the multiple discourses (and identities) that youth bring to and extend through the meaning-making process. It is in contrast to alternative theoretical frames that leverage more fixed and normative accounts of discourse to interpret learner activity.

Gutiérrez's commentary on this section discusses current methodological possibilities and theoretical priorities related to a more holistic accounting of the repertoires of practice of youth, their learning pathways, and ways of understanding their possible futures. She grounds the work in relevant lines of work and points to promising directions for further development as research on learning across settings continues to develop in the future. She importantly highlights how these more detailed accounts might open up more robust possibilities to promote educational equity and transformative outcomes for youth from nondominant communities. The accounts of ecological and purposeful learning in this section collectively inform how educational systems might provide coordinated, redundant supports for learning across multiple settings over longitudinal time.

References

Banks, J. A., Au, K. H., Ball, A. F., Bell, P. Gordon, E. W., Gutiérrez, K. D., et al. (2007). Learning in and out of school in diverse environments. Seattle, WA: The Life Center and the Center for Multicultural Education, University of Washington.

Crowley, K., & Jacobs, M. (2002). Building islands of expertise in everyday family activity. In G. Leinhardt, K. Crowley, & K. Knutson (Eds.), *Learning conversations in museums* (pp. 333–356). Mahwah, NJ: Lawrence Erlbaum Associates.

Duschl, R. A., Schweingruber, H. A., & Shouse, A. W. (2007). *Taking science to school: Learning and teaching science in grades K-8*. Washington, DC: National Academies Press.

Lee, C. D. (2008). The centrality of culture to the scientific study of learning and development: How an ecological framework in education research facilitates civic responsibility. *Educational Researcher, 37*(5), 267–279.

Reeve, S., & Bell, P. (2009). Children's self-documentation and understanding of the concepts 'healthy' and 'unhealthy'. *International Journal of Science Education, 31*(14), 1953–1974.

Suárez-Orozco, C., Suárez-Orozco, M., & Todorova, I. (2008). *Learning a new land: Immigrant students in American society*. Cambridge, MA: The Belknap Press of Harvard University Press.

Chapter 8
Creating Within and Across Life Spaces: The Role of a Computer Clubhouse in a Child's Learning Ecology

Brigid Barron, Susie Wise, and Caitlin K. Martin

It is a quiet Tuesday evening in the fall. A teenage girl sits at a large green table with her head bent over a Venn diagram. A younger boy sits at a nearby computer searching for images on the Internet using Google. This familiar scene of after-school concentration can be found in any number of American community contexts, including a public library, a local school, or a family kitchen. However, in this particular case, we look farther into the room and find more youth at work using high-end technology equipment, including a full digital recording studio, new computers with professional design software, digital animation supplies, and the latest gaming technology. Welcome to the Simmons Computer Clubhouse, part of an international network of over 100 similar informal after-school learning environments where young people work with adult mentors to "explore their own ideas, develop skills, and build confidence in themselves through the use of technology." Luis, a skinny dark-haired 13-year-old boy, sits at a computer station in the back corner of the room with a set of plastic action figures from the X-men comic series, a basket of play dough, and an Intel Digital Blue stop-animation camera. He is producing his latest movie.

Introduction

The clubhouse environment described above has its origins in concerns about equitable access to tools, people, and ideas that support the development of *technological fluency*—defined generally as the capacity to express oneself using a broad

B. Barron (✉)
School of Education, Stanford University, Stanford, CA, USA
e-mail: barronbj@stanford.edu

S. Wise
d.school Institute of Design, Stanford University, Stanford, CA, USA

C.K. Martin
H-STAR, Stanford University, Stanford, CA, USA

B. Bevan et al. (eds.), *LOST Opportunities: Learning in Out-of-School Time*, Explorations of Educational Purpose 23, DOI 10.1007/978-94-007-4304-5_8,

range of computing tools (Resnick & Rusk, 1996) and to adapt technology to advance one's own goals. Digital technologies offer children and adolescents rich opportunities to design and create artwork, movies, games, animations, interactive robots, and other artifacts. Online communities that reflect "cultures of participation" (Jenkins, 2006, 2009) allow creators to share their work, receive feedback, and expand their social networks. Informal collaborative relationships develop as learners share knowledge and codevelop interests. It has been suggested that participation in these informal collectives nurtures important twenty-first-century capacities such as collaboration, knowledge of how to build social networks, manage information, direct one's own learning, engage in design, and capitalize on opportunities for distributed cognition and the building of collective intelligence. Design activities, including information gathering, creative thinking, prototyping, improvisation, and tinkering, are thought to provide potential pathways to these crucial twenty-first-century capacities (Balsamo, 2010).

Although these cultures of participation are becoming more common, they are not equally accessed. Recent research has shown that despite the emerging cultural image of the average youth as constantly connected and technologically savvy, those who can actually create digital media or interactive environments are in the minority (Barron, Walter, Martin, & Schatz, 2010; Ito et al., 2009). More common are socially motivated genres of participation such as social networking and texting. Those that can use technology in more advanced ways have typically been deeply supported by parents, peers, or teachers that have expertise. For example, in a study of technologically sophisticated youth in Silicon Valley, (Barron, Martin, Takeuchi, & Fithian, 2009), we found that parents advanced their children's learning when they collaborated with them, learned from them, brokered outside learning opportunities for them, provided nontechnical support to them, or hired them to do technical work. Parents also played instrumental roles when they shared their technical expertise through informal teaching processes or provided their children with learning resources such as books or new media tools. Studies of school-based access to learning opportunities show that it is the more economically advantaged communities that offer electives focused on advanced topics such as computer science (Goode, 2007; Margolis, Estrella, Goode, Jellison-Holme, & Nao, 2008).

Other studies show that youth who live in communities in areas of lower socioeconomic status (SES) like Luis's often are not using technology at home or at school to do much beyond basic Internet searching, social networking, or typing out a report using Microsoft Word (Barron et al., 2010; Warschauer & Matuchniak, 2010). While these skills are certainly important, there are still stark differences among children and adolescents in access to learning opportunities that will help position them to use computers in ways that can promote their own development (Goode, 2007; Warschauer & Matuchniak, 2010). It is becoming increasingly evident that differences in the types of participation youth engage in will further contribute to inequities along gender, SES, or cultural dimensions.

Theoretical Goals and Methodological Strategies

Community technology centers can provide an important space for youth with less home access, offering multiple opportunities to learn through mentors and material resources (Kafai, Peppler, & Chapman, 2010; Penuel, et al., 2000). In our snapshot of Simmons, youth are diligently working on their stop-motion animation skills, tinkering with the timing of their movie soundtracks, and laying down complex beats in the recording studio. Observations of the creative work emerging from environments like the clubhouse raise a host of questions about the learning activities that take place there, how they evolve over time and place, and who is involved.

The case study analysis we present in this chapter is part of a larger research program investigating conditions and consequences of persistence of engagement in technologically mediated design activities that offer adolescents opportunities for imaginative work.[1] A focus on engagement in research on learning, in contrast to an exclusive focus on knowledge acquisition, is consistent with contemporary theories of learning that conceptualize moments of learning as part of a process of identity development (Beach, 1999; Nasir, 2002; Wenger, 1998). Participatory views of learning draw attention to membership in communities of practice that are defined by affinity groups (Gee, 2000) based on interest-driven activities (Wenger 1998). For newcomers, joint endeavors offer not only opportunities to develop knowledge in a particular domain but also increasing levels of commitment, sense of belonging, and identity as a practitioner that develops and is sustained across time and place. Practice-linked identities typically emerge when learners view their own engagement in the practice as an important part of who they are (Nasir & Hand, 2006) and when this connection is made, self-sustaining strategies of continued learning can often be observed (Barron, 2006). Members of affinity groups come to develop practices and sets of experiences that position them to engage the world in particular ways that offer continual opportunities for learning.

Youth who lived in the community served by the Simmons Computer Clubhouse, on average, had much less access to computing tools at home than their Silicon Valley neighbors whose parents worked in the technology industry (Barron et al., 2009). The goal of the clubhouse study was to better understand how this intentionally designed space provided opportunities for learning and how it intersected with

[1] In this research program, we use both quantitative approaches and ethnographic approaches. Surveys allows us to collect data from large samples in order to compare communities with respect to the breadth and depth of creative production activities that adolescents have experienced (Barron, 2004; Barron et al., 2010). This quantitative approach also allows us to examine the relationships between variables. The ethnographic case study work allows us to more deeply understand the social processes that lead to and sustain engagement. It allows us to see the dynamic nature of a child's learning ecology and how it changes as new learning resources are made available or disappear.

other spaces for learning such as the their own homes, the homes of friends and relatives, schools, libraries, churches, and virtual settings such as online environments. We have selected one clubhouse member, Luis, as a focus for this chapter, because his level of productivity at the clubhouse was relatively high and we were interested to see how his learning history and learning processes compared with those from the community whose parents were employed in the information technology industry.

Our particular approach to case studies involves taking a longitudinal perspective. Interviews and observations are summarized to create portraits of learning about technology in a genre that has been called "technobiography" in recent work (Henwood, Kennedy, & Miller, 2001). A life narrative approach allows us to chart a learning history in terms that go beyond metrics such as numbers of courses taken to include the meaning and attribution behind decision making and narratives of how the learning activities unfolded across time and setting (Bruner, 1994; Elder, 1994; Linde, 1993).

Our representations of Luis's activities in narrative form provide what might be called a wide-angle view of learning, losing direct observation of micro-interactional phenomena but offering a glimpse at the dynamics of learning and interest development over weeks, months, and years (Lemke, 2000). Our methods involve observation, interviews, analysis of the artifacts learners create, and data collection through an occasional questionnaire. To advance our conceptualization of learning over time and settings, we create visualizations that map key learning activities, relationships between activities, where they take place, and the people and resources involved in each activity. In the next section, we provide Luis's technobiography and then present a visualization that maps his activities across time and setting.

Luis's Learning Pathways

Luis was 13 years old and in the last months of seventh grade at the time of our final interview. We had been talking to him and observing him in the clubhouse for two years. He lived in the low-to-middle class primarily Hispanic, Northern California community served by the clubhouse and attended the local public middle school. Luis had been coming to the Boys and Girls Club after school for five years. He shared the one computer at his home with his parents, his sister, and his brother. Luis often used it to play games. Although the family did not have Internet access or a printer, his older brother had an analog video camera. When Luis was 10 years old, his brother showed him how to shoot video and use the animation special effects available within the tool.

Getting Started at the Clubhouse

Luis became a member of the computer clubhouse when it opened on-site at the Boys and Girls Club when he was 11. "I'm not sure how long I've been here at

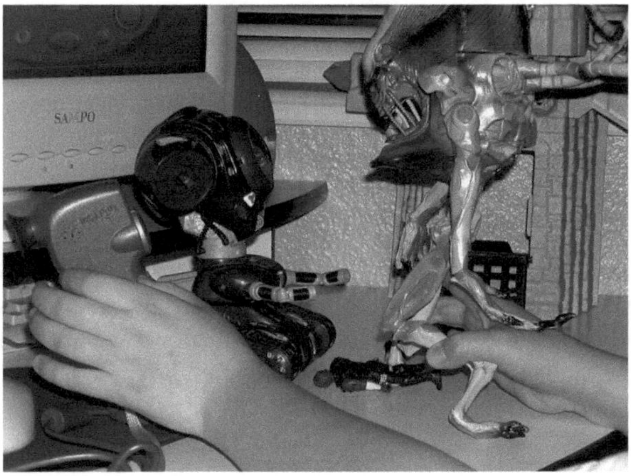

Fig. 8.1 Luis positioning the Intel Digital Blue camera and various action figures for one of his stop-motion animations

the computer clubhouse, but it's been a while. I animate a lot here. I do stop-motion. I also make photos on Photoshop."

Luis noticed the video camera equipment in the space and, using his existing knowledge and interest from working with his brother, set out to make live action movies with his friends, "I just asked [the clubhouse coordinators] what it was and they old me and it was for taping and stuff, so I just started running around taping my friends, trying to do scenes and stuff...." The clubhouse was equipped with Intel Digital Blue cameras and although the clubhouse coordinators had taken a workshop on how to use them, they were not experts in this field. The camera set, which could be used to shoot both video and still images and came with its own editing and special effects software, became Luis's main tool for creating. He returned to the cameras again and again, experimenting with different methods.

> *Luis*: "Well when I first saw the camera I didn't know what it was for, and [the clubhouse coordinators] didn't know either, so my friends and me were just doing stuff, like movies making it look like people were going super fast. Then I started getting ideas about like moving things and then taking pictures."

While he was learning the process of stop-motion, Luis experimented with animating paper drawings. "Just to draw a bunch of animations on paper and see how it worked. ... It worked out pretty good. I drew a picture and then I take a picture of it, then erased it, then drew, just moved it a little, drew it, and erased it." The paper drawings evolved to claymation, from which Luis then developed his own style, a combination of claymation and plastic action figures, like the ones shown in Fig. 8.1.

While Luis was finding his own technical way with the equipment, the clubhouse coordinators recognized and encouraged his incoming interest in film-making and offered new ideas.

Coordinator: "To begin with, I guess he's always been into movies and stuff so we would always take the cameras out and him and his buddies would go out there and shoot these little fight scenes and make little sound effects with their mouths and stuff like that and bring it back in. They made little clips of them fighting and stuff. We had a box of clay that we would bring out … we showed them how to use the stop-animation tool in the software. I think he just started … I know, like his brain just started thinking up all these other ideas. He came up with just a bunch of different scenarios."

The clubhouse coordinators included Luis in field trips connected to his interests, including one to the game design company Electronic Arts. Luis remembered, "That was pretty cool. We learned a lot about its history. We got to try some games that hadn't come out yet." The coordinators also often promoted his work by showing his movies to guests and new members.

Movie-Making Process: Idea Generation, Feedback, and Revision

Luis describes his process in a nutshell as, "think about it, get some supplies to make it, and then do it."

Think About It

For inspiration, Luis drew heavily on contemporary media, from popular cartoons to Hollywood action and Kung Fu movies, using them as touchstones and markers of the kind of movie he wanted to make. He sometimes cited what he saw as a flaw in a narrative as the spark for a movie of his own, such as bringing a favored dead character back to life. He also talked about watching and enjoying other stop-motion films including claymation.

He also was able to generate design process ideas from examples in mass media, such as when he picked up the ideas of developing concept art (drawings to guide digital look and feel) and storyboarding from a special feature in a video game:

Luis: In the video games, there are a few things you can unlock and it's in the extras, it's called extras and it shows you the storyboard of how…um…about the scenes in the game. They look pretty much like a comic and no words. They just like show what they are doing and write down what they do.

Get Some Supplies to Make It

Once Luis has planned out his general idea for a movie, he gathered the materials he needed to produce his work. Initially, this included action figures from home for the characters and play dough and clay from the clubhouse for different effects (such as using red play dough to model spilling blood as shown in Fig. 8.2). As he made more movies and got feedback from himself and others, he looked for new tips and tricks. When his friends complained about seeing the clubhouse

Fig. 8.2 Battle scene created by Luis with red clay used to simulate blood and animation effect added to convey explosion

computer equipment in the background of a scene, Luis began to make his own backgrounds for different shots. He also looked for new ideas and solutions to problems online, especially from a site called zipster.com.

> Sometimes I use boxes as houses. I'm looking forward to getting some fishing lines so I don't have to use my fingers to hold them up. The fishing line, you won't be able to see them.

Then Do It

Luis worked tirelessly to position, shoot, and edit his movies. He added his own sound effects and music, sometimes using the clubhouse sound studio. Once the shots were assembled in the Digital Blue software, Luis added edits and special effects to enhance the narrative.

Luis's workspace at the clubhouse was frequently crowded with boys who were very much a part of the scene, laughing and giggling while making sound effects, and gesturing with the action figures that appear in the movies. They contributed ideas and offered advice in terms of what they liked or did not like. In addition to being audience and critic, they also occasionally contributed sound effects to the films.

Luis's process of revision entailed reflecting on movies he had already made and setting goals for new ones. In some cases, this was self-critical, attempting to fix mistakes he saw in his previous work. Luis described his desire to make more complex and better-structured films.

> *Interviewer*: So how would you do it so it would make more sense?
> *Luis*: Actually have like a story in it instead of just random fighting.
> *Interviewer*: And where did you get that idea?
> *Luis*: Well, I watch a lot of Jet Li movies and I like fighting. So I just tried it. But um the first one was just random punches. The second one is going to be more choreographed.

Boundary Crossing: Connections to Home and School

Although access to camera, computer, and the editing and effects software designated the clubhouse as the main site for his movie production, Luis was able to work on elements of his movies in multiple contexts. His work in the clubhouse was often directly impacted by what was occurring in these other contexts.

Movie-Making and the Context of Home

Despite the fact that his family computer was old and not equipped for video production, home remained an important context for design. Luis played games on his Playstation, exploring the extras, sketched and storyboarded movie frame ideas, received feedback on his work, and procured the resources he needed, such as action figures. His mother and older brother were central figures in the development of his projects.

Interviewer: Does Luis at home sometimes draw things that are for movies that he makes [at the clubhouse]?

Mother: He draws … the same animation he does, he draws his own figures that he has at home from a book or something and he makes like … you know like in the newspaper, the little comic things, he makes little squares and he draws it and puts little comments on them.

Luis's older brother had moved out of the family home, but was close enough to have a significant impact on Luis's learning. In addition to initially introducing him to and teaching him about video production, at his brother's house, Luis used his faster computer and played games with him on his Xbox. He also received important feedback on his video design work.

Like with my brother, he saw [characters in my movies] just like floating around instead of moving their feet. So he would tell me put toothpicks in the hold in them, on their feet, so it looks like they are actually walking.

Luis and his mother frequently talked about subject matter for his movies, with her suggesting that he do something "funny, because he is funny, he has a good sense of humor" and him replying that he "likes monsters and gore." In addition to being an audience for his projects and discussing ideas, Luis's mother also monitored Luis's work on the computer. At the end of seventh grade, Luis's grades in school were falling to the point where he was worried he would need to repeat the grade. His parents attributed some of this to his spending too much time on his extracurricular computer projects, like the computer animation, and sought to limit his work.

Mother: Well, his grades kept coming down and we had a conference and the teacher was concerned because Luis is like a super smart boy and that he was wasting his time on not doing what he's supposed to do. Me and his Dad know that he does those kind of things, the animation on the computer. Me and his Dad don't put him down about it and we, you know, we are standing right by him if this is what he wants and we'll support in every which way like I do with my little girl. But, um ... he needs to know that he needs to keep doing what he needs to do in school and then in the computer clubhouse he needs to do what he needs to do. It's two different worlds. If he doesn't graduate, if he doesn't ... if he's getting bad grades, that's where we know how to punish him, right here because this is his main interest. We had to restrict his activity at the clubhouse because he was not focusing on school.

At the same time, the close attention Luis's mother paid to his work at the clubhouse allowed her to recognize the importance of capitalizing on his outside interests to motivate his production. She does not seem to have the same detailed level of understanding about what is going on in school.

Mother: "I think if ... if they did have this kind of project in school, it would probably keep a lot of children out of trouble from going into the street and trying to find something bad to do. All they have in school is just work, do your homework and recess time. Sometimes I think that during recess they should, you know, the kids that are interested in doing this, give them reasons ... if they want to do something in the computer, let them go do it instead of wasting their time outside fighting and arguing like other little kids. They don't let them go into the computer only when he's in class. I don't think they ... they don't let them use the computers to do fun projects."

Movie-Making and the Context of School

About once a week, Luis used the Internet for research or used the school computers to type out his school assignments. After he started developing movies at the clubhouse, he found ways to use opportunities at school to advance his own learning and production. He talked about having access to some new programs, including an animation program called Sweep, "like a flash animation but you could draw something, like move it around." He found time to work on his clubhouse movies in school, storyboarding scenes during free time. He found an elective class in school where they used iMovie. As part of the class he worked with a group of peers to create a movie, all participants taking on roles of actors, scriptwriters, and directors.

That same year, he brought a CD of work created at the clubhouse to school and showed many of his teachers there.

Luis: I actually showed [some of my video work] to my teacher and she said she really liked it. She learned that I knew how to animate. The computer lab teacher actually, she liked it a lot. I showed her on the DVD of my work and she really liked it...

According to Luis, his teachers were surprised to see what he was able to do, "They were kind of amazed, like they liked it a lot. Their eyes were wide like looking at it." In addition to making him feel "real happy" and "kind of like proud of himself," this opportunity to represent himself through his movie productions led to new opportunities in school.

Luis: Well, the experience. It's been really fun and uh...all the ideas coming up, the opportunities. Like my PE teacher, he is paying me $50 to make a stop-motion animation about the movie War Ball, I mean, the game. He's going to show it to his class this year to learn about how to play it, more like a "how to" movie.

Luis at the End of Seventh Grade

At the time of his final interview at age 13, Luis had firmly established himself as a video production artist. He had over 30 items stored in his folder on the clubhouse server, including movies and sound files. He estimated that he had taught five boys at the clubhouse about stop-animation movies. His films included a tagline for his production company, "I like bacon productions," a phrase that Luis was fond of using from the TV show *The Simpsons*.

He had plans for the future, including learning more about "stunt choreographing, also fighting, more about making movies and stop-motion" and "making longer movies." In the short term, he was trying to get actors together to make a live-action horror movie based on *Resident Evil* and had Halloween as a due date. More long term, he said he planned "to get myself a computer, and I might become a computer programmer or a video game designer."

Summary and Discussion

In this chapter, we shared the learning history of one child in the form of a techno-biography. Our cross-setting focus led us to describe his activities at the computer clubhouse, at home, and at school. Expanding the temporal dimension from days to years allowed us to chart the development of his engagement in movie-making over time, his design process, and the way that his social network evolved to support his learning. His pattern of sustained activity resulted in more stable interests, depth of expertise, and an identity as a stop-motion animator (Hidi & Renninger, 2006). We were able to understand the ongoing design activities Luis engaged in as he came up with new and creative solutions, tinkered with different ideas and tools, and created new words in his head that he translated through developmental design phases into the shareable medium of video. Encouragingly, these activities reflect those that researchers believe can build critical twenty-first-century skills (Balsamo, 2010).

Figure 8.3 maps his activities using a timeline representation, a visualization that allows us to see patterns of learning activities and to quickly compare the pathways and resources of different case learners. Age is represented along the horizontal dimension and setting (home, school, community center) along the vertical dimension. Although Luis played games on a computer as early as age eight (represented by the computer icon for "first used a computer"), his first digital media productions took place at the clubhouse when he was 11 years old (represented by the light bulb icon for "onset of fluency building activities"). At the top of the diagram, people and other resources that supported each of Luis's activities are noted. As this diagram

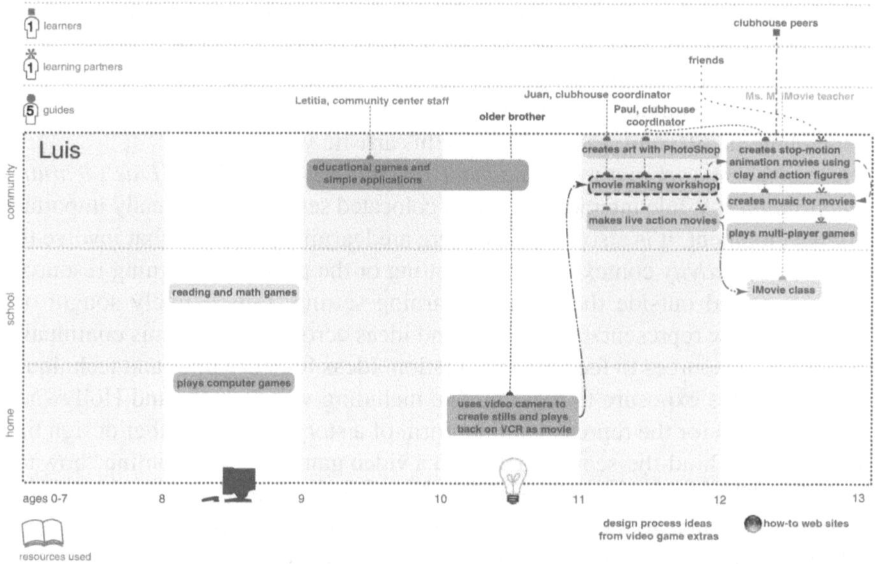

Fig. 8.3 Graphic visualization of Luis's learning pathway over time and across setting[2]

shows, and as the narrative reveals, Luis was able to recruit the help of the clubhouse coordinators, his mother, his brother, teachers at school, and peers in the clubhouse to support his movie-making work. He also drew on Internet-based resources, professional movies, and storyboarding examples within videogames as inspirations and sources of learning.

Luis's timeline makes apparent the centrality of the clubhouse in his learning ecology— it is the main site of his production, providing him with new powerful creative tools, new learning partners, and a dedicated space for daily work on his projects. The timeline also exhibits a path into these activities from his preexisting interest in animation and games, and outcomes from his new skills and knowledge, as he found new opportunities for learning in school. Three important themes emerged in this case that we discuss below.

1. *The clubhouse was critical for the development of Luis's skill and identity as a producer of stop-animation films.* His prior interests and experiences with video and animation from his analog video work with his older brother and his love of video games and cartoons, primed him to take advantage of the digital video opportunities available at the clubhouse. These interests were the basis of his initial inspirations, and the tools and Internet access at the clubhouse enabled Luis to explore further, learn more, and develop his ideas and talents in the area. Specific elements at the clubhouse that supported Luis's interests and allowed

[2] There were no technology learning events to map for Luis before age eight, the age at which he reported he first used a computer.

him to develop them in different ways include access to the tools of production and coordinator recognition and encouragement of his work. The stop-animation camera, the software, his action figures, and clay were basic materials provided to him by the clubhouse and his parents, which he was able to combine and recombine in different ways to produce his artistic visions.

2. *The clubhouse work motivated and made possible diversification of Luis's learning opportunities.* While interactions within colocated settings are critically important for development, it is also clear that there are learning processes that involve the *creation of activity* contexts in a new setting or the pursuit of learning resources that are found outside the primary learning setting. Luis actively sought out feedback, new representational tools, and ideas across settings. Luis continually sought out resources to feed his imagination. Ideas for story lines and techniques came from his exposure to mass media, including video games and Hollywood movies. Ideas for the representational form of a storyboard and other design tips came from behind-the-scenes footage in a video game DVD and online "how to" websites. The continued engagement in film-making created an ongoing cycle of learning for Luis. As he was out in the world viewing stories and professional films, he watched for directors' cuts and angles for shots and their narrative story lines. As he envisioned special effects in his claymation action drama, he generated visual design goals that set in motion the pursuit of learning new techniques that advanced his dual aims of realism and a professional look. At a much broader and perhaps more generative level, his very relationship to film viewing changed such that he was no longer simply watching or being entertained but rather was looking with an eye toward learning in order to expand his repertoire of productive strategies (Gutiérrez & Rogoff, 2003).

His language reflects his attention to aspects of the expressive and designed aspects of what he creates and we hear from him terms such as "choreography," "realistic," and "animate," as he described goals for his future work, marking his membership in the broader community of practice of animators (Lave & Wenger, 1991). This pattern of a practice-linked identity, leading to a persistent pursuit of learning opportunities, is consistent with what we have observed in case studies from more affluent communities and speaks to its generality, at least when a minimum of resources are available (Barron, 2006). His articulation of a possible future self (Markus & Nurius, 1986) as a programmer or game designer provides further evidence of the importance of his cross-setting activity for his identity development.

Across the observations and interviews we found that Luis's learning partners played a variety of roles. Luis's social interactions around his film-making grew as he became more committed to claymation as a medium. As we noted in the introduction, our prior work on parents' roles in their child's learning about computers and technology identified several roles that parents played that directly or indirectly supported knowledge development. These included the roles of learning broker, project collaborator, teacher, employer, resource provider, nontechnical consultant, and learner/audience (Barron et al., 2009). Definitions and examples of these roles are provided in the appendix. In contrast to Lareau's (2003) well-known argument that lower SES families follow a parenting practice of "natural development," Luis's parents engaged his hobbies in important ways. They primarily played roles that did not require expertise with technology, most notably those of resource provider and audience. As he sought out feedback from his family, he developed a sense of the desires of different audiences, and in response to a request for a different genre of story he strengthens his commitment to his own vision. Other pivotal roles were played by the coordinators at the clubhouse (including teacher, project collaborator, learning broker, and resource provider roles), his schoolteachers (including audience and employer roles), his friends at the clubhouse (including audience and project collaborator roles), and his brother (including teacher, nontechnical consultant, and audience roles). In fact, each of the seven roles, originally developed based on research with parents in Silicon Valley, most of whom worked in the technology industry, was filled for Luis by someone in his social network. Luis's parents also played the role of a monitor. In Luis's case, his challenges in school led to restrictions on his club time. His parents believed that access to the club was one way that they could motivate him to work harder on his school assignments. Despite this, increasingly, the growth of his learning network is attributable to his sharing of expertise and the subsequent uptake of his expertise by those at school.

3. *Links between the clubhouse, home, and school were present but could have been stronger.* Despite his slipping grades at school, Luis's work at the clubhouse clearly demonstrated a rich imagination, persistence, attention to detail, and resourcefulness in furthering his own development. These characteristics are markers of the potential to thrive as a learner. Had his teachers been attuned to his expertise development earlier, they may have found ways to build on his excellent out-of-school learning skills to reengage him in academic content. Typically, parents act to coordinate across settings and in Luis's case they intervened to reallocate his time at the clubhouse so that he would spend more time on schoolwork rather than try to bring his talents to the attention of school staff.

   The fact that Luis took the initiative and brought his animations to the attention of his teachers was remarkable and the digital form of his work made it possible. His recounting of the surprise and admiration of his teachers teach an important lesson about the missed opportunities for nurturing a child's development when the school-based work is the primary lens through which teachers come to know their students. The invitation by his PE teacher to create a "how to" video in exchange for financial compensation marked a potentially transformative moment as it sets up yet another design challenge and learning opportunity for Luis while positioning him as a creator worthy of pay. Had the boundaries between the school and the clubhouse been more permeable, mentor–teacher collaborations may have been possible that could have further amplified the possibilities for learning.

In closing, the clubhouse was critical in helping Luis to develop the kinds of technological fluencies that may position him for further learning and creative work, demonstrating that intentionally designed environments can help bridge the digital divides that are of growing concern. As his case study illustrates, understanding the origins and consequences of sustained engagement with content requires research methods that go beyond more commonly assessed near-term knowledge gains (e.g., after the completion of a course). It is important to trace connections between learning activities and to characterize how content-related interests evolve over time and across life settings. As this case study and more broadly the research in this volume suggests, a better understanding of how learning takes place across settings and time, and of the possible synergies and barriers between them, may help educators and parents find ways to supplement school-based or home-based opportunities. The rapid increase in access to information and to novel kinds of technologically mediated learning environments such as online special interest groups, tutorials, or games makes it particularly important to understand how, when, and why adolescents choose to learn and the emergent developmental processes that can arise once interest is sparked. These new opportunities for specialized interests to develop are due to what has been called the "long tail" of learning resources (Anderson, 2006). The Internet, for example, allows for the proliferation of communities of learning that cater to very specific kinds of interests and that are available to anyone who has access to the Internet and the skills to understand them, such that even young learners can develop high levels of competence. It is conceivable that in the future, teachers may be supported to take on the role of learning broker intentionally and that parents who do not perceive the benefit of helping their child pursue their interests or do not find ways to attend to their interest may be encouraged to do so. It is such hobbies and pleasurable pursuits that often provide a sustaining pathway of learning that can pave the way to careers and new ideas about possible selves.

References

Anderson, C. (2006). *The long tail: Why the future of business is selling less of more*. New York: Hyperion.

Balsamo, A. (2010). Design. *The International Journal of Learning and Media, 1*(4), 1–10.

Barron, B. (2004). Learning ecologies for technological fluency: Gender and experience differences. *Journal of Educational Computing Research, 31*(1), 1–36.

Barron, B. (2006). Interest and self-sustained learning as catalysts of development: A learning ecologies perspective. *Human Development, 49*, 193–224.

Barron, B., Martin, C. K., Takeuchi, L., & Fithian, R. (2009). Parents as learning partners in the development of technological fluency. *The International Journal of Learning and Media, 1*, 55–77.

Barron, B., Walter, S., Martin, C. K., & Schatz, C. (2010). Predictors of creative computing participation and profiles of experience in two Silicon Valley middle schools. *Computers in Education, 54*, 178–189.

Beach, K. (1999). Consequential transitions. In A. Iran-Nejad & P. D. Pearso (Eds.), *Review of research in education* (Vol. 24, pp. 101–139). Washington, DC: American Educational Research Association.

Bruner, J. S. (1994). The remembered self. In U. Neisser & R. Fivush (Eds.), *The remembering self: Construction and accuracy in the self-narrative* (pp. 41–54). Cambridge, UK: Cambridge University Press.

Elder, G. (1994). Time, human agency, and social change: Perspectives on the life course. *Social Psychology Quarterly, 57*, 4–15.
Gee, J. P. (2000). Identity as an analytic lens for research in education. *Review of Research in Education, 25*, 99–125.
Goode, J. (2007). If you build teachers, will students come? Professional development for broadening computer science learning for urban youth. *Journal of Educational Computing Research, 36*(1), 65–88.
Gutierrez, K. D., & Rogoff, B. (2003). Cultural ways of learning: Individual traits or repertoires of practice. *Educational Researcher, 32*(5), 19–25.
Henwood, F., Kennedy, H., & Miller, N. (2001). *Cyborg lives? Women's technobiographies*. York, UK: Raw Nerve Books.
Hidi, S., & Renninger, A. (2006). A four-phase model of interest development. *Educational Psychology, 41*, 111–127.
Ito, M., Baumer, S., Bittanti, M., Boyd, D., Cody, R., Herr-Stephenson, B., Horst, H.A., Lange, P.G., Mahendran, D., Martinez, K.Z., Pascoe, C.J., Perkel, D., Robinson, L., Sims, C., & Tripp, L. (2009). *Hanging out, messing around, and geeking out: Kids living and learning with new media*. Boston, MA: MIT Press.
Jenkins, H. (2006). *Convergence culture: Where old and new media collide*. New York: New York University Press.
Jenkins, H. (2009). *Confronting the challenges of participatory culture: Media Education for the 21st century* (White paper prepared for the MacArthur Foundation). Cambridge, MA: MIT Press.
Kafai, Y., Peppler, K., & Chapman, R. (Eds.). (2010). *The computer clubhouse: Constructionism and creativity in youth communities*. New York: Teachers College Press.
Lareau, A. (2003). *Unequal childhoods: Class, race, and family life*. Los Angeles: University of California Press.
Lave, J., & Wenger, E. (1991). *Situated learning: Legitimate peripheral participation*. Cambridge, UK: Cambridge University Press.
Lemke, J. (2000). Across the scales of time: Artifacts, activities, and meanings in ecosocial systems. *Mind, Culture, and Activity, 7*, 273–290.
Linde, C. (1993). *Life stories*. New York: Oxford University Press.
Markus, H., & Nurius, P. (1986). Possible selves. *The American Psychologist, 41*(9), 954–969.
Margolis, J., Estrella, R., Goode, J., Jellison-Holme, J., & Nao, K. (2008). Stuck in the shallow end: Education, race, and computing. Cambridge, MA: MIT Press.
Nasir, N. (2002). Identity, goals, and learning: Mathematics in cultural practice. *Mathematical thinking and learning* (Vol. 4, No. 2&3, pp. 213–248).
Nasir, N. S., & Hand, V. (2006). Exploring sociocultural perspectives on race, culture, and learning. *Review of Educational Research, 76*(4), 449–475.
Penuel, W. R., Korbak, C., Daniels, M., Kim, D., Yarnall, L., Hawkins, J., & Pacpaco, R. (2000). *Community technology centers program findings summary: A review of FY99 grantees' annual performance reports*. Menlo Park, CA: SRI International.
Resnick, M., & Rusk, N. (1996). The computer clubhouse: Preparing for life in a digital world. *IBM Systems Journal, 35*(3 & 4), 431–440.
Warschauer, M., & Matuchniak, T. (2010). New technology and digital worlds: Analyzing evidence of equity in access, use, and outcomes. *Review of Research in Education, 34*(1), 179–225.
Wenger, E. (1998). *Communities of practice*. Cambridge, UK: Cambridge University Press.

Appendix

Notes on Methods

Data Collection

Luis was formally interviewed twice, once in the spring of 2004 and once in the fall of 2005. At the time of the first interview, the survey of Access, Interest, and Experience was conducted. His mother and the two clubhouse coordinators were also interviewed. Observations at the clubhouse took place across 14 months.

Data Representation and Analysis

Our multi-informant interview methods yield reports on learners' histories in the form of conversations between the interviewers, the learners, and their parents. Responses to questions posed by the interviewer include rich information about children's activities, their learning resources, and the ways their parents and peers support their learning, as well as their future goals, attitudes, and interests. We view these as stories or accounts, and realize that retrospective accounts are subject to biases in memory and that the interview situation itself is a social situation that has its own demands. Beyond these informant accounts of learning, the interviews offer a sample of language that can be analyzed with respect to vocabulary, means of expression, and syntax.

In order to maximize the potential for developing new insights from these records, our research team has created a number of intermediate representations that summarize the raw interview data. Each representation highlights unique information contained within the records. These representations include: *narrative texts* that tell a learners' story along a number of set dimensions; *Excel spreadsheets* that tabulate types of learning resources and allow us to code and quantify variables such as the number of people in the child's knowledge network or the number of structured learning contexts a child has participated in; *lists* of the technical terms a learner used while recounting their history or describing a project they created during the Artifact-Based Interview; *formal codes* for parent roles that are applied to turns; *graphs and tables* that present descriptive statistics for each code; *parent participation diagrams* that show which parents played specific roles for each child; and finally, *timeline representations* that locate fluency-building activities across setting and time, depict relations between activities, show the involvement of peers or adults in the activity, and note the types of material resources used for learning. Each type of representation offers us new ways of understanding the activities and learning of our focal learners. For example, the timeline representations offer a quick overview of the onset and duration of activities and where they occur. Bursts of activity and the increasing distribution of learning activity over settings become apparent. Placing these learning maps for different learners side by side has helped us attend to significant variations along dimensions of time and resources.

The first clubhouse was housed in an art museum in Boston, adjacent to MIT. Media Lab graduate students ran the club, collaborating and mentoring youth who came to explore, play, create, and invent with professional computing tools of the day. Visitors to the Boston clubhouse were struck by the energy and productivity of youth in this space and soon funding was made available to replicate this model in the USA and internationally.

Tools for Data Collection

Survey of Access to, Interest in, and Experience with Technology

Several scales from a survey developed for use in previous studies were administered (Barron, 2004; Barron, Walter, Martin, & Schatz, 2010). Both Likert-response items and checklist format questions were posed. The questions were designed to tap into four main areas: (1) students' access to technology at home and school; (2) history of technology use across communicative, entertainment, learning, and fluency-building activities; (3) students' use of formal and informal learning resources; and (4) motivational aspects of learning about technology including interest, confidence, and valuing of technology as a subject and potential career.

Learning Ecologies Interview

This interview is designed to give us a portrait of how the child is learning to use technology across the contexts of home, school, community (e.g., church, libraries, clubhouse, camp), and through distributed resources such as books, tutorials, and magazines. This interview also gets at the child's sense of what it takes to be good with computers, their plans for learning, and how they see themselves in relation to technology. A simple diagram illustrating different settings is used to help focus the child and interviewer's attention.

Artifact-Based Interview

This is a semi-structured interview that is designed to provide a focused look at what kinds of projects youth are doing and how they learned, how the projects came to occur (pathway), and the opportunities for fluency building within different projects. We asked interviewees to select one project to show, but often they would share more than one. Questions focused on the story of creation and their learning, although, when appropriate, we asked interviewees to define terms or share technical knowledge. This interview was video-recorded with the camera focused on the screen and keyboard to capture the visual referent of the interviewee.

Parent Interview

The goal of the parent interview was to obtain a developmental history that would help confirm the information provided by the focal participants and to better understand parent perspectives on their child's activities and how they saw their role in helping their children learn. We were also interested in understanding the parent's own experiences with technology and so we began the interview with a request for them to tell the story of their family and technology.

Learning Partner Interview

These interviews focused on how the adult clubhouse mentors recalled working with the focal case learners and their history of coactivity with them.

Coding Categories for Parent Roles in Learning

	Description	Examples
Teacher (T)	Parent has taught child how to do something on the computer over some period of time, which can be either high- or low-fluency in nature (word processing to programming). The parent possesses more knowledge about the subject than the child	*Father*: And back then the scanner we had was not very good. So that's when, I think, I started showing [child] Photoshop. So he could draw on Photoshop and include it into his reports. I showed him some basic things and he took off from there *Father*: I think [child] picked that up mostly on her own. I explained for her what the concept is though and how you need to define it so that it can be shown anywhere. I told her that
Project Collaborator (PC)	Parent has collaborated with child on a project. The parent may or may not know more about the subject than the child. The project is a shared learning experience	*Mother*: I know that [child] and [father] have a pretty close relationship. He will write macros for him. I know that there has been a lot of collaborative work that is way over my head. They will sit and discuss things. That is a learning process like a work place
Learning Broker (LB)	Parent seeks learning opportunities for child by networking, the Internet, parents, and other information sources. Signs child up and provides necessary support for endeavor	*Father*: The only thing he did get help with is there are always the tidbits of educational information you will not get out of the book, and we sat with one of my MIT buddies. It took him and Caleb about 8 h to get the final bug out of this. *Child*: My Mom signed up for the school newsletter and she said they were saying they were short a few members [on the Robotics team] and does anyone want to sign up
Resource Provider (RP)	Resources provided to the child beyond the family computer (e.g., books, video equipment, software, accounts, etc.) in support of their technology learning Resources can be those owned by the parent and used by the child or purchased specifically for the child	*Child*: Yeah. I make music, too. But I don't have the program right now. My other dad has it and I asked him to bring it next time *Child*: I'm not sure, but I think six, seven. I don't know. And then [my dad] got me an HTML book also so I started learning HTML. I got into websites *Mother*: Yeah, we got him Macromedia Dreamweaver, that is when he started learning how to do the web pages

(continued)

(continued)

	Description	Examples
Nontechnical Support (NTC)	Parent provides information/ advice to child on nontechnical issues such as business or artistic design. This role also covers when a parent gives advice about continuing in his/her learning such as project management, learning organization, or basic encouragement	*Father*: I kind of knew that [child] would be like: 'Oh, we can do this and this and that.' But I wanted him to focus and understand the business side of things. You get a task and you are told to do it this way. I was trying to make him concentrate on the assignment *Child*: In terms of like charging money for [my IT services], I think that's just cause I want to have money for things. My Mom also, she has her own, like a business, just so she can do work for people and take deductions and stuff. So I talked to her about it
Employer (EMP)	Parent employs child for technical services rendered. This role can include a formal paid position or more informal activities such as technical support for a home computer	*Interviewer:* So [mother] has got her own computer? *Child*: Yeah, but I take care of it and that sort of thing, you know what I mean. All the updating and stuff *Parent*: No actually it was because I was working for this new company and I decided it would be good to get [child] to help me, I wanted to bring her something that would be interesting so just told her, heck, why don't you take this piece of software and find really good bugs and I think that I paid her something like $25
Learner/ Audience (LRN)	Parent learns technical skills/content from child or looks at child's work on the computer	*Child*: [Father] didn't even know … he doesn't even … like he always asks me to give him like a…how to do that, like you know, like a tutorial on how to do it *Child*: Well, like I said, I help my Dad when he has questions about his work
Monitor (MON)	Parent imposes rules or limits on child's technology use (time, activities, websites, etc.) out of concerns for child's safety, identity, balance, academic performance in other areas, health, etc.	*Mother*: The amount of time we have allowed him to be on the computer has increased over time. When he was very young if he was spending more than an hour we would try to get him out of there *Mother*: I feel kind of weird with the concept of [child] playing a game with someone that he has no idea who they are. We let him do it online but with someone he knows

Chapter 9
Discovering and Supporting Successful Learning Pathways of Youth In and Out of School: Accounting for the Development of Everyday Expertise Across Settings

Philip Bell, Leah Bricker, Suzanne Reeve, Heather Toomey Zimmerman, and Carrie Tzou

Documenting the Cultural Learning Pathways of STEM Expertise Development

A fifth-grade girl, born in Haiti and adopted into a Seattle family, talked at home about how she wanted to be a chemist or a paleontologist when she grew up. For six months, she spent portions of her Saturdays mixing perfumes, as a chemist might, with her mother. But her public schoolteacher, who is a seasoned professional with sophisticated teaching expertise, did not believe the girl always put forth her best effort and was surprised to see her become highly excited about and engaged in a science curriculum unit at the end of the year that the girl counted as "real science." A fourth-grade boy in the same school was often moved to the back of the classroom because he was "off task" and "resistant" to the school curriculum. He spent periods of his time in the back of the room mentally deconstructing the physical environment around him, "thinking in structures" as he put it. Unbeknownst to his teachers, the boy had been deepening his participation in a hobby—an elective vocation—since

P. Bell (✉)
Learning Sciences & Human Development, College of Education,
University of Washington, Seattle, WA, USA
e-mail: pbell@u.washington.edu

L. Bricker
School of Education, University of Michigan, Ann Arbor, MI, USA

S. Reeve
Bellevue School District, Bellevue, WA, USA

H.T. Zimmerman
College of Education, Pennsylvania State University, University Park, PA, USA

C. Tzou
Education Program, University of Washington, Bothell, WA, USA

B. Bevan et al. (eds.), *LOST Opportunities: Learning in Out-of-School Time*, Explorations of Educational Purpose 23, DOI 10.1007/978-94-007-4304-5_9,

attending a summer design program at a local park when he was six years old. Outside of school he engaged in sophisticated design, construction, and building projects with all manner of physical and technological objects. It would be three more years before he came to understand that there is such a field as engineering and that it might be a good match for his interests. By that point it would be much more difficult to make his way along the typical academic path. To simply say that these youth may be "at risk" for making their way along academic pathways ignores the depth of their academic-related interests and developing expertise. It skirts the evaluation and positioning of them that occurred in different contexts based on a partial understanding of who they were at the time and who they wanted to become, and it severely discounts the complexities associated with them productively pursuing and becoming who they might wish to become. We argue that we need to discover and then support the successful learning pathways of youth across social settings over developmental time so that we can promote the development of interests and expertise that may lead to both academic and personal success.

Learners navigate a range of diverse social, material, and discursive contexts everyday—from the classroom to home, after-school programs, informal education institutions, and out into their communities—with a variety of purposes and value systems in place (Banks et al., 2007). Learning is accomplished across these diverse pathways of participation in activity and affiliation with cultural groups in ways that the field of education barely understands. Our empirical literatures tend to focus on the details of learning that occur (or fail to) within specific contexts (e.g., instructional-units-taught classrooms, programs offered by museums, or moments of elective activity in family life). We need more complex empirically informed theoretical models for how learning is accomplished and impeded across sociocultural contexts throughout the diverse social niches and networks of activity in society. We lack theoretical ways of accounting for the learning processes involved with extended pathways of deepening participation and expertise development across physical settings and social groups along developmental timelines. Toward these ends, this chapter describes and reports on a longitudinal study of child development as it occurs across the breadth of contexts present in the lives of diverse youth.

US society is becoming increasingly diverse in terms of ethnic and racial group membership, immigrant status, and linguistic variation. Some schools are now serving significant numbers of nonwhite youth or immigrant youth for the first time. Yet, as evidenced by a long history of inequitable outcomes in science education, few teachers or university professors are equipped to work effectively with all students. In our research, we give central attention to individuals and groups from historically nondominant communities, in order to expand scientific accounts of learning related to how all people learn. We continue to be in need of theoretical accounts of the learning worlds of nondominant groups, in order to understand the normal variation present in the circumstances of learning found in society writ large. Such an approach also helps to expand and develop our theoretical accounts in ways that have more direct benefits for society (Flyvbjerg, 2001).

Cultural and ecological perspectives are increasingly understood to be central in the scientific understanding of learning and development, and they have strong implications for educational practice (Banks et al., 2007; Bell, Lewenstein, Shouse, & Feder, 2009; Gutiérrez & Rogoff, 2003; Lee, 2008; Moll & Gonzalez, 2004; Rosebery, Warren, Ballenger, & Ogonowski, 2005). Scholars have made significant progress in describing how science learning is influenced by the cultural histories, practices, and values of learners and communities (Barton, Ermer, & Burkett, 2003; Bell, Bricker, Lee, Reeve, & Zimmerman, 2006; Lee & Luykx, 2007; Medin & Atran, 2004). In response to the complexities associated with taking a holistic view of how people learn across settings and the cultural variation in human activity present in society, we agree with Lee (2008) that a major program of interdisciplinary research in the field should focus on better understanding the multiple pathways associated with socially significant learning and development of youth.

Over the past several years, we have been developing a theoretical framework focused on everyday expertise development that seeks to account for the social and material dimensions of sophisticated domain learning as it relates to the interests and practices of individuals and their communities. More specifically, our empirical project has involved: (a) documenting the range of expertise that youth develop and apply in their lives that is personally consequential and meaningful to them; (b) understanding the learning pathways associated with the development of that expertise and the myriad sociocultural forces that give shape to those pathways; and (c) aiding in the systemic coordination of successful learning pathways that are meaningful to youth, their families, and their communities and studying the effects of those efforts. We are ultimately trying to understand the extended learning pathways of youth at this historical moment in order to shift and stabilize those pathways by recognizing and leveraging their developing competencies across the range of informal and formal learning environments in which they participate.

In this paper, we describe our efforts to engage the driving question: How do everyday moments—experienced across settings, pursuits, and social groups—result in expertise, sophisticated understanding, and expert identification? We have focused on theory development related to this question as a result of specific gaps in our literatures on learning and expertise development and given present opportunities for interdisciplinary research and synthesis on how, why, and where people learn science across settings over developmental timescales (Bell et al., 2006). As Bransford and Schwartz (2009) argue, "it takes expertise to make expertise," acknowledging that the social processes that support expertise development are understudied and undertheorized. The ultimate explanatory goal of our effort is to better understand the extended learning pathways (e.g., related to the accomplishment of expertise development in science and other domains) that are culturally architected through complex sequences of contingent interaction and activity that occur across the breadth of everyday life.

Cultural learning pathways are conceptualized as a series of linked actions where individuals are positioned—or position themselves—in ways that deepen their participation in a practice amidst a myriad, and often competing, set of different systems of competency. These systems of competency operate throughout cultural experiences taking place across the breadth of social groups and settings in a learner's life. These are complex processes. We pay heightened attention to the various social processes that shape learning, how stylistic forms of talk occur within and across different settings and influence learning, the affordances and constraints of material resources that help us understand the accomplishment and evaluation of situated performance, the multiple cultural meanings that circulate around specific domains of activity, and the linguistic forms of talk that shape and inform learning pathways (e.g., how sense-making gets accomplished by groups across social encounters). Before describing research findings related to this broad effort, we start by summarizing the study.

The Purpose of the Study: Documenting Life-Long, Life-Wide, and Life-Deep Learning Pursuits and Pathways

Our theoretical and empirical accounts of learning need to more directly mirror how people learn as they routinely circulate through the settings, activities, and pursuits of everyday life. Disparate accounts of learning and teaching that exist in balkanized literatures need to be brought together, juxtaposed, coordinated, and synthesized—or actively differentiated. The project described in this chapter is an attempt to more holistically account for human development and learning in cultural and cognitive terms by documenting the myriad of activity systems and consequential decisions that individuals and groups navigate and constitute as a fixture of social life. As we detail below, our conceptualization of *everyday expertise* involves a complex coordination of the personal meanings, cultural practices, identities, motives, underlying ideologies, and the specific learning resources that come to be intertwined, as learning pathways open up in conjunction with the development and application of personally meaningful or locally consequential expertise. The work has broad implications for understanding learning as a cultural and cognitive phenomenon that shapes, and is shaped by, a complex, interacting mixture of social forces associated with the formal, informal, and nonformal educational institutions present in the lives of youth. The approach sheds insight into the contributing and interfering influences of various formal and informal learning environments, and related institutional routines and systems, in the development and application of everyday expertise (Bell et al., 2006).

Building upon the broad conceptualization of learning developed in the consensus study of how people learn in diverse environments (Banks et al., 2007), we orient to the three conceptual dimensions of learning:

1. *Life-long Learning* refers to the acquisition of fundamental competencies and a facility with real-world information over the life course—from infancy to old age. Generally, learners prefer to seek out information and acquire ways of doing things because they are motivated to do so by their interests, needs, curiosity, pleasure, and sense that they have talents that align with certain kinds of tasks and challenges.
2. *Life-wide Learning* shows how learners navigate diverse social ecologies each day as they circulate through everyday activities and settings—from the classroom to

home, after-school programs to informal educational institutions, and into their communities and online spaces. Learning derives, in both opportunistic and patterned ways, from this breadth of human experience and the related supports and occasions for learning—in ways we do not really understand. As a result of the boundary-crossing nature of social life, people need to learn how to navigate the different underlying assumptions and goals associated with education and development across the settings and pursuits they encounter.
3. *Life-deep Learning* embraces religious, moral, ethical, and social values that guide what people believe, how they act, and how they judge themselves and others. In these ways, learning, development, and education are tied deeply to value systems—although frequently implicitly.

In the empirical and theoretical aspects of our project, we give primacy to "recovering persons" as causal agents in their own learning. In arguing for using ethnographic approaches to better detail human development, Jessor (1996) makes the development of ethnographic cases a scientific priority. This chapter theoretically frames, argues for, and empirically showcases how personally consequential science and technology learning is accomplished across the social ecologies of everyday life by youth and families within an urban, multicultural community.

Conceptual Themes of the Study

Over the course of five years, using a team-based ethnography approach, we have conducted a longitudinal study of youth development and learning across the social settings of their lives. We have employed a mixture of ethnographic and experimental methods to help us navigate into the social lives of these youth, their families and friends, and their classmates and teachers in order to identify successful and unsuccessful learning pathways. We consider how specific pathways and associated outcomes can be viewed as successful (or not, or indeterminate) from both member-driven (emic) and analyst-driven (etic) perspectives.

To bind and focus the work, we have focused on four conceptual themes—or topical spaces of concentrated data collection and analysis—in this study:

1. *Personally Consequential Biology*: How do youth learn about the living world across social settings and apply that understanding in their own lives? The focus is on consequential topics: personal health, nutrition, and local environmental conditions.
2. *Everyday Argumentation*: What are the forms of argument youth engage with and construct across settings? How do they learn about and through argumentation?
3. *Images of Science and Self:* Based on the various accounts and images they encounter, what do youth count as "science" and why? How do these images influence their own identity formation?
4. *Technological Fluencies*: How do youth learn with and about digital technologies? How are technologies a focus of their learning or bound up in the learning of other domains of interest?

The Everyday Expertise Framework: How Significant Learning Is Accomplished Socially and Materially in Everyday Life

Although the expert/novice literature has shed significant insight into the nature of disciplinary expertise and competence, it has not given enough clarity to the everyday forms of significant competence rooted in social life. In contrast to more traditional mentalistic accounts of expertise, we conceptualize everyday expertise as a social construct that is given meaning and form within specific cultural ecologies of practice (Cole, 1996; Gutiérrez & Rogoff, 2003; Hutchins, 1995; Lave & Wenger, 1991; Saxe & Esmonde, 2005; Scribner, 1984). We give primacy to problem domains of everyday life where situated judgments, with corresponding meanings and consequentialities, are made (cf. Spradley, 1979). In this view, specific aspects of disciplinary domains are viewed not as end goals in the development of expertise, but as composite elements that serve to make up what are taken to be successful solutions to problems from the perspective of the learners and those in the local contexts in which they participate.

In contrast to a reductionist theoretical accounting, we are actively striving to understand the "buzzing complexity" of social life associated with learning pathways as they get architected, navigated, and renegotiated. In contrast to experimental traditions that might seek to develop *corridors of (parsimonious) explanation* across multiple levels associated with a phenomena (e.g., cognitive system neuroscience, perceptual/sensorimotor, cognitive behavioral), we actively seek to develop a scientific accounting of the *blankets of contextual explanation* that render the complex systems and interacting phenomena and features of social life associated with successful and unsuccessful learning pathways.

The Strands of Domain Proficiency: A Multifaceted Approach for Understanding Expertise Development

At the core of our framework for the development of everyday expertise, we focus on how people develop means of participating in science, technology, engineering, and math (STEM) domains in increasingly sophisticated ways. We leverage recent consensus reports from the National Research Council that summarize research on science learning and define six strands of science expertise development (or disciplinary proficiencies) (Bell et al., 2009; Duschl, Schweingruber, & Shouse, 2007), and we add a seventh, navigation knowledge. In situated moments of activity, the seven dimensions are intertwined in complex ways (e.g., symbolic knowledge is frequently learned and marshaled through social sense-making routines like argumentation; knowledge is leveraged and manifested in judgments and moments of material practices). But, each strand also represents an important and unique aspect of what is being learned associated with STEM practices. The seven strands, taken together, define the "outcome space" associated with sophisticated STEM learning (see the center of Fig. 9.1 below).

Fig. 9.1 Bridges and barriers in everyday expertise development in relation to the strands of domain proficiency

The seven strands of STEM proficiency and expertise we focus on are (1) personal interest in the domain, (2) social sense-making routines (e.g., forms of reasoning, explanation, or argumentation), (3) social and material practices (i.e., specialized ways of talking and acting), (4) symbolic knowledge (i.e., disciplinary facts, concepts, models, and explanations), (5) navigation knowledge (i.e., how people learn to support their own learning with resources and experiences), (6) knowledge of the enterprise (i.e., what counts as disciplinary work, how it relates to everyday life and society), and (7) a domain-linked identity (i.e., coming to think of oneself as someone who knows about, uses, and sometimes contributes to science). We take these seven strands as the focus of "what people develop" during STEM expertise development. Next we specify social and material influences on the development of these seven strands.

Social and Material Supports for Extended Learning: Bridges and Barriers in Everyday Expertise Development Across Encounters

What are the social, material, and cultural processes that shape the learning of these intertwined strands of proficiency? Our theoretical stance on expertise development builds upon the social, cultural, and material perspectives associated with situated perspectives on learning (cf., distributed cognition (Hutchins, 1995), situated learning (Lave & Wenger, 1991), the agency-identity framework (Holland, Lachiocotte, Skinner, & Cain, 1998), and critical feminist perspectives (Barton et al., 2003; Suchman, 2007)). These perspectives allow us to develop a theoretical and empirical understanding of the social and material influences on what is taken to be sophisticated learning and activity that occurs within and across social contexts. Such situated moments, exhibiting significant cultural variation, are often contested among social actors, and are inequitably available to individuals and groups. The Everyday Expertise framework allows for an accounting of how moments of situated meaning and activity (e.g., how a child is positioned to have relevant expertise for an immediate task) are contingently related across a series of encounters influenced by multiple actor-networks operating with multiple systems of competency at a given moment (e.g., formal instruction shaped by high-stakes accountability pressures relative to actions related to peer youth culture). Within the range of efforts that focus on the cultural and material accomplishment of complex disciplinary activity, our approach resonates heavily with the actor-network theory view (Latour, 1987; Law, 1999). This perspective postulates that activities are best understood by examining how actors-in-activity create the operating material networks in which they are situated. In terms of equity dimensions of the analysis, we focus on how (in)equalities in participation and recognition are discursively manufactured and regulated in these situated moments and we highlight the broader, arranged actor-networks that influence such dynamics (e.g., how educational accountability systems shape the evaluation of what teachers count as relevant expertise in the classroom—and what expertise gets marginalized). Expertise is then taken to develop along extended cultural learning pathways that get architected across social encounters over the course of developmental time (Bell et al., 2006). Learning is afforded or constrained across settings through the material resources that are available to shape actions, the value systems that are operating to evaluate actions, and the specific bridge-and-barrier mechanisms that are in place to explicitly or implicitly connect the meanings and actions from one moment to prior one.

Methods and Data

The data utilized in our work were collected as part of a four-year team ethnography. Researchers followed the same youth across the settings of their lives to study how these youth learn about science and technology, as well as develop various areas of expertise (Bell et al., 2006). In the spring of 2005, researchers formed a partnership with a local elementary school (pseudonym Granite), which caters to a student body that is diverse with respect to ethnicity, nationality, languages spoken, and socioeconomic status. In the fall of 2005, researchers began recruiting families into the ethnographic study. Thirteen families agreed to participate and the sample of focal participants from each of those families was balanced for age (six youth were in fourth grade and seven were in fifth grade at the beginning of the study) and gender (seven boys and six girls). Besides the focal participants and their immediate family members, extended family members (e.g., grandmothers, cousins), teachers, and peers consented to participate in the study.

This team ethnography charted the learning of 123 people—including 99 youth, 13 in great depth— over multiple years from an urban, multicultural, multilingual community with significant levels of poverty. We conducted thousands of hours of fieldwork over the first three years and have followed up with many of the participants at a lower level of fieldwork over subsequent years. After we developed a saturated accounting of our conceptual themes (and families participated in the study for one to four years based on their circumstances), primary data collection was pared back and analysis has been expanded in the latter years. Periodic visits to the homes of many families are still being conducted. Fieldwork in the school remains at a high level, although it has increasingly focused on research surrounding collaborative curriculum design research in science.

A guiding methodological principle of this research was to follow the same people as they moved across settings. The majority of the observations of the focal participants took place in school and at home. However, focal participants were observed participating in activities and interacting with others in many additional settings, such as religious institutions, after-school clubs, museums, sporting events, camping excursions/vacations, neighborhoods, and parks. Across these settings, data collection methods included (a) observation and participant observation; (b) interviews (both ethnographic and clinical); (c) self-documentation techniques, where focal participants were given digital cameras and asked to document various objects and phenomena (e.g., argumentation) and then answer questions about their photographs; and (d) document collection. Two surveys, designed to gather information about socioeconomic status and ethnic identity and participation in science respectively, were administered. Researchers also conducted analyses of public census tract data for the neighborhoods in which families lived.

The resulting data corpus was constructed from over 2000 hours of in situ video-recording and field-noting across dozens of social settings (homes, classrooms, neighborhoods, etc.). Data sources included (a) *field notes* of observations, interviews, participant self-documentation assignments, and documents collected; (b) *video- and audio-tape* of observations and interviews (when in settings that allowed video and/or audio taping); (c) *digital photographs* taken during observations and interviews; (d) *video and/or digital photographs* taken by participants as part of their self-documentation assignments; (e) *documents* collected during family visits (e.g., magazines, school work, writing samples from clinical interviews, written survey responses); and (f) *survey results*.

Lines of Research

The compiled data set of people engaged in everyday expertise development supports a broad variety of analyses. We are currently pursuing the following lines of analysis and theory development in conjunction with the methodological approaches described.

Bridges and Barriers in the Learning Pathways of Everyday Expertise Development

Current analytical efforts are documenting *the multitude of cultural learning pathways* associated with expertise development that come into existence through a coordination of social and material influences across settings and over extended timescales of activity as people come to more deeply participate in a set of personally consequential social practices (Dreier, 2009). In collaboration with colleagues in the interdisciplinary Learning in Informal and Formal Environments (LIFE) Center, we have been identifying a set of socially occurring bridges and barriers that influence the learning along extended learning pathways of expertise development. Studies of early expertise development highlight the *multiple roles of learning partners* to cultivate expertise of individuals (Barron, Martin, Takeuchi, & Fithian, 2009; Bell et al., 2006; Crowley & Jacobs, 2002). Among these important roles is the recognition of early interest in the domain of the learner (e.g., by a parent) and ongoing efforts to *sustain interest by mediating and architecting subsequent choices*. For example, parents provide material resources to learners; they broker access to future learning experiences; and they arrange for more expert-others to teach their children how to improve their practice. Learning is also accomplished in situated moments of activity through an *exploitation of flexible learning arrangements* found in particular contexts—the leveraging of social and material resources to accomplish sophisticated action (Stevens et al., 2008). We have also discovered that learning is accomplished across settings through *interdiscursive uses of language*—specific linguistic terms and styles of talk that connect multiple encounters. We have been able to analytically connect sense-making in one situated moment to that in prior moments in the developmental history of learners by attending to the linguistic details of participant talk.

Through a series of encounters with situated activity, learners often develop *social reputations* for these interests and subsequently as "developing experts" in the domain. Such reputations serve both as a marker and a maker of expertise. That is, social reputations denote developing expertise and also provide an entrée to subsequent related learning experiences (e.g., providing a youth with a social reputation as an expert in the Halo videogame can put him/her in more challenging gaming scenarios with other experts). Such reputations and opportunities to learn are strongly influenced by the local *positioning dynamics* (Harré, 2008) constructed through talk and action that assign and regulate the expertise-related rights and responsibilities of individuals within particular moments of activity (e.g., whether or not a young person is positioned as having relevant science knowledge related to classroom instruction). Often, these positions are influenced by *cultural stereotypes of domains* (or storylines) that circulate in the culture more broadly in relation to domain context and specific demographic groups of learners (e.g., whether women excel at doing science and whether girls should be encouraged in learning science). Such stereotypes and supportive positioning dynamics have a strong influence on whether learners come to personally identify with the domain. We have documented how negative positioning dynamics can keep significant STEM-related expertise from being recognized in specific learning environments, although it is rhetorically of interest (e.g., how youth with significant material competencies can be seen as not having relevant expertise for science instruction in school; Bricker & Bell, in review).

Documenting Children's Understanding of Health

In the context of this team ethnography, we have documented the focal participants' health-related beliefs and behaviors through the use of photodocumentation tasks, semi-structured interviews, and two case-study analyses (Reeve, 2009). Given the far-reaching consequences of health-related decisions and recent increases in childhood obesity, type II diabetes, and other serious conditions, everyday management of personal health is an area of expertise that must be better understood (Reeve & Bell, 2009).

Youth Understandings of "Healthy" and "Unhealthy"

We asked each participant to document in words and photos the range of everyday materials and activities he/she believed was healthy or unhealthy (see Clark-Ibanez, 2004 for a background on self-documentation as a general method). In ethnographic interviews debriefing this activity, young people expressed a surprising breadth of meanings for the concept of health, including weight gain or loss; mental and emotional health; environmental health; organic or "natural" foods; health as determined by growth, strength, or color; cleanliness; and elements of the natural environment that help to sustain human life (e.g., trees that produce oxygen, air for people to breathe). Each participant also described health from multiple perspectives, often giving explanations that incorporated different definitions of health for the same object, or that described complex and nuanced ideas. The responses of the youth also revealed meanings that served specific functions for them and their families and were rooted in recurring home activities.

Semi-Structured Interviews About Health and Nutrition

We also interviewed each focal participant about his or her understandings of five areas related to health and nutrition: (1) staying healthy; (2) sickness and wellness; (3) questions the youth had about health or food-related topics; (4) images of medical careers; and (5) what food is and why people need it.

Again, these youth had multifaceted ideas about the five areas, showing that young people can simultaneously hold multiple ideas about scientific processes (cf. diSessa, 1988). Their responses also illustrated that explanations rooted in folk traditions or everyday experience do not necessarily signal the absence of more accepted understandings. For example, seven of the 13 youth suggested that illness can be related both to transmission of germs and to temperature- and weather-related factors (e.g., playing in the rain without a jacket). Although Western science typically recognizes only the former explanation, science educators have a great opportunity to investigate young people's multiple ideas through discussing recent research on this topic (e.g., Johnson & Eccles, 2005; Lowen, Mubareka, Steel, & Palese, 2007) and through helping students think about different ways to evaluate evidence and the dynamic nature of scientific knowledge.

Young people's questions about health and nutrition, another area investigated in the interviews, largely focused not on *what* to do to stay healthy, but on *how* and *why* such behaviors work, as well as topics they had heard about recently or that were relevant to practices in their own families (e.g., where do diseases [bird flu, AIDS, etc.] come from? How does calcium help to build muscles?). Their questions suggest significant thought and curiosity about health-related topics, even at this relatively young age; current curricula and instruction would do well to listen to and address such complex questions that reflect young people's personal areas of interest.

Case Studies of Health Practices: Everyday Health Expertise and Cross-Cultural Forms of Health Care

Two case studies represent different kinds of everyday interactions with health and nutrition (cf. Flyvbjerg, 2001). Because of his mother's work as a home-based distributor for two health-related products, one boy's home became a unique learning environment that provided him with instrumental knowledge relative to managing his own health and shaping his future career goals. Bob (pseudonym) learned about health through hearing and taking up distinct types of discourse (e.g., sales claims); reading and hearing print, audio, and visual media; personal experience with serious illness (e.g., chronic food allergies); and the modeled behavior of his mother and her business associates. Despite his deep knowledge and experience, however, we rarely saw Bob make connections (either at home or at school) between classroom curriculum and his health- and nutrition-related home activities. Bob's case suggests opportunities for science curriculum and instruction to help young people see relationships between scientific content and issues that are important to them and their families (Banks et al., 2007; Korpan, Bisanz, Bisanz, Boehme, & Lynch, 1997).

A second boy, who immigrated with his family to the USA from the Philippines at the age of six, grew up in a context of transnational health-care use (across the USA, Vietnam, and Canada). His family flexibly used professional health-care providers, home or over-the-counter remedies, and traditional folk treatments as they made health care decisions (cf. Chrisman & Kleinman, 1983). Luke's (pseudonym) home interactions with health also occurred in contexts of social, economic, and personal significance, such as a grandmother's serious illness. In sharp contrast to Luke's experiences, however, his formal science and health education focused only on Western systems of knowledge and largely separated out the social and economic factors that were closely intertwined with his family's everyday decisions.

These data lay the foundation for increasingly relevant, health-related science curriculum and pedagogy, and underscore the importance of taking nonschool experiences into account when designing and delivering health-related instruction, especially for vulnerable and historically marginalized populations who experience increasing health disparities (Agency for Healthcare Research and Quality, 2008; Lee, 2002; US Department of Minority Health Care and Health Disparities, n.d.). By incorporating health topics into instruction, science education has a golden opportunity to help young people make sound health decisions and increase their long-term quality of life.

Documenting the Everyday Argumentation of Youth

With respect to argumentation, we examined the argumentative practices youth utilize in their activity across settings and over time (see Bricker, 2008). We examined youth everyday argumentation within activity to better understand the learning affordances of this discourse practice and also to dialogue with the science education community, which currently proposes that youth in science classrooms should learn how to argue scientifically in order to mimic actual scientific practices in which argumentation plays a central role in knowledge construction (e.g., Duschl et al., 2007). Designs of learning environments meant to engage youth in school science with what it means to argue scientifically have to date not attended to the existing argumentation practices of youth,[1] although the field is strongly oriented to utilizing students' prior knowledge in instruction (e.g., Bransford, Brown, & Cocking, 2000; Linn & Songer, 1991). We have argued that curricula and instruction designed to engage youth with what it means to argue scientifically could be much better informed by youth's everyday argumentative competencies (e.g., Bricker & Bell, 2008).

We know that youth bring a rich set of argumentative practices to formal education (cf. Corsaro, 2003; Corsaro & Maynard, 1996; Goodwin & Goodwin, 1987; Kyratzis, 2004; Ochs, Taylor, Rudolph, & Smith, 1992). They routinely interpret

[1] For an exception, see the work of the Chèche Konnen Center at the Technical Education Research Centers (TERC) and publications from those Centers, such as Hudicourt-Barnes (2003).

and produce arguments as they navigate the social settings and activities of their lives but rarely, if ever, are these practices acknowledged and utilized by those designing argumentative learning experiences. To guide our investigations of youth everyday argumentation in order to add to the literature base and possibly inform the design of learning environments, we asked the following research questions: (1) What meanings do youth associate with argumentation and how do they describe aspects of their argumentation practices? (2) How do youth report learning how to argue and do youth argumentation practices help us understand how youth learn? (3) What are the relationships between youth, family, and community culture and argumentation?

Youth Understanding of "Argument"

What do youth associate with the word "argument" and how do they characterize their own argumentative practices? Findings indicate there is enormous variety with respect to youth ideas about argumentation and their accounts of their practices. Furthermore, youth appear quite capable of explicating the fine-grained details of their argumentative practices, some of which are quite sophisticated. We found, however, that without asking youth about their argumentative practices as associated with *specific* activity in *specific* settings, youth tend to associate the word "argument" with fighting, yelling, and inappropriate behavior in general, which has implications for engaging youth in school science with what it means to argue scientifically.

Cultural Grounding of Youth Argumentation

What are the relationships between argument, learning, and culture? Youth utilize culturally influenced frames associated with argumentative practices, such as argument as decision-making and/or problem-solving or argument as social/political protest, in order to make sense of those practices within activity. Findings also show that some youth identify argumentation as a learning practice (e.g., Billig, 1987/1996), highlighting its similarity to critique and its role in helping to make ideas visible so that others can learn from those ideas. This has important implications for utilizing aspects of and details about youth argumentative practices in curricular and instructional design.

Forms of Argumentation That Cross Settings

How do youth linguistically construct arguments that invoke life experiences from different settings and over time? Findings indicate that youth use linguistic elements (both verbal and nonverbal), such as discourse markers, evidentials, and indexicals when bringing evidence to bear on their claims (cf., Aikhenvald, 2004; Schiffrin, 1987).

Furthermore, findings show that some of these linguistic elements mark sources of evidence and are helpful in identifying when youth learn something in one setting and transfer it to another setting. Determining what aspects of youths' linguistic competencies are useful for curricular and instructional purposes and how those identified aspects should be utilized as curricular and instructional tools are important areas of future study.

Youth Perspectives on Argumentation in Science

Lastly, how do youth perceive the role of argumentation in the sciences and what are their thoughts about being asked to argue as part of their school science experiences? While many youth understand that argumentation is a critical practice in the sciences, many conclude that such efforts in science education are "strange" given that argumentation is not an activity they code as appropriate in school settings, save for specific exceptions (e.g., persuasive writing in English/language arts classes). Findings indicate that the culturally influenced frames associated with certain in-school activities, such as science class, for example, might inhibit youth from employing their argumentative practices during those activities, even when they routinely employ them as part of other activities across the settings of their lives.

Who Counts What as Science?

To explore the conceptual theme related to images of science and self, Zimmerman (2008) analyzed youths' ideas about science and their science-related talk and activities across school, home, and neighborhood settings. This line of work has two goals: how youth define science in consequential moments of their lives and how this definitional work relates to how youth participate in science-related practices across settings (McDermott & Webber, 1998; Stevens, 2000). Through this work, we empirically documented developmental trajectories that began to distance and disenfranchise youth from science and those that brought youth closer to science.

Youth Images of Science

Because of concerns about the decreasing interest that children show toward science as they move into middle school (e.g., Zacharia & Barton, 2004) and because so much of the research imposes an external framework which judges children's views on science (cf. Driver, Leach, Millar, & Scott, 1996), an analysis was conducted to give voice to how the youth perceive scientific practices in their daily activities. To accomplish this, we developed a game-like task, called the Science Activity Task (SAT) where the focal participants rated the frequency of the activities that they did and then reflected how these activities connected to scientific knowledge, practices, and tools (Zimmerman and Bell, in review). The analysis of the SAT found that youth participated in scientific practices and saw science in their homes and community activities as well as in school. We identified design principles such as *build science activities from the learner's preexisting connections to science* (i.e., mixing things together, conducting informal experiments at home, and understanding how iPod® and like devices work) rather than traditional home–school connections often featured in curriculum. For example, the youth, as a group, did not see science connections to building with Lego® or to sports.

Youth Participation and Identification with Science

Developmental biographical accounts of learning showed how science practices were embedded in the activities and practices of two young women, Penelope and Raven, and how these crossed multiple social settings. These accounts examined how youth performed science practices within activities and how youth crafted different pathways toward science for personal goals. Both girls reported scientific people as doing certain scientific practices like observing, teaching, and measuring. For Penelope, science was when she was engaged with content around nature, technology, or school science. For Raven, science was when something changed from one thing into another (i.e., when a seed grows into a plant) and when a person discovered something as in an archeologist. In both cases, a tension was present between participating in science and having their participation not be seen as negative by peers or having a negative impact on time to be spent on other personal goals.

Raven and Penelope were recognized for their science work in elementary school, yet they both stopped participating in science-related out-of-school programs in middle school. Raven found her academic enrichment program overwhelming and prohibitive of her nonacademic pursuits. She projected that if she remained in the enrichment program during the sixth-grade school year, it would adversely affect her ability to make honor roll in middle school—her personal goal. Penelope participated in a science after-school club during the academic year offered by a sixth-grade science teacher, yet she also ultimately disengaged from this club. Penelope expressed concerns about enrichment programs as having too much work. Penelope and her mother Eve agreed that school is important, but they stressed that a balance is needed; getting good grades without time for fun is not a fair exchange.

Social Supports for Science Learning

In looking at who youth tapped as learning partners, Raven and Penelope were assisted in science by people that would not be normally classified as scientific. Their social networks included a guardian's golfing partner, nurse's aides, owners of home businesses, pet shop workers, godparents, farming grandparents, former teachers, peers, and more.

Results (Zimmerman, 2008) have implications for the assumptions we make about youth and their development, to learning theory, and to the development of design principles for program developers of informal spaces and formal education curriculum. First, the youth in this study had complex practices in their homes related to science in multiple domains as well as school. Second, children's interests in science were not always aligned to the school science content, pedagogy, or school structures for participation, yet youth like Raven and Penelope found ways to engage with science despite these differences—through crafting multiple pathways into science. A positive outcome was that the youth who did not connect to science at school found a space at home to participate in science in their hobbies and other personal pursuits. Third, urban parents were active supporters of STEM-related learning environments through brokering access to social and material resources. The brokering involved a full deployment across the social network, bringing in fellow church parishioners, family members, godparents, retail workers, and friend within and outside of formal science connections. A final result was that the natural world was a relevant context for urban youth to learn about science, albeit in nontraditional ways. The connections with houseplants and animals as pets provided opportunities to integrate cultural understandings, build competencies with scientific practices, and develop expertise relevant to peer and community groups.

Connecting Repertoires of Practice Across Home, Community, and School Boundaries: The Micros and Me Science Curriculum

For many students, learning science in school is like learning another culture (Aikenhead, 1996). Students may not see themselves or their ways of knowing reflected in the practices of science in school. We argue for the need to diversify the images of science that students encounter in school so they may come to see themselves as people who can do science, based on images that reflect actual scientific practices and their own culturally based ways of knowing. Gutiérrez and Rogoff (2003) argue that we need to see individual students as having their own experiences and histories that are influenced rather than dictated by their membership in certain cultural groups. In this way, we can start to see students as coming to the classroom with various repertoires of practice, or areas of everyday expertise, that stem from their membership in multiple groups—peer, cultural, community, and family, just to name a few.

This section describes a design-based research effort aimed at constructing learning pathways between students' culturally based *repertoires of practice* (Gutiérrez & Rogoff, 2003) around health and school science. We asked two questions: (1) How can we *elicit and make visible* students' everyday expertise around health in science instruction? (2) How can we *deeply connect* this expertise to authentic scientific practices? In the unit, we used the self-documentation technique described earlier in this paper to elicit students' repertoires of practice and leverage them in classroom science instruction. Our findings showed that self-documentation shows promise in both eliciting and complicating culturally based practices in classroom instruction.

In this study, we designed a seven-week curricular intervention for fifth grade that was studied across four enactments called *Micros and Me* (Tzou & Bell, 2010).

This curriculum attempted to (a) make science more personally consequential to students' lives, and (b) connect authentic scientific practices deeply with students' areas of everyday expertise. In this unit, we attempted to elicit and leverage students' repertoires of practice around health in order to motivate their study of microbiology and the connection between microbiology and health. Through this series of design experiments, we explored a set of interlocking design principles for culturally responsive instruction.

Overlapping Science Curriculum with the Lives of Youth

Youth should be engaged in classroom science investigations heavily focused on the social practices they participate in as part of family and community life outside of school. We extend a classic Deweyan ideal (Dewey, 1902) by leveraging instructional approaches (e.g., youth documentation of everyday life) to systematically overlap the curriculum with the social practices of the youth and their communities (McDermott & Webber, 1998). In this way, we attempt to bridge youths' social practices from their informal environments into the reflective context of formal instruction with the hopes of better understanding and, perhaps, informing family and community practices.

Building Upon Prior Interests and Identity

Agency in learning should be supported as a coordination of the interests of youth (and their communities) and the goals of science education as they relate to providing equitable access to capital in society. We pursue agency as a relational construct that is developed and regulated between the learner and other actors in the learning context (Holland et al., 1998). Instructional strategies need to intentionally position youth as having interests and identities relevant to the societal roles of science (Tzou & Bell, 2010). We do this, in part, by supporting youth in developing a capacity to interpret and conduct research focused on the interests of their community.

Supporting Extended Learning Pathways by Building on Developing Expertise

The social and material capacities being developed by youth should be sanctioned and leveraged in instruction. Extending the research base on culturally responsive instruction to issues of science learning (e.g., Bell et al., 2009; Nasir, Rosebery, Warren, & Lee, 2006), our current work focuses on surfacing and leveraging the social and material capacities of youth in relation to the goals of the unit. This involves leveraging the sense-making routines (e.g., around argumentation) associated with specific cultural group membership.

Conclusions

People routinely learn—or fail to successfully learn—across the breadth of their life experiences in ways that we barely understand (Bell et al., 2009). As Lemke (2000) has noted, it is important to engage in a documentation of how people learn across multiple settings over extended time periods. Many research efforts have served as direct influences and provided theoretical inspiration to the effort reported on in this chapter. We have oriented to the theoretical perspectives associated with this prior work and have worked to develop a culturally and cognitively oriented theoretical framework that is specifically tailored to our scientific purposes and commitments. We are seeking to parcel out the social and cultural influences that shape the development of locally meaningful and personally consequential expertise. The research summarized in this chapter highlights the variegated pathways of human development that exist in diverse communities and the range of bridges and barriers associated with extended learning pathways. Current analytical work continues to document the barriers and bridges associated with the extended learning pathways of everyday expertise development. Current design research is attempting to architect successful pathways as students position themselves and are positioned to participate in activity across the breadth of their life experiences and settings. We hope that such forms of culturally responsive instruction will help engage all youth in meaningful learning experiences and promote more equitable access to desirable futures.

Acknowledgments This work has been conducted by members of the Everyday Science & Technology Group as part of the Learning in Informal and Formal Environments (LIFE) Science of Learning Center (http://life-slc.org/). We extend deep gratitude to the families participating in the study. We also gratefully acknowledge the intellectual influence of our LIFE colleagues and wish to thank the National Science Foundation for the opportunity to conduct this research through awards SBE-0354453 and SBE-0835854.

References

Agency for Healthcare Research and Quality. (2008). *2007 National healthcare disparities report* (AHRQ Publication No. 08–0041). Rockville, MD: U.S. Department of Health and Human Services, Agency for Healthcare Research and Quality.

Aikenhead, G. S. (1996). Science education: Border crossing into the subculture of science. *Studies in Science Education, 27*, 1–52.

Aikhenvald, A. Y. (2004). *Evidentiality*. Oxford, UK: Oxford University Press.

Banks, J. A., Au, K. H., Ball, A. F., Bell, P., Gordon, E. W., Gutierrez, K., Heath, S. B., Lee, C. D., Lee, Y., Mahiri, J., Nasir, N. S., Valdes, G., & Zhou, M. (2007). *Learning in and out of school in diverse environments: Life-long, life-wide, life-deep*. Seattle, WA: UW Center for Multicultural Education.

Barron, B., Martin, C., Takeuchi, L., & Fithian, R. (2009). Parents as learning partners and the development of technological fluency. *The International Journal of Learning and Media, 1*(2).

Barton, A. C., Ermer, J. L., & Burkett, T. A. (2003). *Teaching science for social justice*. New York: Teachers College Press.

Bell, P., Bricker, L. A., Lee, T. R., Reeve, S., & Zimmerman, H. T. (2006). Understanding the cultural foundations of children's biological knowledge: Insights from everyday cognition

research. In *Proceedings of the seventh International Conference of the Learning Sciences (ICLS)* (pp. 1029–1035). Mahwah, NJ: Lawrence Erlbaum.
Bell, P., Lewenstein, B., Shouse, A. W., & Feder, M. A. (Eds.). (2009). *Learning science in informal environments: People, places, and pursuits*. Washington, DC: National Academies Press.
Billig, M. (1987/1996). *Arguing and thinking: A rhetorical approach to social psychology* (new ed.). Cambridge, UK: Cambridge University Press.
Bransford, J. D., Brown, A. L., & Cocking, R. R. (Eds.). (2000). *How people learn: Brain, mind, experience, and school*. Washington, DC: National Academy Press.
Bransford, J., & Schwartz, D. (2009). It takes expertise to make expertise: Some thoughts about why and how. In K. A. Ericsson (Ed.), *Development of professional expertise: Toward measurement of expert performance and design of optimal learning environments* (pp. 432–448). New York: Cambridge University Press.
Bricker, L. A. (2008). *A sociocultural historical examination of youth argumentation across the settings of their lives: Implications for science education*. Unpublished doctoral dissertation, University of Washington, Seattle, WA.
Bricker, L. A., & Bell, P. (2008). Conceptualizations of argumentation from science studies and the learning sciences and their implications for the practices of science education. *Science Education, 92*(3), 473–498.
Bricker, L. A., & Bell, P. (in review). "*I want to be an engineer*"*: Network, framing, and positioning dynamics associated with youth STEM learning and expertise development in and out of school.*
Chrisman, N. J., & Kleinman, A. (1983). Popular health care, social networks, and cultural meanings: The orientation of medical anthropology. In D. Mechanic (Ed.), *Health, healthcare, and the health professions* (pp. 569–590). New York: Macmillan.
Clark-Ibanez, M. (2004). Framing the social work with photo-elicitation interviews. *The American Behavioral Scientist, 47*(12), 1507–1527.
Cole, M. (1996). *Cultural psychology: A once and future discipline*. Cambridge, MA: Belknap.
Corsaro, W. A. (2003). *We're friends right? Inside kids' culture*. Washington, DC: Joseph Henry Press.
Corsaro, W. A., & Maynard, D. W. (1996). Format tying in discussion and argumentation among Italian and American children. In D. I. Slobin, J. Gerhardt, A. Kyratzis, & J. Guo (Eds.), *Social interaction, social context, and language: Essays in honor of susan Ervin-Tripp* (pp. 157–174). Mahwah, NJ: Erlbaum.
Crowley, K., & Jacobs, M. (2002). Building islands of expertise in everyday family activity. In G. Leinhardt, K. Crowley, & K. Knutson (Eds.), *Learning conversations in museums* (pp. 333–356). Mahwah, NJ: Lawrence Erlbaum Associates.
Dewey, J. (1902). *The child and the curriculum*. Chicago: University of Chicago Press.
diSessa, A. A. (1988). Knowledge in pieces. In G. Forman & P. B. Pufall (Eds.), *Constructivism in the computer age* (pp. 49–70). Hillsdale, NJ: Lawrence Erlbaum Associates.
Dreier, O. (2009). Persons in structures of social practice. *Theory & Psychology, 19*(2), 193–212.
Driver, R., Leach, J., Millar, R., & Scott, P. (1996). *Young people's images of science*. Buckingham, UK: Open University Press.
Duschl, R. A., Schweingruber, H. A., & Shouse, A. W. (2007). *Taking science to school: Learning and teaching science in grades K-8*. Washington, DC: National Academies Press.
Flyvbjerg, B. (2001). *Making social science matter: Why social inquiry fails and how it can succeed again*. Cambridge, UK: Cambridge University Press.
Goodwin, M. H., & Goodwin, C. (1987). Children's arguing. In S. U. Phillips, S. Steele, & C. Tanz (Eds.), *Language, gender, and sex in comparative perspective* (pp. 200–248). Cambridge, UK: Cambridge University Press.
Gutiérrez, K. D., & Rogoff, B. (2003). Cultural ways of learning: Individual traits or repertoires of practice. *Educational Researcher, 32*(5), 19–25.
Harré, R. (2008). Positioning theory. *Self-Care, Dependent Care & Nursing, 16*(1), 28–32.
Holland, D., Lachiocotte, W., Skinner, D., & Cain, C. (1998). *Identity and agency in cultural worlds*. Cambridge, MA: Harvard University Press.

Hudicourt-Barnes, J. (2003). The use of argumentation in Haitian Creole science classrooms. *Harvard Educational Review, 73*(1), 73–93.

Hutchins, E. (1995). *Cognition in the wild.* Cambridge, MA: MIT Press.

Jessor, R. (1996). Ethnographic methods in contemporary perspective. In R. Jessor, A. Colby, & R. A. Shweder (Eds.), *Ethnography and human development: Context and meaning in social inquiry* (pp. 3–14). Chicago: University of Chicago Press.

Johnson, C., & Eccles, R. (2005). Acute cooling of the feet and the onset of common cold symptoms. *Family Practice, 22*, 608–613.

Korpan, C. A., Bisanz, G. L., Bisanz, J., Boehme, C., & Lynch, M. A. (1997). What did you learn outside of school today? Using structured interviews to document home and community activities relating to science. *Science Education, 81*(6), 651–662.

Kyratzis, A. (2004). Talk and interaction among children and the co-construction of peer groups and peer culture. *Annual Reviews of Anthropology, 33*, 625–649.

Latour, B. (1987). *Science in action: How to follow scientists and engineers through society.* Cambridge, MA: Harvard University Press.

Lave, J., & Wenger, E. (1991). *Situated learning: Legitimate peripheral participation.* Cambridge, MA: Cambridge University Press.

Law, J. (1999). After ANT: Complexity, naming and topology. In J. Law & J. Hassard (Eds.), *Actor network theory and after* (pp. 1–14). Hoboken, NJ: Wiley-Blackwell.

Lee, C. (2002). Environmental justice: Building a unified vision of health and the environment. *Environmental Health Perspectives, 110*(Suppl. 2), 141–144.

Lee, C. D. (2008). The centrality of culture to the scientific study of learning and development: How an ecological framework in education research facilitates civic responsibility. *Educational Researcher, 37*(5), 267–279.

Lee, O., & Luykx, A. (2007). Science education and student diversity: Race/ethnicity, language, culture, and socioeconomic status. In S. K. Abell & N. G. Lederman (Eds.), *Handbook of research on science education* (pp. 171–197). Mahwah, NJ: Lawrence Erlbaum.

Lemke, J. L. (2000). Across the scales of time: Artifacts, activities, and meanings in ecosocial systems. *Mind, Culture, and Activity, 7*(4), 273–290.

Linn, M. C., & Songer, N. B. (1991). Cognitive and conceptual change in adolescence. *American Journal of Education, 99*(4), 379–417.

Lowen, A. C., Mubareka, S., Steel, J., & Palese, P. (2007). Influenza virus transmission is dependent on relative humidity and temperature. *Public Library of Science (PLoS) Pathogens, 3*(10), 1470–1476.

McDermott, R., & Webber, V. (1998). When is math or science? In J. G. Greeno & S. V. Goldman (Eds.), *Thinking practices in mathematics and science learning*. Mahwah, NJ: Erlbaum.

Medin, D., & Atran, S. (2004). The native mind: Biological categorization, reasoning, and decision making in development and across cultures. *Psychological Review, 111*(4), 960–983.

Moll, L. C., & Gonzalez, N. (2004). Engaging life: A funds-of-knowledge approach to multicultural education. In J. A. Banks & C. A. M. Banks (Eds.), *Handbook of research in multicultural education* (2nd ed., pp. 699–715). San Francisco: Jossey-Bass.

Nasir, N. S., Rosebery, A. S., Warren, B., & Lee, C. D. (2006). Learning as a cultural process: Achieving equity through diversity. In R. K. Sawyer (Ed.), *The Cambridge handbook of the learning sciences* (pp. 489–504). New York: Cambridge University Press.

Ochs, E., Taylor, C., Rudolph, D., & Smith, R. (1992). Storytelling as a theory-building activity. *Discourse Processes, 15*, 37–72.

Reeve, S. (2009). *Health beliefs and practices of young people in a multicultural community: Findings from a child-centered ethnography.* Unpublished dissertation, University of Washington, Seattle, WA.

Reeve, S., & Bell, P. (2009). Children's self-documentation and understanding of the concepts 'healthy' and 'unhealthy'. *International Journal of Science Education, 31*(14), 1953–1974.

Rosebery, A., Warren, B., Ballenger, C., & Ogonowski, M. (2005). The generative potential of students' everyday knowledge in learning science. In T. Romberg, T. Carpenter, & F. Dremock (Eds.), *Understanding mathematics and science matters* (pp. 55–80). Mahwah, NJ: Erlbaum.

Saxe, G., & Esmonde, I. (2005). Studying cognition in flux: A historical treatment of "fu" in the shifting structure of Oksapmin mathematics. *Mind, Culture, and Activity, 12*(3&4), 171–225.
Schiffrin, D. (1987). *Discourse markers*. Cambridge, UK: Cambridge University Press.
Scribner, S. (1984). Cognitive studies of work. *The Quarterly Newsletter of the Laboratory of Comparative Human Cognition, 6*(1 & 2), 1–42.
Spradley, J. P. (1979). *The ethnographic interview*. Belmont, CA: Wadsworth Group/Thomson Learning.
Stevens, R. (2000). What counts as math? Emergent and assigned mathematics problems in a project-based classroom. In J. Boaler (Ed.), *Multiple perspectives in mathematics teaching and learning* (pp. 105–143). New York: Elsevier.
Stevens, R., Satwicz, T., & McCarthy, L. (2008). In-game, In-room, In-world: Reconnecting video game play to the rest of kids' lives. In K. Salen (Ed.), *The ecology of games: Connecting youth, games, and learning* (The John D. and Catherine T. MacArthur Foundation Series on Digital Media and Learning (pp. 41–66)). Cambridge, MA: MIT Press.
Suchman, L. (2007). Agencies in technology design: Feminist reconfigurations. In *Human-machine reconfigurations: Plans and situated actions*. New York: Cambridge University Press.
Tzou, C., & Bell, P. (2010). Micros and me: Leveraging home and community practices in formal science instruction. In K. Gomez, L. Lyons, & J. Radinsky (Eds.), *Learning in the disciplines: Proceedings of the 9th International Conference of the Learning Sciences* (Vol. 1, pp. 1127–1134). Chicago: International Society of the Learning Sciences.
U.S. Department of Minority Health and Health Disparities. (n.d.). *Eliminating racial and health disparities*. Retrieved June 30, 2009, from http://www.cdc.gov/omhd/About/disparities.htm
Zacharia, Z., & Barton, A. C. (2004). Urban middle-school students' attitudes toward a defined science. *Science Education, 88*(2), 197–222.
Zimmerman, H. T. (2008). *Everyday science and science every day: Science-related talk and activities across settings*. Unpublished dissertation, University of Washington, Seattle, WA.
Zimmerman, H. T., & Bell, P. (in review). *Where children see science: Youth's activities as connected to science.*

Chapter 10
Doing Science with Others at Preschool and at Home: A Comparison of Contextually Situated Interactional Configurations and Their Implications for Learning

Siri Mehus, Reed Stevens, and Linda Grigholm

Introduction

In this chapter, we compare aspects of the home and preschool contexts in terms of how they afford particular types of scientific exploration and inquiry in early childhood. Our goal is to identify the opportunities for learning that are available in one environment but unlikely to occur in others. We ask, in other words, what makes a place *special* as an environment for learning? As such, rather than surveying the entire range of activities in which children may engage in each of these environments, we focus on two types of interactional configuration, each chosen because we have observed it occurring in one of these settings and because, we argue, there are particular aspects of the environment that make it likely to occur there.

The first is an adult-guided mode of interaction in which the adult closely monitors the child's attention and action and accordingly orchestrates the emergence of opportunities for scientific observation, exploration, and knowledge construction. Like others (e.g., Crowley & Jacobs, 2002; Goodwin, 2007), we have observed this occurring in family interactions. The other configuration is a type of peer play that occurs frequently within "free-play" periods at preschool, in which children collaboratively explore the physical properties of the world around them—making

S. Mehus (✉) • L. Grigholm
LIFE Science of Learning Center, University of Washington, Seattle, WA, USA
e-mail: sirimehus@gmail.com

R. Stevens
School of Education and Social Policy, Northwestern University,
Evanston, IL, USA

B. Bevan et al. (eds.), *LOST Opportunities: Learning in Out-of-School Time*,
Explorations of Educational Purpose 23, DOI 10.1007/978-94-007-4304-5_10,

observations, developing hypotheses, and even conducting informal experiments in short-lived social groupings of varying sizes. In both contexts, the type of science learning we focus on in this chapter can be termed emergent, incidental, or serendipitous (Bell, Lewenstein, Shouse, & Feder, 2009; Stevens, 2000) in that it occurs in the process of everyday living and playing with others, rather than within events that are explicitly designed for science learning.

In the pages that follow, we analyze interactions engaged in by two 5-year-old girls. These include home interactions in which children and parents discuss such topics as the plants growing in a garden and a child's rock collection, as well as preschool-based peer interactions such as one in which a child and her peers use a tape measure to calculate the length of various objects in their classroom and then discover how to use it as a projectile. Through microanalysis, we compare these activities in terms of the *learning arrangements* that occur within (Stevens, Satwicz, & McCarthy, 2007), considering how these interactional arrangements allow for or constrain children's entry into scientific activities as well as the particular opportunities for learning they may afford.

Background

This study responds to previous calls for research investigating learning across school and out-of-school experiences (Stevens et al., 2005). While much educational research has focused on children's school experience at the expense of what happens in the home, research on informal science learning in early childhood has focused primarily on the family. Such research has shown that children have opportunities to learn scientific concepts in everyday life and that adults in their lives have practices for supporting them in this process (e.g., Callanan & Oakes, 1992; Tizard & Hughes, 1984). Other researchers have pointed to the value of learning from peers in early childhood (Rogoff, 1990; Williams, 2001), even documenting a relationship between frequency of peer interaction and academic performance (Hanish et al., 2007). Our observations suggest that informal play activities create a rich interactional environment for such peer learning. Interestingly, a primary site for such informal peer interactions seems to be the quasi-formal context of the preschool.

We approach our investigation of young children's home and preschool interactions from the theoretical and methodological frame of microethnography, i.e., close analysis of talk and action as situated in particular social, cultural, and material contexts (Stevens & Hall, 1998; see also Jordan & Henderson, 1995; Streeck & Mehus, 2005). Through moment-by-moment analysis of video-recorded interaction, we demonstrate possibilities for learning in the activities in which children participate. This form of qualitative inquiry allows for the generalization of findings, not to populations, but rather to practices and activities based on principles derived from the analysis of structures of action.

Study Description

We report here on one piece of a larger study examining young children's learning as it occurs in the multiple environments of their everyday lives. By observing and recording young children's interactions in multiple contexts (such as preschool, home, and playgroups) and among different configurations of co-interactants (including peers, older and younger children, teachers and caregivers), we seek to capture the complexity and variety of the social environments in and through and across which young children learn.

The first phase of this study involved weekly video-recording in a childcare classroom for approximately five months. In addition, two of the children in the classroom were observed and recorded in their home environments. In a second phase of this study, we video-recorded activities in six rooms at two preschools for a total of nine months. In addition, out-of-school activities of eight children were recorded.

In this chapter, we draw on data collected during the first phase of the study, in which we observed and recorded home and school activities of children in the "Rocket Room," a kindergarten-readiness classroom (four and five years old) in a nonprofit NAEYC-accredited childcare center. The center is located in a middle-class neighborhood of a midsized Western city. There are 20 children in the Rocket Room, supervised by a lead teacher and an assistant teacher. Our observations took place in the mornings, during which time the children participated in free play, breakfast, art activities, and circle time. We describe learning interactions experienced by two focal children in the study: Molly and Darcy. Molly is five years old and spends three days per week in the Rocket Room. She is an only child. Her mother is a freelance artist who stays home with Molly two days per week; her father works full-time. Darcy is also five years old and has an 18-month-old sister. Both Darcy and her sister attend the childcare center full-time. Darcy's parents both work in high-tech occupations. First, we provide a view of learning arrangements and interactions in the home settings of these two focal children. We then provide a view of the learning arrangements and interactions for the same children in the preschool setting.

Parent-Guided Learning Interactions at Home

When video-recording children in home environments, we frequently captured interactions that were intensely mediated by parents. In the section that follows, we describe some such interactions that arguably could have relevance for the development of scientific skills and interest, and analyze the particular opportunities for learning they afford.

Key Characteristics of the Learning Arrangement

What do we mean by "parent-guided learning interactions"? These are events that may be planned in advance, but even if not they are continuously organized for children's learning as they proceed. Within these activities, children's talk and actions are closely monitored by adults and frequently commented upon (usually praised). Children's actions and other aspects of the environment are frequently used as a starting point for elaboration and exploration, within which there are clear attempts to guide children's perception toward particular aspects of the scene and to provide labels for what children are experiencing (Stevens & Hall, 1998). The segment below exemplifies several of these features. This occurs in the backyard of Darcy's home. Darcy is splashing in an inflatable swimming pool while her mother and father watch. Her mother gets up to take a look at their garden and calls Darcy's attention to what she sees.

Example#1: Sunflower Sprouts

1 Mom: Here you know what (.) Darce? It looks like the (.) the: uh sweet peas
2 (over here) need water. (That soil still) looks dry: doesn't it.
3 *((Darcy starts watering where her mom pointed))*
4 Mom: Do you need some help with that? *((gets up))*
5 It's probably on the far side so it's kinda (.) kinda rough huh.
6 Mom: You want me [to
7 Darcy: [I can smell the sweet peas.
8 Mom: You can smell the sweet peas? What do they smell like.
9 Do they smell (.) sweet?
10 Darcy: *((grunts))*
11 Mom: Woah. (.)
((Darcy is rocking watering can back and forth))
12 That's kinda clever.
((Darcy steps up onto side of bed))
13 Mom: You're like (.) rocket-shipping.
((Darcy reaches out with watering can over bed, wobbles))
14 Oh there you go.
15 Here I'll hang on to you: *((Mom holds Darcy around torso. Darcy*
16 *holds watering can))*
17 Mom: Hang on to you: (.)
18 That's kinda heavy huh. *((takes watering can))*
19 °Here you go. °
20 (can I) do these?
21 Darcy: Mm hm. *((turns away from garden, towards pool))*
22 *((Darcy gets into pool, Mom continues watering))*
… [01:10 exchange between parents and researcher not transcribed]
23 Mom: Darcy! (.) Did you no:tice?

24 Dad: Oh there's some sprouts coming up?
25 Mom: Check it ou::t!
26 Holy cow that's fa:st. We just put this in last week.
27 Look!
28 What are tho::se?
29 Darcy: Sprouts.
30 Mom: What kinda sprouts.
31 Darcy: Sunflower sprouts
32 Mom: =sunflower sprouts. [Holy cow::!
33 Dad: [°Sunflower sprou::ts.°
34 Darcy: I've done (.) some (.) very good watering
35 Mom: =You've done some very good watering.
36 That is instant gratification. (.) [Right there.
37 Dad: [Oh. Educational moment.
38 Res.: Yeah.

This segment exemplifies several of the features that characterize adult-child learning interactions in the homes that we studied. First, these *events are managed for children's learning*. In some cases, they are preplanned, involving the preparation of materials and the use of predesignated roles and actions. Even when they are not, the activities are managed by adults in the moment, with slots for children's participation being provided by the parents. While the particular event above was not planned at the level of a classroom lesson, it was purposefully brought into being for educational purposes by the parents. Darcy's mother told us that they decided to plant a garden specifically for Darcy's benefit (as a means to "inspire little Darcy about growing things"; they also hoped she might become more interested in eating vegetables). Within the moment, Darcy's mother enlists her participation in familiar interactional routines. For example, in lines 28–32, Darcy's mother engages her in a school-like initiation-response sequence in which the mother asks Darcy "known-answer" questions (Cazden, 1988; Mehan, 1979).

Within these events, *children's actions and talk are monitored*, commented on, praised, and elaborated upon. For instance, in line 12, Darcy's mother verbally calls attention to Darcy's action and assesses it as "kinda clever," further describing it in line 13 as being "like rocket-shipping" (presumably referring to the way she is propelling water from the can). *Phenomena or objects in the environment*, in many cases *natural* phenomena, are also commented upon, even used as opportunities for "occasioned knowledge exploration" (Goodwin, 2007). We see this in lines 1–3 above, when Darcy's mother first calls her attention to the garden's need for water, as well as in lines 23–32, when she discovers new growth and subsequently guides Darcy in specifying what they are seeing. Parents orient to these as teaching relevant moments (though not usually as explicitly as does Darcy's father in line 37 above).

Opportunities for Learning in Parent-Guided Interactions

There are several types of learning opportunities that children encounter in these types of interaction. For instance, a prominent source of potential learning in the interaction above comes from the mothers' directing of her daughter's attention and the guiding or disciplining[1] of her perception (Stevens & Hall, 1998). We see this in lines 1–2, in which Darcy's mother not only calls her attention to a specific aspect of the environment but guides her in how to *see* it—i.e., to recognize the dirt as soil that "looks dry."

We can also see this in the way Darcy's mother calls attention to Darcy's actions. For example, Darcy's mother takes notice of an action being performed by the child and frames this action as being both purposeful and ingenious (lines 12 and 13).

This is not a unique occurrence. Darcy's parents do this at other times and it happens in other children's interactions with their parents, as well. In the segment below, Molly (the other focal child) is engaged in a project of making thank-you cards. This project is highly arranged and planned: Molly's mother has gathered the materials, designated a space for Molly to work, explained the purpose of the activity, and provided advice about how to do it. This is a practical activity, but Molly's mother demonstrates throughout that she is attentive to the learning opportunities that are embedded within it.

Example #2: Linen

1 Mom: [You want-
2 Molly: [Do:ggies *((picking up sheet of stickers))*
3 Mom: Ye:::s do:ggies. *((leans forward to pick up additional stickers))*
4 Molly: I think Morgan likes doggies so I'll give her one.
((Molly places sticker in her hand on edge of fabric box))
5 Mom: Oh::kay. (.)
6 That's a good idea Molly.
7 Is to stick them on something that they won't stick too fa- (.) too much
8 to.
((Fingers sticker on the edge of box))
9 This is called linen *((rubs fabric on box))*.
((Molly's gaze is on stickers she is holding.))
10 It's not like paper where it really sticks tight. (.)
11 *((Mom picks up some more sticker sheets, puts them in a pile))*
12 So: yeah we have kind of a mess right here.=
((Molly reaches over and places another sticker on the edge of the box.))

[1] The construct of "disciplining perception" was originally used in the context of disciplinary activities (e.g., mathematics, engineering), and thus "discipline" took on a double meaning. Since the activities analyzed here are not intentionally disciplinary, we will henceforth use the more general phrase "guiding perception," but the interactional mechanism identified previously is the same.

13 Molly: =Linen? *((Still sticking the sticker to box))*
14 Mom: Linen. See the fabric? *((Mom rubs fabric near where Molly is placing sticker.))*
15 It doesn't fee:l- it's cloth. It's not (.) uh paper.
((Molly rubs the fabric in the same area.))
16 Molly: I know=
17 Mom: =Yeah.
18 This is kind of a plastic paper that's why it doesn't stick to that.
((Moves a sticker sheet to the top of the box.))
19 Should we throw these away these old paint chips?

In line 6, Molly's mother notices an action of her daughter's that reveals some attention to the properties of the materials she is working with. She takes this opportunity to articulate the practical distinction that Molly has made between the characteristics of the materials. She provides labels for the materials and models a way for Molly to further explore their tactile features.

Note that in each case, an explicit positive assessment is made, linked to the child's action with the deictic pronoun "that's" ("That's a good idea," Linen, line 6; "That's kinda clever," Sunflower Sprouts, line 12). The parent then immediately provides elaboration of *why* the child's action seems like a good idea, in a way that points to what the child is achieving through her action. In Example #2, Molly's mother adds, "Is to stick them on something that they won't stick too fa- (.) too much to." (lines 7–8). In Example #1, Darcy's mother comments "you're like (.) rocket-shipping." (line 13). At this moment, Darcy is standing on the edge of the garden bed and rocking the watering can back and forth, which causes the water to slosh out, reaching further into the bed than it would if she were simply pouring it out. Darcy's mother's "rocket-ship" analogy conveys the notion that through her action, Darcy is causing something to be propelled through the air.

In the segments above, parents not only provide labels and guide children's perception of their own actions but they also guide children's perception of objects in the world around them. For example, in Example #2, Molly's mother provides a label for the material that she has referred to as "something [the stickers] won't stick too much to," while rubbing the fabric box with her finger (lines 8–9). However, Molly is looking at the stickers in her own hand and does not shift her gaze toward her mother's finger. (See Fig. 10.1 below.) As Molly places another sticker on the fabric box, she repeats with a questioning intonation "Linen?" (Fig. 10.2). Molly's mother responds with a confirmatory repetition: "Linen." She then verbally directs Molly's visual attention, saying "See the fabric?" As she does this, she moves her finger back to the box, not to the spot she rubbed previously, but to another spot adjacent to where Molly is currently attaching the sticker and directly in Molly's gaze (Fig. 10.3). However, Molly's mother is not just guiding Molly's visual attention, as she does not merely point at the fabric. Rather, she rubs it, suggesting another modality through which Molly can perceive the fabric. Molly moves her finger slightly to the left, closer to her mother's, and also rubs the linen box.

Fig. 10.1 Molly's mother rubs the fabric box

Fig. 10.2 "Linen?"

This segment provides an example of an adult engaging in tactical work (monitoring the child's gaze and positioning her hands accordingly) to guide a child's attention to particular features of an object in the environment. Notably, she does this by modeling a tactile mode of engagement with the material, even as she verbally instructs her to use a visual mode.

In the following example, Molly's mother also models a way of exploring the characteristics of an object. On this day, Molly has proposed showing the researcher her rock collection. The segment begins as Molly extracts the first rock from her bowl.

Fig. 10.3 "See the fabric?"

Example #3: Tiger's Eye

1		*((Molly takes a rock out of bowl and holds it up toward the camera))*
2	Mom:	What kind of rock is that.
3	Molly:	A cry:stal.
4	Mom:	Uh huh::
5		Did we open a rock to get into it?
		((Molly pulls another rock out of bowl))
6	Molly:	Oh. I found the ti:ger's eye.
7	Mom:	Yeah:.
8		Is it a ge:ode? Did we open a geode or a-
9	Molly:	*((in "baby talk"))* Oh der h- de:rs ↑Misty: *((addressing cat))*
		[02:30 minutes not transcribed, during which Molly chases the cat and then returns to the rocks]
16	Molly:	I think this is a blue tiger's eye but I don't know. *((holding up rock towards Mom))*
17	Mom:	Uh hah I think it is a tiger's eye: *((takes rock))*
18	Mom:	Oh well no it's not tig- well I don't know:
19		Let's look at it and compare.
		((sets it down on the table))
20		(.)
		((reaches for tiger's eye taken out previously))
21		It has [kind of (.) [shiny stripes *((Mom holds new rock next to other on table))*
22	Molly:	[b- [b-
23		Because lookit because lookit- [(.) it's dark. *((fingering rock))*
24	Mom:	[yea::h

25 Mom: It's uh it's redder. *((Molly removes hand))*
26 This one's more (.) yellowy-brown:: *((Mom sets rocks next to one another again))*
27 That would be a good question for somebody who know:s (.) geology.
28 =Geology is the study of rocks and formation of rocks.
29 And there's people who make- who study it.
30 It's like mama is an artist that's what they do for their work. (.)
31 *((Molly sips tea))*

When Molly raises a question about a rock in line 16—whether it is a tiger's eye—Molly's mother responds by suggesting a way of answering that question: "Let's look at it and compare" (line 19). She then enacts this process with her body, her words, and the objects before her. As she fingers the two rocks, she provides verbal descriptors of the one in question. Molly participates by doing the same (line 23). In lines 27–30, Molly's mother presents another means of answering Molly's question: asking a geologist. By doing so, she links their activity to a scientific discipline and introduces the notion of consulting an expert as a way of answering a scientific question.

Learning within such adult-guided activity is hardly a passive process for children. Rather, children take advantage of this interactional format to initiate and further their own learning experiences. One way in which they do this is by asking questions or otherwise indicating uncertainty. For example, in Example #2, Molly responds to her mother's earlier labeling of the material with a questioning repeat: "Linen?" (line 12). Molly's mother then confirms the label, elaborates on the characteristics of the material, draws her visual attention to the material, and encourages its multimodal exploration. All of this happens after Molly reinitiates the topic of the material with her question. Similarly, in Example #3, above, Molly initiates the rock-comparing sequence by proposing a candidate identification of a rock, then expressing uncertainty: "I think this is a blue tiger's eye but I don't know" (line 16). She further solicits her mother's assistance by holding the rock toward her mother when she says this.

In a context in which the child's talk and actions are being closely monitored by the parent, an assertion can be as effective as a question in eliciting a parent's support for learning. In the segment above, Darcy twice issues declarative statements that draw responses from her mother. For instance, in line 7 of Example #1, Darcy asserts that she can smell the sweet peas. Her mother responds with the question "what do they smell like?" then proposes a candidate answer: "Do they smell sweet?" Darcy's mother could be guiding her olfactory perception here; however, Darcy has become preoccupied with the task of the watering the garden and does not respond. In line 34, in response to her mother's enthusiastic pointing out of the new sprouts, Darcy comments that she (Darcy) has "done some very good watering." The timing of Darcy's assertion suggests that she is making a connection between the sprouting of the sunflower seeds and her previous watering actions. By making her assertion in the context of an interaction with her mother, Darcy makes it available for confirmation, which her mother provides unambiguously.

Discussion: Adult-Child Interaction in the Home as Learning Arrangement

Adult-child talk in the home can provide many learning opportunities for children: parents can guide children's attention, deliver comments and questions that are tightly coupled with children's activities, and design and orchestrate interactions around material objects with children's learning in mind. Children's initiations of learning opportunities can be taken up readily and enthusiastically. The interactional demands on children are low: these adults do much work to invite children's participation and ensure that these activities work as learning arrangements. These do not represent the totality of children's interactions in any home; surely, there are times when the children do not have the parents' full attention and even when they do, parents cannot take advantage of every learning opportunity. It should also be noted that other types of interaction occur in these homes and other homes and that these offer different types of learning opportunities (such as guided participation in adult activity, intent observation of adult or peer activity, and peer learning in sibling play, to name only a few). We focus on this type both because we observed it frequently and because particular features of the interactions make them rich with learning opportunities, as described above.

We propose that multiple structural and cultural aspects of the home environment make these types of interaction more likely to occur in homes than in preschools, chiefly the respective ratios of adults to children. In preschool, the preponderance of children means that teachers must frequently shift attention in order to monitor and care for everyone. While adults in the home may also have other demands on their attention, they are much more likely to be able to carry out extended interactions in which they devote significant amounts of attention to the talk and action of an individual child. This does not mean that similar interactions cannot occur in settings such as preschools, and sometimes they do. It does suggest, however, that the unique features of the preschool environment may give rise to *other* interaction patterns, which can in turn be analyzed for the ways in which they allow science learning to occur.

Learning Through Peer Interaction in the Preschool Classroom

In this section, by way of comparison, we analyze the learning opportunities inherent in an interaction structure that we found frequently in our preschool observations and rarely in homes: small group activities organized and guided by children. Though this type of activity might occur from time to time in other contexts, there are features of the preschool environment that make it more likely. For instance, during free-play periods at school, teachers rarely spend long periods of time with a single child or group of children. This means that while they do set up activities for children and step in to mediate children's interactions, teachers infrequently engage in the intense moment-to-moment guidance typical of what we observed in homes. As such, rather than having interaction initiated and maintained for them, children engage one another.

Key Characteristics of the Learning Arrangement

The type of activity analyzed in this section is characterized by (1) including three or more children (more children may come in and out of the activity); (2) being initiated, guided, and maintained by the children, rather than by adults; (3) enduring over a relatively long stretch of time (e.g., the measuring activity analyzed in this chapter endured for nearly 45 min in total); and (4) being organized toward some purpose (although individual goals may not be shared by all participants and may shift during the activity). For the purposes of this chapter, we analyze one extended activity in which all four characteristics are present. Though the activity we describe is not explicitly undertaken as a "science" or "math" lesson, there are opportunities for scientific and mathematical exploration and discovery embedded within it.

Measuring Tape

The activity we analyze in this section occurs over an extended period of time during which a core group of three children, joined occasionally by other children in the room, play with a tape measure that one of the children has brought from home.

It is free-play time in the preschool classroom. One of the children, Anna, takes a drawing she has just finished to her cubbie, and pulls out a tape measure. "I brought this from home:::!" she says to an entering parent. Anna then brings the tape measure over to Nancy, one of the teachers, who compliments her on the idea to bring a tape measure from home and asks her what she wants to measure. Anna begins to measure a bookshelf. Though Nancy soon becomes occupied with other children, Anna is able to regain her attention long enough to get some help with the measuring task she has begun. Specifically, Nancy provides instruction in reading numbers from the tape ("A three and a seven is thirty-seven.") and with the unit of measurement ("An inch is about that long," gesturing with thumb and forefinger). She also provides a different type of guidance: in the form of a question, she suggests to Anna ways of linking her current measuring activity with adult professional practices ("Are you a carpenter? A mathematician? A scientist?"). After this point, the children's measuring activity proceeds with no significant input from any of the adults in the room.

Perhaps taking her cue from Nancy's question, Anna then initiates a game with some other children, in which Anna plays the role of carpenter. That game is short-lived, and when it ends, Anna and one of the children begin stretching out the measuring tape and letting it go so it retracts into the case. They are then joined by one of the girls who will make up the core group: Susan. Susan suggests that they measure people, and though Anna initially resists this idea, when we find them again (after a one minute gap in which they are out of camera range), Susan is measuring Anna. Susan struggles with the challenge of extending the tape the length of Anna's body and also being able to see the numbers at the top.

It is at this point that Darcy (one of our focal children) joins them. She approaches the two girls, takes the end of the measuring tape, stretches it to the top of Anna's head, and holds it there. This allows the girls to move to the next step of the measuring task, which is to read the number. Anna and Susan then measure Darcy, and the three then move on to measuring a large cardboard tree. The three girls continue to work together as a team, measuring objects all over the preschool classroom.

For most of this period, they measure preexisting objects, such as tables, the refrigerator, chairs, bookshelves, the elevated stage, room dividers, and a large block. During this time, they negotiate a set of roles and a turn-taking system. The set of roles evolves and shifts over the course of the activity; however, some core "jobs" emerge, such as holding (and hooking) the tab end of the tape measure; extending the tape measure case; looking at, reading, and shouting out the measurements (numbers); being the one to yell "Let go!"; and releasing the tape measure case (so that it springs back to the hooked tab end as the tape retracts into the case).

In the last minutes of the activity, the girls begin to put blocks together to construct a low wall to measure. When the block line gets all the way to a low stage on one side of the room, they discover that when they let the tape case go, it will slide along the top of the block line and then jump up onto the stage. This effect is greeted with celebratory whoops and jumps and repeated several times.

There is then an interruption of the dominant activity while the girls join some of their classmates in a different activity with one of the teachers. When the girls leave the line of blocks, other children immediately approach it and begin to appropriate the blocks for their own purposes. The girls enlist the help of a teacher (Nancy) in reestablishing the exclusive use of the block line for measuring activities. One of the children, Tom, sticks around and gains access to the group by assigning himself a role ("I'm the guy who watches") in the activity. It is at this point that our focal child, Darcy, exits the group to engage in another activity.

Opportunities for Learning in a Preschool Setting

Over the course of this activity, there are multiple opportunities for the children to learn from (and teach) each other. We focus on the experience of our focal child, Darcy, and find types of learning afforded by the activity can be separated into two rough categories. The first is more directly related to measuring as ordinarily carried out and relates to a set of concepts that include quantification and number identification, dimension, and comparison (determining whether one thing is longer or shorter than another thing). The other category is not directly tied to the activity of measuring but rather to the physical affordances and constraints of the tape measure as an object. This set includes learning about the limitations of the material—both in terms of its length (the tape can only be stretched a certain distance) and its strength (the tape will collapse if it is stretched too far without any external support). Especially interesting for Darcy and the other girls is a particular affordance of the tape measure tool: that the tape automatically retracts into the case and that this mechanism will also cause the case end to spring toward the tape end when that end is held (or hooked on something).

Over the course of the activity, we see the children engaging in negotiation and discussion about how to do things, trying out terminology with one another, making discoveries, issuing predictions about the behavior of physical objects (informal hypotheses), and watching to see if their predictions are borne out.

Learning About Measuring

One of the basic subtasks involved in measuring is *identifying and reading the numerals* printed on the tape. Anna receives direct instruction in this from the teacher (as described above). After Darcy joins the group, the girls continue to negotiate how do to this. In the example below, the girls are measuring their sixth object together and still working out how to read numbers from the tape. Also under discussion and available for learning are issues of *dimension*, i.e., that there are different terms to describe measurements made in different directions (lines 13 and 15) and that this is a distinction worth noting.

Example #4: Table-Measuring
1 Susan: *((jumps off stage and bounces over to Anna))* How about the ta::ble.
2 Anna: (oew:::)
3 *((girls run off-screen))*
4 Anna: I know.
5 How about we measure this table.
6 Susan: Okay.
7 Anna: Okay. so.
8 Um Susan [we're measuring to thi::s
9 (Darcy:) [(we're measuring the table)
10 Susan: Okay.
11 Anna: You know::?
12 Okay um
13 Susan: You mean you're measuring how long it is.
14 Anna: Yes.
15 Susan: Not how tall it is.
16 Anna: Yeah.
17 Susan: Hey. When it (springs) back it might (.) pinch your fingers.
18 Darcy: Yeah. I know that.
19 Anna: =How many [is it?

20 (Susan:) [(so if it falls (.) back it might)
21 Darcy: It's seventy-five feet.
22 Susan: Lemme see! Actually (.) it's (.) fifty-seven feet.
23 Anna: OKAY:: [( )
24 Susan: [Actually actually it's (.) seventy-five feet.
25 Anna: Okay. STA:ND BACK EVERYONE:
26 *((sound of tape measure case scooting over table and dropping to floor))*
27 Girl: YES!
28 Anna: How about the fridge?
29 Susan: Yeah we'll mea:sure the fri::dge!

Learning About the Constraints and Affordances of the Object

Over the course of their play activity, the girls encounter and explore various properties of the tape measure itself. For instance, they display awareness that the tape is limited in length, and concern themselves with limiting the length of their block wall accordingly. Earlier on, they discover that the strength, or the ability of the tape measure to support its own weight when extended, is also limited—e.g., when measuring a tall cardboard tree, the measuring tape bends over on itself and Anna yells, "It's too high!". However, it is a curious affordance of the tool that captures their interest the most. That is the self-propelling action of the tape measure—that the tape end automatically retracts into the case, but also the case end can spring back to the tabbed end if the tabbed end is held in place (see lines 17–18 and 25–27 above).

That this is a learning experience for Darcy is evidenced at a later moment. The girls are measuring a bookcase by placing the tape measure case on the floor and extending the tape toward the top of the bookcase. Darcy is holding the tabbed end at the top of the bookcase. When Anna lets go of the tape measure case, which is resting on the floor, Darcy abruptly pulls her hand back and jumps away from the tabbed end (Figs. 10.4 and 10.5).

Evidently, Darcy expects the case to spring up toward the tabbed end. Darcy's embodied prediction represents learning in progress. Her earlier experiences with the tape measure have led her to form the expectation that the case will spring toward the tabbed end when released. It is fairly certain that she would not have expected a metal case to rise into the air if she had not just had this experience with the tape measure. That the tape measure does not do what she expects provides an opportunity for Darcy to revise her conceptualization of the properties of this object, potentially gaining a more complex (though unarticulated) understanding of the interaction between gravity and the pull of the spring retracting the tape measure into the case.

Later, the girls discover that when the tape measure is released so that it runs along the block wall, it will act as a projectile and pop up onto the stage (beyond the spot at which the end was hooked). The first time this happens, the girls react with

Fig. 10.4 Darcy holds the tabbed end of the measuring tape

Fig. 10.5 When Anna lets go of the tape measure, Darcy jumps back, anticipating that the case will spring up towards her hand

excitement and try it again. The second try is less successful and meets with less enthusiasm. The third and fourth tries are shown (in boxes) below.

Example #5: Block-Measuring

1 Anna: OK<u>AY.</u> Now:::: *((nasal))* I wanna do (.) as many ( ) *((picks up*
2 *blue block and puts at end of wall, looks back to Darcy*
3 *and Susan))*
4 *((Darcy stretches measuring tape along the length of the wall))*
5 Anna: *((nods)) ((puts hands on hips))* I think (it'll) stop soon. Because it does
6 stop.
7 *((Darcy gets to end of block wall--back end of tape measure is lined up*
8 *with end of wall))*
9 Anna: Okay. *((looks at end of tape measure, looks at Darcy))*
10 (Darcy:) Okay?

11 Susan: Okay.
12 *((Both girls look at Susan, holding tab end))*
13 (?): (All right. Let's go. Looks goo::d.)
14 *((Susan lets go of hooked end and moves away from block wall.))*
15 Anna: Okay [( )
16 [*((Darcy lets go of tape measure))*
17 *((tape measure pops up onto stage))*
18 Anna: Waaa[::oh *((runs toward tape measure))*
19 Susan: [Wu Hoo::=
20 Darcy: =Wuhoh *((jumps, walks to end of wall))*
21 *((Susan jumps, walks toward group at table.))*
22 Anna: Okay 'm gonna I'm gonna
23 *((Susan moves chair))*
24 Anna: *((calling out))* Okay. Put ano- another block the::re.
25 *((Susan picks up yellow block, starts to set it down at end of wall))*
26 Anna: No.
27 *((Susan sets it down))*
28 Anna: Okay Darcy's gonna ( )
29 *((Darcy skips to stage end of block line))*
30 *((Susan runs to end, Anna starts stretching tape along wall.))*
31 Susan: Oh yeah. yeah.
32 *((Anna reaches the end of wall. Points to end. Susan looks.))*
33 Anna: ( ) read it.
34 *((Susan looks at the number,))*
35 Susan: One (.) hundred (.) and ni:::ne
36 *((Susan points at end of block line (may say something inaudible).*
37 *Darcy moves away.))*
38 Anna: Okay::::
39 Let's see::: *((she lets go of the case))*
40 *((Tape case jumps again onto stage. Girls scream and run toward the*
41 *tape. Anna hands it to Susan.))*

Four times in all the girls let go of the tape measure and it pops onto the stage. The girls display excitement about this when it happens (in the form of whoops, jumps, and screams), but they do not make verbal reference to it. This activity has several experiment-like qualities. For instance, Anna's comment as she lets go in line 39 ("let's see:::") suggests that she is orienting to her action as a "trying out" of something. It is also experiment-like in that it consists of multiple repetitions of a procedure with slight modifications made each time. These multiple repetitions, in which the girls take different roles, allow them to see the phenomenon from different perspectives and potentially observe which aspects remain stable and which aspects change. We should note, however, that there are aspects of the activity that do not clearly fit the label of "experiment." For instance, there is no evidence that the modifications undertaken with each repetition are designed to produce the effect of making the tape measure jump when released. To the extent that there is a particular purpose indicated, it is to add blocks of the right number and length such that the tape measure will still extend to the end of the block wall (lines 1, 5–6, 22–26). Otherwise, the majority of the talk surrounding this activity is about the turn-taking process itself and the appropriate performance of roles. Close consideration of the girls' talk and action yields the impression that the repetitive activity is driven more by the social process of collaboration itself than by an effort to investigate the phenomena.

Interactional Challenges of the Activity

As discussed above, in parent-child interaction, the management that parents do minimizes the interactional work required of children in order to participate in the activity. The same is not true of peer play. Rather, we find that creating or entering into a play group, sustaining the activity of a group, and influencing the activity of the group in the direction one prefers all require substantial effort, even skill. The segments below are chosen to focus on some of the interactional achievements of one of our focal children, Darcy, and compare with difficulties experienced by Darcy and our other focal child, Molly.

Getting In

Darcy first joins the activity when Susan is in the process of measuring Anna. Darcy wordlessly steps in to perform a needed task: pulling up and holding the end of the tape measure so the number can be read.

Example #6: People-Measuring

1 *((Darcy moves across the room toward Susan and Anna, pauses half-way, and continues.))*
2 *((Susan's arm is extended with case end of tape measure near Anna's head. Anna reaches up with her right hand to take the tape measure case.))*
3 Susan: Can you read that?
4 *((Anna looks at tape.))*
5 *((Darcy briefly pauses in front of them, then moves forward as she lifts her arms and takes tape measure.))*
6 Anna: Um:::
7 *((Darcy stretches tape measure up to top of Anna's head.))*
8 Susan: What number does it say?
9 *((Anna steps out and looks at tape, while Darcy holds it.))*
10 Anna: Twe::lve (.)
11 Okay um (twenty).
12 It's a two and a fou:r

13 Darcy: So it's it's forty.
14 Susan: I think ( )
15 Darcy: It's your ( )
16 Anna: No::: *((turns to face Darcy))* I am twe::lve (.) inches.
17 (you didn't make a)
18 *((hands on hips))*

In line 1, Darcy comes into view of the camera on the opposite side of the room from Anna and Susan. As Susan continues to try to stretch the tape, Darcy notices the girls' activity and moves closer, continuing to observe them. In line 2, Susan seems to have realized that she cannot possibly stretch the tape herself all the way. She hands the tape Anna. However, Anna does not pull the tape to the top of her head, nor could she read the tape if she did.

As an observer from across the room, Darcy most likely sees Susan's struggle and is able to join in the activity simply by doing the task that the girls cannot do themselves: pulling the tape up to the top of Anna's head. Darcy does this wordlessly. The girls do not object to Darcy's entry. They do not talk about it, nor do they invite her in. They accept and acknowledge her task and presence by continuing with the activity. Susan's question in line 8, and the fact that they simply go on with the activity, evidences Darcy's acceptance into the group.

We can compare Darcy's experience to that of Molly. When Molly does something very similar to what Darcy did earlier—i.e., stepping in to hold the edge of the tape measure—she is forcefully reprimanded by the existing members of the group. The segment below occurs when the girls are measuring objects around the room. Molly is playing with another girl in the dress-up area. Darcy, Anna, and Susan approach the low wall that divides the dress-up area from the block area and start to measure it. This attracts Molly's attention, and she walks over to the divider and places her hand on the edge of the measuring tape (line 13). As the girls finish measuring, Molly removes her hand and walks around the divider. Susan walks around to meet her.

Example #7: Dividing Wall
1 Darcy: Anna! *((Gestures from block area for her to come))*
2 Anna: *((Runs over to dividing wall where Darcy is standing))* Let's measure
3 this! *((Anna, looks at Susan, points at wall, and jumps up and down))*
4 Susan: *((Walks over to Anna and Darcy))* Okay, but this time=
5 *((Anna puts end of tape at top of wall))*
6 Susan: =let me pull it down *((pulls tape down, kneels on floor))*
7 Anna Okay and I'll hold it.
8 *((Darcy looks up and down tape measure))*
9 *((Anna looks at Darcy))*
10 Darcy: Let go *((puts arms up slightly))*
11 *((Anna takes a step back))*
12 Darcy: <u>Uh:</u>
13 *((Molly places her hand on measuring tape from other side of wall.))*
14 Susan: Okay *((kneels down to grab tape))*

Anna: Okay *((kneels down slightly and then up again))*
((Susan guides measuring tape to top of wall with hand))
((Both Susan and Molly walk around divider to table))
Susan: You guy::s you're not supposed to touch the measuring tape. *((hands on hips))*
Anna: Because you might pinch your fingers and we're measuring everything. *((leans forward))*
Susan: Yeah
Darcy: And don't (.) put your fingers on that again cuz you might (pinch) your fingers. *((leans forward and folds arms))* Okay?
Anna: Okay I'm gonna show you what kinda part you can't put your fingers on. *((taps end of measuring tape))*
Darcy: And it's my turn to pull it. Okay?
Anna: Yeah.
((Girls walk away. Molly turns back to other girl.))

Having finished measuring the wall, Susan steps out first to talk to Molly (line 18). Her tone is sharp, and her expression and body posture display disapproval. She tells Molly what her transgression was: "you guys" are not supposed to touch the tape. In line 20, Anna builds on Susan's comment by making it more specific and providing reasons for the girls not to touch the tape. The first reason is for Molly's own safety ("you might pinch your fingers"). Anna has used this reason with other children who sought access to the tape. She gives Molly another reason, which is their activity. The group is "measuring everything"; in other words, they are not just playing with it but using it for an important project.

In line 22, Susan backs up Anna's comments with "Yeah." At this point, Darcy also contributes by reiterating what has been said in the form of a warning for the future: "Don't put your fingers there again." Darcy adopts Susan's tone and body posture.

Overall, the impact of the group is impressive. Molly watches them and says nothing. They walk off together, almost marching, to continue their project. Molly's action has sparked an interaction in which the girls make it clear that the group is not open to everyone and that they will decide who gets to participate.

It is impossible for us to say with certainty why Darcy's attempt to participate was successful and Molly's was not. We can note, however, differences in the social and material configurations of participants at the moment at which each girl places her fingers on the end of the tape. When Darcy arrived, there were only two girls, one measuring the other with the tape. When Molly encounters the group, there are three girls, the tape, and a dividing wall. The presence of another set of hands and a wall with an edge that affords hooking of the tabbed tape end may render Molly's contribution unnecessary and unwanted. In any case, because her effort at participation is rejected, learning opportunities that Darcy experiences are not available to Molly.

Influencing the Course of Events

Getting into the group is not the only interactional issue with which the girls must contend. Directing the group's activity in the way she desires is also a challenge for Darcy. Anna, as owner of the tape measure, repeatedly asserts the right to make decisions for the group. In the segment below, Darcy employs two tactics for influencing the group's activity. The girls are standing on a stage on one side of the room and measuring a bookshelf at the back of the stage.

Example #8: Stage-Measuring

1 Anna: Okay (inaudible) *((moves onto stage, followed by Darcy and Susan))*
2 I wanna (start it) to be my turn.
3 Okay.
4 *((Anna and Susan hook tape end on top of shelf, extend case downward.))*
5 *((Susan holds top; allows it to dip down a bit.))*
6 Anna: No no no no no no no
7 *((Girls get tape hooked on top shelf and case extended to bottom. Both*
8 *kneel down to look at case end.))*
9 *((Susan unhooks end and brings to floor. She then starts stretching tape to top of shelf.))*
10 Darcy: (inaudible) *((to Anna))*
11 Anna: No::. I'm gonna roll down it down (inaudible) and you might have a
12 you might have this part and you might pinch your fingers.
13 Susan: (inaudible)
14 Darcy: If you don't let me have a turn (I won't invite you to my birthday
15 party) *((hands on hip))*
16 Anna: (inaudible)
17 Susan: (inaudible) *((seems to be reading number))*
18 Darcy: How bout we measure- how about *((runs to right of stage and back,*
19 *pointing at floor))*
20 *((Anna unhooks top of measuring tape; it retracts into case))*
21 Anna: (Cool!)
22 *((Susan does little jump.))*
23 Darcy: *((gets down on her knees))* Now let's measure the stage.
24 *((Susan and Anna look down at her))*
25 Susan: Yeah!! ((jumps off of stage))
26 Anna: *((walks over to end of stage and hooks tape end over side))* Now I'm
27 gonna go like the::is.
28 Girls: (inaudible)
29 *((Anna stretching out tape))*
30 *((Darcy backs up, watching. Susan moves over to tape end and holds*
31 *down))*
32 Anna: *((gets to end))* Okay!!
33 *((Susan leaves tape end and comes over to Anna; Darcy also gathers*

34 *round, looking at tape))*
35 Susan: Okay:
36 (Darcy:) (fifty-nine)
37 *((Susan walks away toward the end of the tape, then jumps off the*
38 *stage))*
39 *((Anna lets go of tape case and it scoots across stage, hitting end))*
40 Darcy and Susan (in chorus): Coo:::l:::!!
41 *((Darcy looks at Anna, smiling))*
42 *((Susan laughs))*
43 *((Darcy laughs))*
44 *((Susan picks up tape measure))*

Darcy's utterance in line 10 is not audible, but from Anna's response, we can surmise that it is a request to participate in some way. Anna rejects this request by asserting her own plans and explaining the risk to Darcy's fingers. Darcy follows up with a threat in lines 14–15, "If you don't let me have a turn I won't invite you to my birthday party." Anna continues with what she is doing; the threat does not appear to have its desired effect.

Darcy then suggests a new object to measure: the stage itself (lines 18–19 and 23). This suggestion is taken up by Susan in line 25 and by Anna in lines 26–27. Although it does not immediately result in Darcy having a turn with the tape measure, she does become more actively involved in the measuring project. It also results in the happy occasion of Darcy and Susan first witnessing the tape case retract toward the tape end, which is greeted with smiles and laughter.

As with Darcy's initial move at getting into the activity, it seems that a good way to ensure involvement is to contribute in ways that keep the activity going. In this case, task skills and interactional skills are one and the same.

Discussion: Peer-Regulated Group Activity as Learning Arrangement

First, and perhaps most importantly, this analysis shows that young children can and do engage in science-relevant (or protoscientific) activities in peer groups *without* adult guidance. We can recognize several advantages of this peer-regulated form of interaction as a learning arrangement for young children. First, there appears to be strong social motivation for participating in these activities. Even when their participation is not solicited nor reliably rewarded by other children—as it tends to be by parents in the home environment—children actively seek out and work hard to gain entrance into these activities. We might say that the activities exhibit something like a gravitational pull on children.

Furthermore, these activities afford different types of learning opportunities. Children negotiate with each other in guiding and regulating the activity themselves—as such, they have increased opportunity to both exercise agency and to develop skills at collaborating with others. Children can position themselves as both learners and teachers in such interaction. The lack of stable and explicit goal orientation gives children a great deal of freedom to explore the aspects of a phenomenon that most interest them (Rogoff, 1990). The iterative nature of this activity is also important, as each repetition, even without an articulated purpose for it, provides a chance to see the phenomenon unfold in a slightly different way.

These peer-organized learning arrangements also seem to offer enhanced opportunities for full-fledged embodied participation and experiential learning, as children move about the room with little constraint, drawing into their activity a variety of different types of artifacts and built structures.

There also are disadvantages when compared with adult-mediated interaction. These have to do with the interactional challenges of the activity and the lack of adult support in developing an activity in the direction of disciplinary knowledge. In child-regulated peer group activity, the interactional demands fall squarely on the shoulders of the children themselves. Our analysis of this activity shows that gaining entrance is itself a difficult maneuver—while one our focal children succeeds, the other attempts to participate but is unsuccessful. Similarly, the children must maintain the activity and direct it in fruitful and enjoyable directions. Here, the children employ skilled practice in this regard, but young children may not always be able to do this effectively.

Children also do not experience the same level of interactional support for learning in peer activities as they do in parent-guided activities. Unlike parents, peers do not necessarily monitor situations for opportunities to support other children's learning. To the extent that children's activities are monitored, we have observed that peers are more likely to see shortcomings and offer correction than recognize innovation or respond with elaboration. These qualities of the interactional format provide opportunities to develop skills at argumentation and negotiation—surely part of the collaborative practice of scientific work. However, for the child who gets left out or shut down, learning opportunities will be missed.

Another possibility is that rich opportunities for learning may occur but not get developed in the direction of disciplinary knowledge. For instance, in the block-measuring activities described above, there is no investigation of *why* the tape measure flies. Over the course of the repeated tries, girls encounter "information" about the physical workings of the tool; however, they do not seem to engage in purposeful or principled experimentation in order to find out more. Without the involvement of knowledgeable adults or peers, there is little opportunity to link this activity to disciplinary principles or practices. We can draw a contrast with Example #4 ("Tiger's Eye") above, in which Molly's mother, after modeling a comparative process through which Molly might determine whether the rock was a Tiger's Eye, suggests asking a geologist. In doing so, she both links their current activity to a scientific discipline and introduces the notion of an expert as someone who can be consulted when one wants to know more about a topic.

Conclusion

The distinction between formal and informal learning takes on a peculiar status when applied to the lives of very young children. Though formal education reaches into the mostly informal lives of young children in many ways, "school" does not neatly map onto "formal" and "home" does not neatly map onto "informal." In fact, due to the particular social configurations of preschools, they seem to offer *more* opportunity for the highly informal learning arrangement of uninterrupted, child-organized group play than do middle-class homes.

Another distinction that is blurred through our analysis is that between doing (and learning) science and doing (and learning) social interaction. We see that in both home and preschool, the opportunity to participate in social activity can create opportunities for learning. Scientific inquiry and exploration are not incidental to the social activity but are an integral part of it (and vice versa), because keeping the activity going and making it interesting are socially beneficial. For example, Darcy has more success in exerting agency within the activity when she comes up with good ideas for measuring things than when she threatens her friends with the loss of social favor (i.e., not being invited to her birthday party). Our analysis thus far suggests, furthermore, that while adult-guided interaction may be a more reliable learning arrangement for introducing scientific content, group peer play may have certain benefits for developing skills at doing science as a collaborative activity—such as negotiation, argumentation, and the ability to maintain the cohesion of a group while having a say in the direction of its activity.

Our comparison of peer-organized play in preschool and adult-organized interactions in the home suggests that both arrangements may contribute fruitfully to children's early science learning. The analysis in the first section reveals some of the mechanisms through which parents can support early STEM-related domain learning by exploiting everyday experiences with natural phenomena and materials (see also Goodwin, 2007). By guiding children's perception and categorization of objects in the environment and linking children's actions to the practices of disciplinary "experts" (such as geologists), parents create conditions for the early development of science-relevant interests and knowledge (Crowley & Jacobs, 2002). Informal peer-group interactions, as we have seen, can also provide opportunities for children to encounter, discuss, and physically interact with science-relevant phenomena. Further, children in these activities display levels of affective engagement and self-motivation unmatched in the adult-child interactions in our data. However, it is less clear what role participation in this type of activity might play in a trajectory of "becoming a scientist" or even succeeding in science learning later in school, as the learning that occurs is less likely to be linked to disciplinary knowledge or practices.

Our analysis thus supports pedagogical practices that take advantage of the special opportunities for peer learning afforded by the preschool classroom setting by allowing children the freedom to self-organize into groups and pursue activities of their choice. But it also suggests that strategic adult support, provided at key moments in children's peer play, could enhance the benefits of these activities by providing disciplinary connections and promoting deeper learning of a kind that may not happen when children play together without intervention.

Acknowledgements This work has been conducted as part of the Learning in Informal and Formal Environments (LIFE) Science of Learning Center (http://life-slc.org/). We wish to thank the National Science Foundation for the opportunity to conduct this research through awards SBE-0354453 and SBE-0835854.

References

Bell, P., Lewenstein, B., Shouse, A., & Feder, M. (Eds.). (2009). *Learning science in informal environments: People, places, and pursuits*. Washington, DC: National Academies Press.

Callanan, M. A., & Oakes, L. (1992). Preschoolers' questions and parents' explanations: Causal thinking in everyday activity. *Cognitive Development, 7*, 213–253.

Cazden, C. (1988). *Classroom discourse: The language of teaching and learning*. Portsmouth, NH: Heinemann.

Crowley, K., & Jacobs, M. (2002). Islands of expertise and the development of family scientific literacy. In G. Leinhardt, K. Crowley, & K. Knutson (Eds.), *Learning conversations in museums*. Mahwah, NJ: Lawrence Erlbaum Associates.

Goodwin, M. H. (2007). Occasioned knowledge exploration in family interaction. *Discourse & Society, 18*(1), 93–110.

Hanish, L., Barcelo, H., Martin, C., Fabes, R., Holmwall, J., & Palermo, F. (2007). Using the Q-connectivity method to study frequency of interaction with multiple peer triads: Do preschoolers' peer group interactions at school relate to academic skills? *New Directions for Child and Adolescent Development, 118*, 9–24.

Jordan, B., & Henderson, A. (1995). Interaction analysis: Foundations and practice. *The Journal of the Learning Sciences, 4*(1), 39–103.

Mehan, H. (1979). *Learning lessons: Social organization in the classroom*. Cambridge, MA: Harvard University Press.

Rogoff, B. (1990). *Apprenticeship in thinking: Cognitive development in social context*. New York: Oxford University Press.

Stevens, R. (2000). Who counts what as math: Emergent and assigned mathematics problems in a project-based classroom. In J. Boaler (Ed.), *Multiple perspectives on mathematical teaching and learning* (pp. 105–144). New York: Elsevier.

Stevens, R., & Hall, R. (1998). Disciplined perception: Learning to see in technoscience. In L. Magdalene & M. L. Blunk (Eds.), *Talking mathematics in school: Studies of teaching and learning*. Cambridge, MA: Cambridge University Press.

Stevens, R., Satwicz, T., & McCarthy, L. (2007). In game, in room, in world: Reconnecting video game play to the rest of kids' lives. In K. Salen (Ed.), *Ecology of games: MacArthur Foundation series on digital media and learning*. Cambridge, MA: The MIT Press.

Stevens, R., Wineburg, S., Herrenkohl, L. R., & Bell, P. (2005). The comparative understanding of school subjects: Past, Present, and Future Research Agenda. *Review of Educational Research, 75*(2), 125–157.

Streeck, J., & Mehus, S. (2005). Microethnography: The study of practices. In K. L. Fitch & R. E. Sanders (Eds.), *Handbook of language and social interaction*. Mahwah, NJ: Lawrence Erlbaum.

Tizard, B., & Hughes, M. (1984). *Young children learning*. Cambridge, MA: Harvard University Press.

Williams, P. (2001). Preschool routines, peer learning and participation. *Scandinavian Journal of Educational Research, 45*(4), 317–339.

Chapter 11
Shifting Languages, Spaces, and Learner Identities: Learning Mathematics After School

Aria Razfar

Introduction

Over the last 30 years, there has been growing interest in how children learn in informal and "out-of-school" contexts especially through the prism of discourse analysis of interaction. One of the primary rationales for this development is the rigidity of formal, classroom discourse structures. This is further complicated when restrictive language policies limit the use of bilingual children's native languages and cultures, especially for Latinas/os. While there has been growing recognition within the mathematics education community for the need to consider discourse in the learning of mathematics and "mathematization" (Rogers, Mosley, Hui, & O'Garro-Joseph, 2005), there has been a dearth of research on how children engage in mathematical discourses in informal and "out-of-school" spaces. Given the well-documented achievement gap between Latinas/os and their dominant counterparts, documenting learning in informal contexts such as an after-school club provides us with a unique opportunity to enhance our understanding of how children mathematize using multimodal mediational tools.

Following the primarily literacy-driven work of the *Fifth Dimension* (Cole & The Distributed Literacy Consortium 2006; Gutiérrez, Baquedano-López, & Tejeda, 1999; Vasquez, 2003) and the theoretical principles of cultural historical activity theory (CHAT) that guide it, we set out to create a similar activity system, whereby bilingual, Latina/o children could more fully draw on their linguistic and cultural toolkits to develop mathematical thinking and discourse practices through participating in community-based activities as well as games (Khisty, 2001). While the after-school club is a single setting, it is an activity system that has been designed to

A. Razfar (✉)
Department of Curriculum and Instruction, University of Illinois at Chicago, Chicago, IL, USA
e-mail: arazfar@uic.edu

B. Bevan et al. (eds.), *LOST Opportunities: Learning in Out-of-School Time*, Explorations of Educational Purpose 23, DOI 10.1007/978-94-007-4304-5_11,

foster cross-contextual, multigenerational, and multimodal interactions with varied ends including mathematical learning, academic discourse development, and building community relations. Given the inherent hybrid design of the after-school club, meaning making, whether mathematical or not, is often contested and filled with tension. These interactional moments are critical to the learning and development of all the participants: children, graduate students, preservice teachers, parents, and the faculty investigators as other chapters in this volume show. All participants interact with each other not only for the purposes of the after-school activities but also as a part of belonging to the community. Thus, the interactions observed within the after-school club are a microcosm of a broader discourse with cognitive, sociocultural, and ideological implications.

Given our focus on mathematical learning, for the purpose of this analysis, I defined mathematical discourse as those practices that are primarily used to discuss and understand quantities, shapes, spatial relations, and deductive/inductive reasoning (Sfard, 2002). Similar to other *Fifth Dimension* sites, play and community-based activities were central to learning and development where they serve to create hybrid spaces for learning through the use of multiple modalities, languages, and other mediational tools (both material artifacts and cultural models). The after-school club could not be considered a purely elective learning space and nor could it be considered a formal instructional setting with compulsory learning expectations; it was simultaneously in school and out of school and formal and informal. Students were encouraged to engage in specific mathematical problem-solving contexts, but they were free to choose how they participate and were encouraged to leverage their cultural knowledge and linguistic repertoires in the process. In this chapter, I use discourse analysis to illustrate, from a cultural historical point of view, how bilingual, Latina/o children engaged in mathematical practices within game-play activity. In particular, I show how mathematical strategies related to probability evolved in the course of multigenerational and multilingual game play. Furthermore, I show how bilingual children over time displayed evidence of learning through shifts in participation as they negotiate multiple and sometimes conflicting goals with adults. This analysis draws on data taken from *Counters* play in the *Los Rayos* after-school club, a game designed to build understanding of probability reasoning.

Context: The Counters Game and the Fifth Dimension

The *Counters* game was situated within an urban area with a predominant Latina/o population. Not surprisingly, there was variation with respect to bilingualism, immigration status, and length of residency in the United States. *Counters* was one of several activities that children played that required the development of advanced probability strategies in order to be considered an expert player. The activity took place in an informal, after-school mathematics club called *Los Rayos* (the rays of CEMELA), located in a bilingual, urban public school. *Los Rayos* participants met twice a week for approximately 1.5 hours each session. The student participants were

14–20 bilingual Spanish/English male and female fourth and fifth graders, most of whom have participated in *Los Rayos* since third grade. The facilitators were bilingual undergraduates, some of whom were preservice teachers, as well as graduate fellows. Groups of three to five students, with one or two facilitators in each group, engaged in activities designed to foster mathematical thinking and interactions. Within the sessions, students were allowed many freedoms to choose their activities and to dictate the course a project will take. This design was deliberately chosen by researchers to create a drastically different environment than what is typically found in the traditional classroom. The activities were designed to foster embodied meaning making practices and higher-order thinking.

All of the games in the after-school club were designed to include written task cards in multiple languages as mediating artifacts for orienting participants to the goals of the activities. Technology- and computer-mediated communications are integrated within the *Fifth Dimension* after-school clubs. It is part of the multimodal design that informs the social organization of learning at the *Los Rayos* after-school club. In *Los Rayos,* technology-mediated learning happens in two prominent ways: (1) communication with a cyber entity known as *El Maga* and (2) composition of digital stories using *iMovie* at the end of the year. Participants used computers to play games and communicated with an imaginary cyber entity known to the participants as *El Maga. El Maga* had regular online communication with the participants about the immediate task and created another context for assistance and dialogic inquiry. As part of their daily routine, participants communicated electronically with *El Maga* at the beginning and end of each session. *El Maga* was purposely designed to be ambiguous and somewhat ubiquitous so that each student could freely engage with him/her/it on their own terms and in the mode they were most comfortable with. Of course, *El Maga* often spoke in Spanish as part of the overall goal of repositioning Spanish as a viable language for mathematics. *El Maga* positions students as active learners and agents of knowledge construction. At the end of the year, the children composed digital stories based on their experience in the after-school club. For example, Rafael (to be discussed later) told the story (in Spanish) of a hybrid board game he created based on a variety of games he played throughout the year (including the *Counters* game) and another probability game his mother introduced called *the Game of Five Numbers.*

Like all activities, the *Counters* game is governed by explicit and implicit rules of participation. For example, the undergraduate and graduate facilitators, for the most part, are expected to ask leading questions (e.g., "why?" and "how?") as the primary form of assistance and mediation rather than provide answers or "next step" types of assistance. They also try to promote meaning making and "third spaces" by encouraging debate, contestation, inquiry, and *zones of proximal development*[1] (Gutiérrez, 2008; see also Chap. 15 by Baker et al., this volume, for a description).

[1] Vygotsky (1978) defined the zone of proximal development (ZPD) as the difference between what a learner can do individually and what they can accomplish with assistance.

However, as I will show later, there was quite a bit of variation in terms of how undergraduate and graduate facilitators scaffolded children's learning, especially if primary languages and discourses were used or not used.

There was an explicit orientation to privilege and encourage the use of Spanish during the activities. Another expectation that organized the activities was the fact that participants worked in small cohorts with the same facilitators for the duration of the activity, thus sustaining deep relations with one another as well as fidelity to the goals of the activity. In addition, participants were explicitly encouraged to draw on multiple modes of knowing and assume dynamic identities as learners with respect to expertise and epistemic authority. Participants were expected to draw on their community knowledge and use all modes of communication they knew and were comfortable with. For example, in the context of the *Counters* game, some students may talk about "blowing on the dice" or the "dice being dead" during game play. Facilitators were expected to draw on students' home-based expertise; however, this was much more likely to happen with the presence of parents. Parents engaged the children and mediated their learning in the after-school activities in ways that other facilitators by themselves were unable to do. They were able to draw on the students' household knowledge, or *funds of knowledge* (Gonzalez, Moll, & Amanti, 2005), in more visible ways (e.g., the use of Spanish, drawing on home mathematical activities, and problem-solving strategies). In addition, parents often used Spanish terms of address that are more affectionate and grounded in students' primary discourse (e.g., "mija" meaning daughter). They challenged one another with respect to the rules of the game and generated *third spaces* for building "formal" mathematical concepts through "informal" means (Gutiérrez, 2008). Ultimately, in creating an interactive third space, these activities served to socialize the students to new mathematical identities and foster a repositioning and refinement of their cultural and linguistic toolkit—developed in other settings and activity systems in their life but applicable to sensemaking in specific ways within this after-school context.

The *Counters* game was one of several planned activities designed to develop the concept of probability. It requires that players create their own game board, consisting of a rectangular strip of colored construction paper with hand drawn boxes spread from left to right, with one of each of the numbers from 2 through 12 written in each space (Fig. 11.1). Each player receives 12 "counters" (small plastic cubes) and places them on their own game board, arranging them according to their own preference. For example, players may put several counters on the numbers 5, 6, and 7 while leaving other numbered spaces unoccupied. All other arrangements of the counters are allowed as long as all are placed on a numbered space on the game board. A pair of dice is passed between each player who then rolls them to get a number between 2 and 12. Each player then checks their game board to see if any of their counters are on that number. If so, one piece is removed and discarded. Then, another player takes a turn rolling the dice. Players take turns rolling the dice until one person has removed all of his or her counters to win the game.

Next, players begin another round and once again arrange their counters. In this activity, we were interested in observing how each student arranged his or her counters.

2	3	4	5	6	7	8	9	10	11	12
X	X	X	X	X	X	X	X	X	X	X

X=Plastic Cubes

Fig. 11.1 Sample counters game board

Specifically, one might ask, "Do students place their counters using strategies formulated by observing the frequency of particular numbers appearing during random throws of dice?" For example, a student might notice that the numbers 6 and 7 appear more frequently than other numbers such as 2 and 12. As a result, the student's rearrangement of game board counters may indicate strategies that the student uses to figure out how to maximize chances of winning the game.

Conceptual Frameworks

Cultural Historical Activity Theory and Learning Through Participation Shifts

Cultural historical activity theory (CHAT) as an *in situ* theory of learning and development is a useful framework for understanding learning across multiple contexts and timescales. This is particularly relevant when one considers learning across what might typically be considered "formal" and/or "informal" settings or in contexts such as the after-school club where traditional borders of age (parents and children), domains of expertise (traditional school subjects, community knowledge), and communicative modalities (e.g., use of formal language, everyday vernacular, Spanish, English) were deliberately blurred. CHAT emphasizes learning as fundamentally social, continuous, and mediated by signs, symbols, material artifacts, and the interactions with more capable others. Scholars working from the CHAT tradition also have a powerful legacy of highlighting the role of play and community-based activities in human learning and development. This body of work shows how children engage with more capable peers and other competent members of a "community of practice" in order to achieve concrete goals (Engeström, Miettinen, & Punamäki, 1999; Lave & Wenger, 1991; Nicolopoulou, 1993; Vygotsky, 1978). Within socially defined activities, participants interact to make meaning and solve problems. It is important to note that all human activities are rule-governed and goal-directed; participants, through the process of mediation, engage in joint activity to accomplish the goals of an activity (Wertsch, 1985). Another significant feature of the CHAT framework is the centrality of history (both synchronic and diachronic; Cole, 1996) toward understanding how participants shift and appropriate the discursive tools to participate in culturally appropriate ways (Rogoff, 1991, 2003).

From a cultural historical perspective, learning can be examined as shifts in participation and discourse and the appropriation of new *discursive identities* (Gee & Green, 1998; Rogoff, 2003; Wertsch, 1998). While human beings undergo a lifelong, life-wide, and life-deep process of language socialization, not all discourses are equivalent both in terms of the process and purpose of appropriation. According to Banks et al., (2007), the learning and learners should be understood in terms of the following four principles:

1. Learning is situated in broad socio-economic and historical contexts and is mediated by local cultural practices and perspectives.
2. Learning takes place not only in school but also in the multiple contexts and valued practices of everyday lives across the life span.
3. All learners need multiple sources of support from a variety of institutions to promote their personal and intellectual development.
4. Learning is facilitated when learners are encouraged to use their home and community language resources as a basis for expanding their linguistic repertoires.

Learning Through Participatory Shifts: Primary and Secondary Discourses and Third Space

Given the continuum and interconnections between "informal" and "formal" contexts of learning, it is useful to delineate discourses in terms of *primary* and *secondary* discourses (Gee, 1996). Primary discourses "are those to which people are apprenticed early in life during their primary socialization as members of particular families within their socio-cultural setting" (Gee, 1996, p. 137), and secondary discourses are "those to which people are apprenticed as part of their socialization within various local, state and national groups and institutions outside early and peer group socialization, for example, churches, schools, etc." (Gee, 1996, p. 133). Secondary discourses have the properties of a more generalizable cultural model, are more explicitly taught, value precision, and are less dependent on the immediate situation for access by a larger audience. Secondary discourses serve to mediate problem-solving in novel situations. The more abstract the literate mathematical discourse, the greater its potential as a widely generalizable problem-solving tool (Sfard, 2002). The activities in the after-school club can be viewed as ones that socialize learners to secondary discourses (in this case probability) with such a generalizable potential while leveraging their primary discourses (i.e., Spanish and community knowledge). From a cultural historical view of learning, secondary discourses develop through mediation and interaction with more capable peers or adults (i.e., "experts" of the secondary discourse).

As people participate in social activities, the use of primary and secondary discourses are usually enabled or constrained by the epistemic and language ideologies of participants (Razfar, 2005). Language ideologies are the beliefs of speakers with respect to the function and purpose of language use, and epistemic ideologies reflect people's beliefs about what counts as legitimate knowledge and ways of knowing. For example, in a traditional classroom where mathematics is done primarily in English, students may come to see "Spanish" as inappropriate for mathematical practices (Vomvoridi-Ivanovic, 2009). In the context of the *Los Rayos* after-school club, Spanish was purposefully repositioned as a central language for learning and mathematization something bilingual students do not typically experience in school. Epistemic ideologies also impact how individuals in various contexts are positioned as "experts" or "novices," and hence, it impacts their ability to make claims within a particular domain of knowledge.

Discourses always embody the language and epistemic ideologies of the participants; furthermore, an ideological view of discourse practices helps us understand why some practices are more valued, privileged, and attributed greater legitimacy than others. Identities and ideologies become foregrounded in the analysis of talk, text, and the use of mathematical symbols especially as participants move between formal and informal social spaces (Street, Baker, & Tomlin, 2005). In the *Counters* game, consistent tensions between participant language choice (English or Spanish) and the use of formal or informal mathematical terms were observed. This interplay is further complicated by the broader epistemic question of what the participants through their discursive actions count as legitimate mathematics and which language is more "mathematical." More significantly, the findings show that these tensions were very productive and the hybrid participation structures of each activity were essential for creating thirds spaces through which advanced meaning making could emerge.

Given the heterogeneous organization of the after-school club especially with respect to age, mathematical content knowledge, and language proficiency, we have observed points of tension as participants struggle to make sense across the various factors impacting the interactions (Razfar, Sutton, & Radosavljevic, 2009). For example, one of the most prevalent tensions was students' use of informal mathematics register and the graduate students' goal of socializing them to formal mathematical terms. Another important tension was the use of Spanish and/or English in relation to mathematical problem-solving. Of course, this was purposefully anticipated as the interest in "meaning making tensions" is borne out of the sociocultural premise that learning, development, and the construction of knowledge are fundamentally mediated through situated spaces that are filled with contestation and cognitive dissonance (Gutiérrez et al., 1999; Moje et al., 2004). As the heterogeneity of interactions increases so do the opportunities for the contestation and construction of knowledge. It is precisely these moments in interaction that lead to greater capacities to solve problems. Learners assume new epistemic identities and show this through their ability to contest and utilize a fuller range of mediational tools available to them especially informal mathematical registers and the use of Spanish. According to Gutiérrez (2008), it is important to conceptualize the zone of proximal development beyond the narrow views of productive adult-children scaffolding and assistance strategies; furthermore, the transformation of everyday concepts (primary discourses) into "scientific or school-based concepts" (secondary discourses) can be conceptualized as movement through third space:

> First, we can document in Third Spaces a reorganization—a movement, if you will—of everyday concepts into "scientific" (Vygotsky, 1978) or school-based concepts. Second, leading activities significant to individuals' subsequent development, specifically play and the imaginary situation, learning, and affiliation, reorganize everyday functioning-the movement-in the Third Space. (p. 152)

Analysis

This analysis is based on an in-depth examination of three consecutive weeks of *Counters* game discourse (Razfar, 2012). The data corpus consists of 220 minutes of actual game play whereby the research team coded for language choice (Spanish/English), use of formal and informal mathematical talk, disagreements between participants in relation to game-play strategies, and explicit references to probability. One of the important features of this game is that in order for participants to successfully negotiate the game, they need to develop strategies that are grounded in understanding of higher-level math concepts surrounding probability. The game structure requires every member to pay attention to each turn of the game with each player receiving control of the dice. In terms of procedure, each player has equal status, an identical role, and identical operations to perform in order for the game to proceed. The game provides for multiple levels of participants, both in terms of generation (parents, students, fellows, undergraduates "UGs") and expertise.

The analysis showed children and adults engaged in multilayered and multimodal discourse practices as they played the *Counters* game. With respect to language choice, participants regularly switched between English and Spanish and in purposeful ways with Spanish used about 60% of the time and English 40%. Given the dynamic participation structure, code-switching occurred systematically for the purposes of assistance, making tasks more comprehensible, asking questions, inclusion, and sometimes exclusion of central and peripheral participants. Children often indexed their awareness of speakers and nonspeakers of a particular code by switching to accommodate understanding. The amount of Spanish use was more than English, and this ratio increased over time as participants became more comfortable with each other and with the game. We also examined numerous instances of "cognitive" and "discursive" tension (25% of the time). These were moments where participants overtly disagreed, challenged, and/or countered each other's strategy for negotiating the game. These episodes suggest how meaning making, especially in informal contexts, is less absolute or fixed and more dialogic, contested terrain.

As expected, learner identities within the *Counters* game were quite fluid and dynamic with learners shifting in and out of expert/novice roles depending on the situation. In the remainder of this chapter, I analyze three different groups of participants as they negotiate the *Counters* game. The first group was a group that interacted in English only, and no parents were present. In this group, the undergraduate and graduate facilitators made numerous attempts to scaffold and mediate a more advanced probability strategy; however, this goal was never realized. In contrast, the second and third groups consisted of parents, and the interactions were characterized by extensive use of Spanish and primary discourses. In these cases, there was much evidence of a "probability" secondary discourse developing. In addition, these interactions exhibited a kind of mediation that leads to meaning making in the third space and greater invocations of an explicit probability strategy.

Fixed on "Luck": Constrained Mediation and Impediments for Third Space

There was quite a bit of variation in how the game play unfolded depending on the participants and the types of mediation available. In the following vignette, the undergraduate and graduate facilitators never used Spanish, there was visible tension between the goals of the facilitators and the children, and rather than moving toward a third space, there was stagnation. The graduate facilitators were non-Spanish speakers. The children simply wanted to play the game and win, while the facilitators made overt moves to scaffold thinking about the mathematics involved. However, a third space where transformative learning could occur never emerged. The children continuously invoked "luck" and "prayers" despite the incessant efforts of the undergraduate and graduate facilitators to move them toward the idea of "probability." The participants consisted of one Latino undergraduate Felipe (F), who considers himself fluent in Spanish but admittedly rarely uses it in "school-like" settings; Erato (E), a female graduate student who does not speak Spanish; Pavle (P), a non-Spanish-speaking graduate student; and Juana (J) and Gustavo (G), two bilingual Latina/o children.

Throughout the interactions, Erato repeatedly attempts to push the children to think about probability (lines 1, 3)[2]:

1. E: Juana. Can you tell me how you decided to put them that way?
2. J: Because that. Because it's hard to get the easy numbers?
3. E: Yeah? Can you tell me why?
4. G: Hey! You got. You don't have two.

Juana (line 2) speaks in everyday terms (primary discourse) saying "it's hard to get easy numbers," and Gustavo (line 4) never affirms Erato's question of "why?" (line 3).

[2] All names are pseudonyms.

He was only concerned with playing the game (line 4). Erato persists (lines 5–8) to push Gustavo's reasoning (lines 5, 7) by trying to get him to respond to a hypothetical situation using an "if" clause (lines 5, 7), but the scaffolding gets minimal uptake from Gustavo (lines 6, 8):

5. E: If she still had one but it was at a different number like at seven. Gustavo?
6. G: Huh?
7. E: If she had still one but it was at a different number and it was seven would she still be the one that would win for sure?
8. G: Yes?

The children continued to play and never rationalized the results of their play in terms of probability nor did they move toward a probability strategy. In fact, throughout this episode of play, the children continue to attribute the results to "dead dice," "magic," "luck," and even pray for better outcomes. Spanish was rarely used except for one instance of "behavioral management" when Felipe tells the other kids, "Ok. Sientate vamos a seguir jugando (sit down, we're gonna keep playing)." Regardless of the minimal acknowledgments, Erato persists in her attempts to mediate an awareness of probability (lines 10, 12, 14, 16) and gets Juana to recount the high incidences of "four" (line 18):

9. F: But it's kind of the same.
10. E: Isn't it similar, yea, no? Four?
11. J: This.
12. E: And why is there…
13. J: Because…they…show more fours.
14. E: More fours than which number?
15. J: Ten?
16. E: Yea?…How do you know that?
17. J: When we were playing I saw that they were showing a lot of fours.
18. E: Ah ha.

Throughout this interaction, Pavle and Erato occupied the more official "first" space of the dialogue with their goal of trying to mediate a more advanced understanding of probability with the questions they posed; the students were acting within their own "second" social space which was oriented toward winning the game and mediated by their discourse of luck and randomness. The tension between first and second spaces is a natural precursor through which third spaces and zones of proximal development can emerge. In this case, a possible third space would constitute a convergence of the seemingly conflicting goals. This would entail the invocation of an explicit probability strategy that the students view as serving their goal of winning and the graduate/undergraduate facilitators view as more mathematically viable. This group of participants never quite achieved this type of third space. One reason could be the visible constraints in the types of meditational tools available to the participants, especially the undergraduate and graduate facilitators. The following lines (19–32) show several more attempts by Erato (lines 19, 22, 24) to mediate the emergence of a third space through which they can develop an explicit probability strategy; however, Gustavo and Felipe were fixated on "luck" (lines 25–26), "feeling good" (line 28), and prayer (lines 31–32)[3]:

19. E: I don't remember how many things came out. I would remember if I wrote them down.
20. F: //Yea,we should have// written them down.
21. G: //Hey it's//
22. E: What do you guys think?
23. G: Eight.
24. E: Felipe is winning! How did <u>that</u> happen?
25. F: I don't know. I guess I got lucky.
26. G: Ten! You're lucky. You're lucky ten (inaudible phrase) ten.
27. J: //<u>Ten</u>?//
28. F: //I woke up feeling good today.
29. G: He's gonna win! Give me.
30. F: Oooh, pretty close there.
31. J: (inaudible phrase) [Gustavo makes prayer gesture with dice then blows on them while shaking his body. Pavle picks up the dice and wipes them off as they all laugh.]
32. F: Wait, no prayer in school.

Making the Probability Strategy Explicit Through Third Space

In contrast to the vignette presented above, the case of a Latino third grader named Rafael demonstrated how an advanced probability strategy emerges through third spaces mediated by parents and a Spanish-speaking undergraduate facilitator. The qualitative difference lies not simply in the fact that Spanish was used, but rather Rafael's mom was able to bring expertise and resources grounded in a probability game played at home (primary discourse) to leverage the development of an explicit probability strategy (secondary discourse). During the early stages of playing *Counters*, Rafael participated in a group with a student, Alfredo, and an undergraduate facilitator, Mateo. Like the other children, Rafael was more interested in playing and "winning" the game than thinking about the mathematics, so he adopted a strategy based on "luck." While playing the game Mateo overtly recommended that the students keep track of the resulting sums each time the dice were rolled and prompted them to notice if there were any sums that came up more or less often. Thus, Mateo explicitly scaffolded Rafael's attention toward the frequency of events (dice rolls). Mateo's recommendation served to mediate Rafael's novice understanding, and as a result, he noticed that the sums 2 and 12 came up the least often. He explained that 2 and 12 are sums that were the least likely to occur because there was only one combination of numbers that would result in each of

[3] Transcript Conventions: // //= Overlapping talk; [*italics*]=English Translation; **BOLD**=emphasis

the sums 2 and 12, respectively. After the session, Rafael wrote the following message to *El Maga*:

> Subject: Counter game. We played with Mateo and Alfredo. I put my numbers in the numbers 5, 6, 7, 9, 9, 10, 10, 11, 11, 8, 8, 7. Because it is less possible to get 2 1 and 2 6. Mateo won all of the games.

In his message to *El Maga*, Rafael first described how he placed his counters on his game board and then justified why he placed them in such a way. By writing "I put my numbers in the numbers 5, 6, 7, 9, 9, 10, 10, 11, 11, 8, 8, 7," Rafael suggested that he placed his counters, which he referred to as "numbers," on specific positions on the game board, which he also referred to as "numbers." Rafael explained that he placed one counter on a position on his game board that corresponded to the sum of 5, one counter on 6, one counter on 7, two counters on 9, two counters on 10, two counters on 11, two counters on 8, and one counter on 7. Rafael justified placing his counters in such a manner by explaining that "it is less possible to get 2 1 and 2 6," meaning that it is less possible to roll two dice and get two ones or two sixes, and therefore, he did not place any counters on the positions on his game board that correspond to the sums of 2 and 12. Not only did Rafael realize that the game was not based on "pure luck," he also realized that there was a mathematical explanation for why some sums were less probable than others. This illustrates the nascent stages of his understanding of theoretical probability. He has developed an awareness of how each rolling of the die constituted an independent event and the likelihood of an event was related to the number of combinations that would lead to that event. The event of rolling a 6 and a 5 and the event of rolling a 5 and a 6 were separate events; the sums 2 and 12 were the least likely to occur as demonstrated above when he wrote that it was less possible to get two ones and two sixes.

Later when Rafael played the *Counters* game with two other students (Marcela and Jesus), his mother Olivia, two more mothers (of different children), and an undergraduate facilitator, Pablo, his understanding of probability developed further. Of course, this learning or "movement" did not happen without tension and frustration (lines 1–15):

1. J: Yay…ya gané…gané…gané…[*Yay...I won already...I won...I won.*]
2. R: Aww.
3. O: Me quedaron tres.[*I had three left.*]
4. J: Cheater…[looking at Rafael].
5. R: Ha mí también. [*Me too.*]
6. M: Ha mí me quedó uno. [*I had one left.*]
7. J: Ha mí me quedó cero. [*I had zero left.*]
8. O: Hiciste trampas. [*You cheated (in a game); you "outwitted."*]
9. Pt1: Acuérdate que aquí el que gana es como si perdiera tú estás jugando el mío. [*Remember, here, whoever wins, it's like they lost. You're playing mine.*]
10. J: Uh uh.
11. Pt1: Tú estás jugando mi juego, y yo juego el tuyo. [*You're playing my game, and I'm playing yours.*]
12. J: Uh uh.

13. O: ¿Por qué tú ya llevas dos ganadas? [*Why do you have two wins?*]
14. P: ¿O alguién ganó? [*Oh! Someone won?*]
15. O: Porque tú ya llevas dos ganadas ya quedas descalificado. [*Because you (Rafael) have two wins, you're disqualified.*]

Throughout the game play, the participants were focused on winning with one student (Jesus) saying things like "hay que ganarle a Rafael [*we have to beat Rafael*]" because he had won consecutive rounds. Rafael clearly had some sense of a probability strategy to arrange the counters which was giving him the advantage. Finally, Jesus pronounced victory loudly (line 1), Rafael was frustrated (line 2), and then Jesus proceeded to call Rafael a "cheater" (line 4) presumably for winning the previous rounds. The parents were equally engaged in the competitive spirit. Only the undergraduate facilitator was apathetic toward winning or losing given his role was to facilitate the development of formal mathematics and in this case an explicit probability strategy. While everybody was engaged in a verbal banter about who should or should not play, Pablo realized late that the game was over (line 14). Olivia, Rafael's mom, playfully accused Jesus of cheating (line 8) and disqualified her own son from playing anymore (line 15).

Rafael's play clearly showed that he recognized that the sum of seven had the highest probability of occurring and used the strategy to win several rounds. After Pablo belatedly realizes the game is over, he attempts to redirect the conversation toward the mathematics instead of winning and losing. However, unlike the previous vignette where the attempts at a third space did not materialize, in this case, a third space emerges where an explicit discussion about probability strategy takes place. Pablo prompts the group in Spanish to explain his strategy for the rest of the group (lines 16–24):

16. P: ¿Por qué tú crees que algunos salen más que otros? [*Why do you think that some come out more than others?*]
17. St1: Yo no sé. [*I don't know.*]
18. R: No sale mucho, porque, porque no hay mucha probabilidad para hacer uno y uno con el dos, y también con el doce. [*They don't come out, because, because, there isn't much probability to make one with one for the two, and the same with twelve.*]
19. P: ¿Y el siete porque sale tanto? [*And the seven, why does it come out so much?*]
20. R: Porque hay más posibilidades para sacarlo. [*Because there is more possibilities of getting it.*]
21. P: ¿Como qué? [*Like what?*]
22. R: Como el cinco y el dos, cuatro y tres. [*Like five and two, four and three.*]
23. St 1: Go! O sí, me toca a mí. [*Go! O yeah, it's my turn.*]
24. P: Oh! Ok. Me convenciste [Oh! Ok, *you convinced me.*] (laughs).

Thus, Rafael (line 18) assumed an "expert" role in relation to the other students (even though he was younger) and explicitly talked about "probability" (line 18) and "possibilities" (line 20) for the first time. It was important to notice that his assessment went beyond his experience of seeing the sum of seven occur most frequently. Rafael's explanation that "there is more possibilities of getting it (the seven)" because it can occur from rolling a "five and two" or a "four and three" was based on his realization that seven is the sum with the greatest number of combinations. In other words, Rafael's thinking was not based on an experiential model (experimental probability); rather, it was based on mathematics (theoretical probability). Rafael's choice to place more counters on 7 was not based on the realization that after playing the game many times is the sum that occurs most often is 7. By explaining that there is a variety of combinations that will result to a sum of seven, "Como el cinco y el dos, cuatro y tres [*Like five and two, four and three*]," he based his reasoning on the fact that there were various combinations that can result in the sum 7, such as rolling the numbers five and two or four and three.

The transformation of Rafael's understanding of probability within the context of *Counters* play was mediated by multigenerational actors using his primary discourse (Spanish). Between the earlier and later stages, there was a clear participatory shift from novice to expert. The significance of leveraging Rafael's primary discourses to develop his understanding of probability became evident at the end of the year while composing a digital story. Rafael's digital story was based on a new probability game he created called *Math Bingo* based on the *Counters* game and a probability game he regularly played at home with his mom called *The Game of Five Numbers*. This example further confirmed how the interactions observed within the after-school club were necessarily linked to contexts outside *Los Rayos*. After one of the planning sessions, Pablo, the undergraduate facilitator, described the game in his field notes:

> Rafael's mom already seemed to be engaged in throwing out ideas based on games we already know in order to think of new ones. She suggested a game that they play a lot at home (but that's why I think Rafael's excellence in math is most definitely attributable to the influence at home and the comfort he feels as a math doer because of that influence) se llama el juego de los cinco números [*It's called the game of the five numbers*]. You generate 5 random numbers with a pair of dice or with some other method like a deck of cards or dominoes, then you try to find different arithmetic operations that will lead the first 4 numbers to equal the fifth. For example you get 4, 10, 3, 2, and 8, so you need to use the first four numbers once each in any order to generate the answer 8. So you might try 4+10=14 then divide by 2=7, minus 3=4, nope try again. Ok, so 4-3=1, 1*10=10, 10-2=8, and we have a solution. The idea is that through trial and error and using all 4 arithmetic operations you all race to see who can get a solution first. I think we need to design an activity with this game, yes it has the feel of drilling or even school, but it is so about "play" and increases comfort with math operations in an informal setting.

In his digital story, Rafael also included images of *Math Bingo,* a game he played with his sisters with the following narration:

> Math Bingo, starring Rafael. I am on the right and my sister, my…the one on the left is my sister and her name is Griselda, and the one on the bottom is my little sister and her name is Veronica. Yo hice mi nuevo juego de Bingo usando matemáticas. Hice cuatro cartas y cada carta tiene diferentes operaciones. También hice una flecha, un…un tablero numérico con una flecha en el medio para girarla. Para ganar, tienes que hacer una línea vertical o horizontal o diagonal. Aquí, todos están tomando turnos girando la flecha y yo, y yo gané como dos veces. Estoy jugando con mi hermana que se llama Griselda y mi otra hermana

que se llama Verónica. Cada quien está tomando turnos girando la flecha y apuntado el número que salga. Fin-The End. [*I made my new Bingo game using mathematics. I made four game pieces and each one has different operations. I also made an arrow, an... a numbered board with an arrow in the middle to spin it. To win, you have to make a vertical or horizontal or diagonal line. Here, everyone is taking turns spinning the arrow and I, and I won like two times. I am playing with my sister whose name is Griselda, and with my other sister whose name is Veronica. Everyone is taking turns spinning the arrow and writing down the number that comes up. The End.*]

Thus, the case of Rafael's probability learning within *Counters* demonstrates how learning the probability secondary discourse not only takes place within the context of the after-school club but also draws on meditational tools that are life-long, life-wide, and life-deep (Banks et al., 2007).

Creating Third Spaces by Leveraging Primary Discourses

The final vignette I present here is another example of how three Latinas move toward a formal probability strategy through primary discourses in the third space. As was the case in the previous examples, the goals of the children (playing and winning) were in sharp contrast to the facilitators (move them toward a formal probability strategy). The participants in this episode were Noelia, Imelda, and Melissa (all fourth graders); parents; Julio (undergraduate); and a Spanish-speaking graduate student. In the early stages of the *Counters* game, it was evident that the children did not share the same goal as the undergraduate and graduate facilitators nor did they have any sense of probability. The following excerpt comes from the initial stages of game play. It is an example of the undergraduate facilitator (Julio) overtly speaking with Noelia (N) about "a strategy" (line 1) and Noelia responds, "who cares!" (line 2):

1. J: Do you guys know of a strategy? That maybe you can get rid of faster?
2. N: Who cares!
3. J: Huh? "Who cares!" Then you're not going to win if you don't know a strategy, you're going to lose.
4. J: You don't care?
5. N: It's just a game.

Like the previous examples, the students were in one social space (playing and winning) and the facilitators in another (pushing toward formal mathematical understandings of probability). The image below (Fig. 11.2) shows that the strategy lacked any sense of differentiated probability for each number using two dice as both children have evenly distributed the pieces across 'their game card.'

However, during later stages of *Counters* play, a more explicit probability strategy emerged as children realized that some numbers were appearing more often. As the image below shows (Fig. 11.3), the pieces were not evenly distributed with more pieces placed on the numbers "6" and "7."

Fig. 11.2 First day, even distribution strategy (*circled*) (Razfar, 2012)

Fig. 11.3 The emergence of a probability strategy (*arrow*) (Razfar, 2012)

The transformative shift evidenced above occurred through specific types of discourse practices that were characteristic of third spaces. The following episode shows how Noelia, Imelda, and Melissa and their parents (Pt) developed their probability strategy through joint participation. Furthermore, the interaction was marked by proposals, counter arguments, marshaling evidence, one-upsmanship, and revision of strategy. These practices were all characteristic of *dialogic* interactions and third space that engender movement in the zone of proximal development (Gutiérrez, 2008; Wells, 1999). This dialogic dance between children and parents was an excellent illustration of how higher-order thinking develops through the use of children's primary discourses. All participants used Spanish and more informal ways of describing what they were observing as evidenced by expressions such as "casi no sale" (lines 6 and 11) or "es muy tarde" (it's very late) to express how the number "11" never came (line 8). The use of primary discourses was critical to extending the conversation and ultimately moving toward a secondary discourse for probability. Initially, Noelia proposed to put more pieces on "7," and Imelda not only concurred (line 2) but also played "one-upsmanship" (line 3) by putting four pieces on seven:

1. N: Ya ahora voy poner tres (fichas) en el siete. [*Now I'm going to put three on seven.*]
2. I: A (inaudible) también. [*(inaudible) too.*]
3. I: **Mejor** lo pongo cuatro. [*Even better I'll put four.*]

Noelia then moved to put two pieces on ten at which one of the parents sighs (line 5). It appears to have opened the floor for Melissa (line 6) to provide a rationale for why it is not a good idea to put two pieces on "10" using a colloquial Spanish expression, "casi no sale el diez" (which literally means "10 rarely leaves" compared with the more formal "diez rara vez viene"). Neither form would be considered a formal, mathematical way of saying it which might be "hay menos posibilidades que salga el diez." All of the participants, including the parents, adopted this style. This expression is later used by one of the parents (line 11) which indicates a very comfortable, interactive discourse style in which the participants have a strong affinity for one another and the object of the game (lines 4–11, Pt = Parent):

4. N: Yo le voy a poner 2 en el diez. [*I'm going to put two on ten.*]
5. Pt2: Hmm?
6. M: **Casi no sale** el diez. [*Ten rarely comes up.*]
7. N: /Entonces dos en el once. [*Then two on eleven.*]
8. I: Casi no sé, no se por qué es muy tarde. [*I really don't know why, it's late.*]
9. M: Aquí para ser once?? [*Here, to make eleven?*]
10. Pt: (offscreen) Sí, para ser once. [*Yes, in order to make eleven.*]
11. Pt2: Yo le voy poner dos en el siete porque **siempre sale** el siete. [*I'm going to put two on the seven because seven always comes up.*]

The discussion continued with parents and kids placing more pieces on "7" (lines 12–15) and one student adding more pieces to "8" and "9" (line 15). They were not using any formal mathematical terms related to "probability," but they were showing a shift in their understanding of probability within a short period of game play:

12. Pt3: Vamos a poner todos en el siete. [*We're going to put them all on seven.*]
13. I: Yo le voy a poner cuatro en el siete. [*I'm going to put four on seven.]*
14. N: Yo voy a poner cuatro en el siete. [*I'm going to put four on seven.*]
15. M: Voy a poner dos en el ocho y dos en el nueve. [*I'm going to put two on eight, and two on nine.*]

This discussion not only showed how the participants' understanding of probability shifted within an interaction but also demonstrated how kids adjusted their strategies. In the ensuing turns, Imelda was observed adjusting her strategy from placing more pieces on "10" to placing less pieces based on the previous discussion (line 17):

16. N: Yo voy a poner dos en el diez. [*I'm going to put two on ten.*]
17. I: Yo puse dos en el diez pero luego lo cambié. [*I put two on ten, but I changed it.*]

In this segment, it is clear that the kids have appropriated a strategy to solve the game that reflected their growing understanding of probability. They developed a secondary discourse related to probability through their primary discourses. This process was primarily mediated by peers, parents, and the use of colloquial Spanish; all tools are made available through the more flexible participation structures of the after-school club. While this context and process may be considered "informal" when compared with traditional school interactions, it does not mean the interactions were not purposeful. Even so-called "informal" social spaces are structured and goal driven, but the goals are less oriented toward discrete learning of subjects (i.e., probability) and more geared toward joint activity and understanding.

Conclusion

This chapter highlights the importance of fostering dynamic interactive spaces that draw on the totality of students' linguistic and cultural tool kits by inviting the community to be agents in the learning and development of their children. The analysis of *Counters* game discourse shows how children moved toward more advanced understandings of probability through their primary discourses, cross-generational interactions, and third spaces. More specifically, the three vignettes discussed in this chapter show how, when parents and primary discourses were absent, the movement toward productive third spaces was impeded. The cases of Rafael, Noelia, Imełda, and Melissa illustrated that parents and other Spanish-speaking facilitators in the after-school club were uniquely positioned to provide access to development across contexts and leverage their primary discourses to develop formal mathematical strategies.

Practitioners and policy makers alike can/must examine or reexamine their assumptions about Latina/o, bilingual learners, especially those who ordinarily struggle in traditional classrooms and on standardized tests because of restrictive language policies and rigid discourse structures that exclude appropriate forms of mediation (Barwell, 2005). By examining learning in nonschool settings such as this one, teachers can learn alternative ways of organizing learning to make it more dialogic and less didactic. If the nature of the mediation is changed and more attention is placed on the context of development rather than individual traits, then the potential for transformative learning and development is enhanced.

While the *Counters* game occurs in an after-school club that is less formal, it is nonetheless structured. It is important to note that this type of learning does not occur spontaneously and must be purposefully and consciously designed in light of who the children are and how they can productively participate to solve problems and use/develop appropriate tools. The *Los Rayos* after-school club as a whole and the *Counters* game were intentionally and explicitly designed on the basis of a language ideology that not only encouraged the use of all available linguistic and cultural tools to the participants but also repositioned Spanish and other nonmainstream cultural tools as central to the learning process. In a sociopolitical context where monolingual norms are practiced, these tools are seriously limited and their use should not be expected to "naturally" emerge.

Acknowledgments This work is based on work conducted with the Center for Mathematics Education of Latinas/os (CEMELA). CEMELA is a Center for Learning and Teaching supported by the National Science Foundation, grant number ESI-0424983. Any opinions, findings, and conclusions or recommendations expressed in this webpage are those of the author(s) and do not necessarily reflect the views of the National Science Foundation.

References

Banks, J., Au, K., Ball, A., Bell, P., Gordon, E., Gutiérrez, K., et al. (2007). *Learning in and out of school in diverse environments: Life-long, life-wide, life-deep* [The Life Center (The Learning in Informal and Formal Environments Center) and the Center for Multicultural Education (pp. 1–36)]. Seattle, WA: University of Washington.

Barwell, R. (2005). Integrating language and content: issues from the mathematics classroom. *Linguistics and Education, 16*(2), 205–218.

Cole, M. (1996). *Cultural psychology: A once and future discipline*. Cambridge, MA: Harvard University Press.

Cole, M. & The Distributed Literacy Consortium. (2006). *The fifth dimension: An after-school program built on diversity*. New York: Russell Sage.

Engeström, Y., Miettinen, R., & Punamäki, R. (1999). *Perspectives on activity theory*. Cambridge, UK: Cambridge University Press.

Gee, J. (1996). *Social linguistics and literacies: Ideology in discourses* (2nd ed.). London: Taylor & Francis.

Gee, J., & Green, J. (1998). Discourse analysis, learning, and social practice: A methodological study. *Review of Research in Education, 23*, 119–169.

Gonzalez, N., Moll, L., & Amanti, C. (Eds.). (2005). *Funds of knowledge*. Mahwah, NJ: Lawrence Erlbaum Associates, Inc.

Gutiérrez, K. (2008). Developing a sociocritical literacy in the third space. *Reading Research Quarterly, 43*(2), 148–164.

Gutierrez, K., Baquedano-López, P., & Tejeda, C. (1999). Rethinking diversity: Hybridity and hybrid language practices in the third space. *Mind, Culture, and Activity, 6*(4), 286–303.

Khisty, L. L. (2001). Effective teachers of second language learners in mathematics. In M. van den Heuvel-Panhuizen (Ed.), *Proceedings of the 25th Conference of the International Group for the Psychology of Mathematics Education* (Vol. 3, pp. 225–232). Utrecht, The Netherlands: The Freudenthal Institute, Utrecht University.

Lave, J., & Wenger, E. (1991). *Situated learning*. Cambridge, MA: Cambridge University Press.

Moje, E., Ciechanowski, K., Kramer, K., Ellis, L., Carrillo, R., & Collazo, T. (2004). Working toward third space in content area literacy: An examination of everyday funds of knowledge and discourse. *Reading Research Quarterly, 39*(1), 38–70.

Nicolopoulou, A. (1993). Play, cognitive development, and the social world: Piaget, Vygotsky, and beyond. *Human Development, 36*(1), 1–23.

Razfar, A. (2005). Language ideologies in practice: Repair and classroom discourse. *Linguistics and Education, 16*(4), 404–424.

Razfar, A. (2012). ! Vamos a jugar counters! Learning probability through bilingual discourse and play. *Bilingual Research Journal, 35*(1), 53–75.

Razfar, A., Sutton, T., & Radosavljevic, A. (2009, April). *Learning mathematics and language: Examining the tensions in two after-school clubs*. Paper presented at the annual meeting of the American Education Research Association, San Diego, CA.

Rogers, R., Mosley, M., Hui, D., & O'Garro-Joseph, G. (2005). Critical discourse analysis in education: A review of the literature. *Review of Educational Research, 75*(3), 365–416.
Rogoff, B. (1991). *Apprenticeship in thinking: Cognitive development in social context*. New York: Oxford University Press.
Rogoff, B. (2003). *The cultural nature of human development*. New York: Oxford University Press.
Sfard, A. (2002). Learning mathematics as developing a discourse. In R. Speiser, C. Maher, & C. Walter (Eds.), *Proceedings of 21st Conference of PME-NA* (pp. 23–44). Columbus, OH: Clearing House for Science, Mathematics, and Environmental Education.
Street, B., Baker, D., & Tomlin, A. (2005). *Navigating numeracies: Home/school numeracy practices*. London: Springer.
Vasquez, O. A. (2003). *La clase magica: Imagining optimal possibilities in a bilingual community of learners*. Mahwah, NJ: Lawrence Erlbaum Associates.
Vomvoridi-Ivanovic, E. (2009). *Latina/o pre-service teachers' use of language and culture while assisting children in mathematics*. Unpublished doctoral dissertation, University of Illinois at Chicago, Chicago.
Vygotsky, L. S. (1978). *Mind in society: The development of higher psychological processes* (M. Cole, V. John-Steiner, S. Scribner, & E. Souberman, Eds.). Cambridge, MA: Harvard University Press.
Wells, G. (1999). *Dialogic inquiry: Towards a sociocultural practice and theory of education*. New York: Cambridge University Press.
Wertsch, J. (Ed.). (1985). *Culture, communication, and cognition*. Cambridge, UK: Cambridge University Press.
Wertsch, J. V. (1998). *Mind as action*. New York: Oxford University Press.

Chapter 12
Expanding Methodologies to Account for Expansive Views of Learning: A Commentary on Part II

Kris D. Gutiérrez

The chapters in this section collectively contribute to a body of work defined by its renewed focus on understanding the contextual and contingent nature of learning—work that also calls for, directly or by implication, a new model of interventionist research designed to improve the educational circumstance of underserved youth, notably students from nondominant communities. I will say more about this point shortly. Of significance, I believe these chapters address conceptual and methodological challenges encountered in carrying out thoughtful empirical work on learning across spaces and contexts.

The conduct of such work has become increasingly complex, as the practices of home, community, and geographical regions that help shape our understanding of race, ethnicity, and culture are shifting in dramatic ways. As Lipsitz (2004) has noted, "the new realities of our time have enacted a fundamental rupture in the relationships linking place, politics, and culture" (p. 3). This new world order calls into question the monocultural and monolingual bias often found in empirical work, as well as a monocular or a singular focus on one context to understand and document individuals' learning trajectories—views in which individuals are understood to be participants in homogeneous, uniform, and bounded practices. Paraphrasing Pavlenko and Blackledge (2004), such conceptions of people and their practices obscure hybrid identities and repertoires of practice, as well as the complex linguistic toolkits of members of nondominant, particularly new immigrant and diasporic, communities. In educational contexts, these new social realities require us to think differently about what counts as learning, as youth move across the myriad and hybrid activity settings of everyday life.

K.D. Gutiérrez (✉)
Inaugural Provost's Chair, Literacy and Learning Sciences, School of Education, University of Colorado Boulder, Boulder, CO, USA
e-mail: kris.gutierrez@colorado.edu

B. Bevan et al. (eds.), *LOST Opportunities: Learning in Out-of-School Time*, Explorations of Educational Purpose 23, DOI 10.1007/978-94-007-4304-5_12,

Youth live and engage in practices that are influenced by the historical and present-moment contexts unique to their experiences. From this perspective, studying youths' "movements" (Gutiérrez, 2008, p. 151) would provide opportunities to understand better how the available tools and resources—historically, culturally, and in the design of learning environments—are subsequently taken up and instantiated in new practices and in new ways.

From the vantage point of my own work, this set of studies of learning in naturally occurring and designed practices, highlights a collective interest in the cultural mediation of human activity and, thus, helps explain the pull toward approaches that expand current conceptions of learning and development, including the constitution of individuals' repertoires of practices across time and space (Gutiérrez & Rogoff, 2003). Within this view, we are interested in what takes hold—that is, what is learned as people and practices and tools travel across activity systems (Gutiérrez, 2008). This shared interest in how participation in everyday practices across informal and designed settings—resulting in expertise, new dispositions, and more expansive forms of learning, as Bell and his colleagues argue—aligns these studies with an ecological and activity theoretical approach to understanding human activity.

Of relevance to the chapters under discussion and the importance of employing new conceptions of learning, as well as tools and methods to capture the complexities of learning across settings, I discuss one approach that resonates with the aims of the present body of work. Third-generation activity theory within a cultural–historical approach provides the theoretical and methodological tools needed to examine local and distal influences on a community and individuals' practices, and to make visible the idea that people are part of multiple activity systems. From this standpoint, the relations among people and tools, as well as the contradictions that exist between activity systems are central to the analysis of learning activity. In short, examining human learning across a minimum of two activity systems (Engeström, 2009) is an essential characteristic of third-generation activity theory—one that distinguishes it from first generation's focus on mediated action (Vygotsky, 1978) and second generation's emphasis on activity as the central unit of analysis of human practice (Leont'ev, 1981). Thus, third-generation activity theory is a particularly robust approach to studying human learning activity, as it is inherently transdisciplinary, multi-methodological, and promotes a form of methodological bricolage needed to capture complexity, border-crossing, and hybridity (Gutiérrez, Bien, & Selland, 2011).

This more expansive focus is instructive to studies of learning interested in how youths' horizontal and vertical forms of expertise are leveraged and supported across formal and informal learning environments. I believe, this is especially important, as accounting for the relation between the horizontal and vertical, the everyday and the scientific, is not the norm in studies of youths' learning activity and trajectories, particularly youth from nondominant communities. For Bell and colleagues, *everyday expertise* involves the coordination of a constellation of things—from the cultural practices in which people participate and the meaning they hold for individuals, to the available resources and tools that come into the mix in the development of learning trajectories and expertise.

As I read the various accounts of learning in the chapters, I looked to the ways each of the authors documented learning, including which of the methodological tools employed would help us capture the ways youth and their learning environments leverage students' repertoires of practice. I was interested in which new arrangements foster learning, and what kind of learning ecology, tools, forms of participation and assistance, and ways of organizing learning would ratchet up development, as well as promote equity and transformative outcomes.

Across the set of chapters, authors challenge notions of communities and their practices as bounded and, instead, take care to attend to the flow and diffusion of youth across settings. As we see in each account, studying what takes hold and what gets leveraged, ignored, and transformed requires new sensibilities and tools, and a new imagination about communities and their practices—where youths' practices are understood as dynamically constituted, rather than as essentialized products of a static community.

In general, the various studies included in this section illustrate the conceptual and methodological shifts required when a dynamic view of culture and more expansive forms of learning are employed; here a first-order assumption is that youths' practices are both culturally informing and culture-producing. Consider Aria Razfar's approach, in which sociohistorical understandings of the language practices of dual language learners provide more accurate and useful descriptions of people's linguistic repertoires, including their genesis and mediating potential. Instead of focusing on students' linguistic "deficiencies," Razfar seeks to know more about students' history of involvement with language and literacy practices and how their enabling and constraining properties influence their learning trajectories (Gutiérrez, 2008). Brigid Barron's and her colleagues' emphasis on understanding engagement in learning activity serves as a means to understand the processes of identity development and their relation to learning. In studying online "cultures of participation" (Jenkins, 2006), these researchers use technobiographies to map learning activity to illustrate how participation in the designed environment, the Simmons clubhouse, with its new tools and forms of assistance, was instrumental in providing youth opportunities that expanded their technological toolkits and afforded new dispositions as learners. In tracing the engagement and shifts in dispositions of one clubhouse member participating in cyber activity, Barron and colleagues make visible possible futures and trajectories for youth like Luis. By following Luis across contexts, we learn of the contradictions and disconnects among the practices and his experiences at the clubhouse, and his home and school practices. One additional and noteworthy contribution of this chapter is its challenge to portrayal of low-income parents' practices, as Luis' parents were engaged in his interest-driven practices in productive ways.

In their chapter, Bell et al. contribute significantly to the development of theory of *everyday expertise*, in which cultural capital or what Yosso calls "community cultural wealth" (2005) provides an important lens to understand the cultural–historical and ecological dimensions of learning. By accounting for what they term, "the material and social dimensions of sophisticated domain learning," their work seeks to understand how domain-specific learning can grow into youths' interest-driven practices. In particular, they examine how leveraging youths' culturally based repertoires of practice in a design-based ecology supports the development of learning pathways around health and school sciences. Using "self-documentation" of participants' repertoires of practice—a method akin to ecological approaches to collecting individuals' reports of daily routines (Gutiérrez, Izquierdo, & Kremer-Sadlik, 2010)—these researchers use students' accounts to leverage their horizontal expertise in classroom learning.

The study by Mehus and colleagues on doing science across contexts focuses on documenting what they call "emergent, incidental, or serendipitous science learning"—forms of learning found in the course of everyday life, practices distinguished from those designed for science learning. We learn again from their study the ways in which the leading activity of play creates zones of proximal development and rich interactional contexts for peer learning (Griffin & Cole, 1984). Here their micro-analytic documentation of children's learning in joint activity across the contexts of home, preschool, and play ensembles, as well in various configurations, captures the routine and variant practices that help shape the social situation of development. I found their elaboration of peer-guided and adult-guided learning interactions particularly useful in understanding the affordances and limitations of various configurations of joint activity in the appropriation of disciplinary knowledge and skills.

Overall, this collection of studies illustrates principled theoretical and methodological treatments of children's repertoires of practice, their learning trajectories, and possible futures. As I have written elsewhere (Gutiérrez, Bien, Selland, & Pierce, 2010), such approaches to documenting and defining what counts as learning illustrate a goodness of fit between relevant theoretical constructs and the complexity of the sociocultural phenomena under investigation. For example, in my own work on the literacy practices of youth from nondominant communities, it is this process of seeking a goodness of fit that allows me to draw in principled ways from expansive theories of learning and development (Cole, 1996; Engeström, 1987, 2001; Rogoff, 2003), critical social theories (Luke, 2003), sociocultural and social practices views of literacy (Barton & Hamilton, 1998), including New Literacy Studies (Gee, 1996, 2005; Street, 1984, 2003), and multi-literacies approaches (Cope & Kalantzis, 2000) to link the particular to the larger social context of development, to document everyday and scientific practices, including their relation, as well as the development of horizontal and vertical forms of expertise. The point here is that studies examining learning across multiple activity systems require robust theories and expansive methods to capture not only what these youth know and do but also who they can become. I believe that the studies represented in this volume will move our work forward as we seek to understand better which social ecologies support—indeed, ratchet up—learning and give meaning for youth, especially those from nondominant communities.

References

Barton, D,. & Hamilton, M. (1998). *Locăl literacies: Reading and writing in one community. London*: Routledge.

Col, M. (1996). Cultural psychology: A once and future displine. Cambridge, MA: Harvard University Press.

Cope, B., & Kalantzis, M. (Eds.). (2000). *Multiliteracies: Literacy learning and the design of social futures*. New York: Routledge.

Engeström, Y. (1987). *Learning by expanding: An activity-theoretical approach to developmental research*. Helsinki, Finland: Orienta-Konsultit Oy.

Engeström, Y. (2001). Expansive learning at work: Toward an activity theoretical reconceptualization. *Journal of Education and Work*, 14, 133–156.

Engeström, Y. (2009). The future of activity theory: A rough draft. In A. Sannino, H. Daniels, & K. Gutiérrez (Eds.), *Learning and expanding with activity theory* (pp. 303–328). New York: Cambridge University Press.

Gee, J. P. (1996). *Social linguistics and literacies: Ideology in discourse*. (2nd ed.). London: Taylor & Francis.

Gee, J. (2005). Semiotic social spaces and affinity spaces from the age of mythology to today's schools. In D. Barton & K. Tusting (Eds.), *Beyond communities of practice: Language, power, and social context* (pp. 214–232). Cambridge, UK: Cambridge University Press.

Griffin, P., & Cole, M. (1984). Current activity for the future. In B. Rogoff & J. Wertsch (Eds.), *Children's learning in the 'zone of proximal Development': New directions for child development* (pp. 45–63). San Francisco: Jossey-Bass.

Gutiérrez, K. (2008). Developing a sociocritical literacy in the third space. *Reading Research Quarterly, 43*(2), 148–164.

Gutiérrez, K., Bien, A., & Selland, M. (2011). Polylingual and polycultural learning ecologies: Mediating emergent academic literacies for dual language learners. *Journal of Early Childhood Literacy, 11*(2), 232–261.

Gutiérrez, K., Bien, A., Selland, M., & Pierce, D. (2010). Syncretic approaches to studying movement and hybridity in literacy practices. In D. Lapp & D. Fisher (Eds.), *Handbook of research on teaching the English language arts* (3rd ed.). New York: Routledge.

Gutiérrez, K., Izquierdo, C., & Kremer-Sadlik, T. (2010). Middle class working families' ideologies and engagement in children's extracurricular activities. *International Journal of Learning, 17*(3), 633–656.

Gutiérrez, K., & Rogoff, B. (2003). Cultural ways of learning: Individual traits or repertoires of practice. *Educational Researcher, 32*(5), 19–25.

Jenkins, H. (2006). *Convergence culture: Where old and new media collide*. New York: New York University Press.

Leont'ev, A. N. (1981). The problem of activity in psychology. In J. V. Wertsch (Ed.), *The concept of activity in Soviet psychology*. White Plains, NY: Sharpe.

Lipsitz, G. (2004). New times and new identities: Solidarities of sameness and dynamics of difference. In G. Gonzalez, R. Fernandez, V. Price, D. Smith, & L. Trinh Vo (Eds.), *Labor versus empire: Race, gender and migration* (pp. 3–16). New York: Routledge.

Luke, A. (2003). Literacy and the other: A sociological approach to literacy research and policy in multilingual societies. *Reading Research Quarterly, 38*(1), 132–141.

Pavlenko, A., & Blackledge, A. (Eds.) (2004). *Negotiation of identities in multilingual contexts* (vol. 45). Clevedon, UK: Multilingual Matters.

Rogoff, B. (2003). *The cultural nature of human development*. New York: Oxford University Press.

Street, B. V. (1984). *Literacy in theory and practice*. London: Cambridge University Press.

Street, B. V. (2003). What's "new" in new literacy studies? Critical approaches to literacy in theory and practice. *Current Issues in Comparative Education, 5*(2), 77–91.

Vygotsky, L. S. (1978). *Mind in society: The development of higher psychological processes*. Cambridge, MA: Harvard University Press.

Yosso, T. (2005). Whose culture has capital? A critical race theory discussion of community cultural wealth. *Race Ethnicity and Education, 8*(1), 69–91.

Part III
STEM in the Organized Out-of-School Time Setting

Chapter 13
Introduction: STEM in the Organized Out-of-School Time Setting

Bronwyn Bevan

The collection of chapters in this section considers organized out-of-school time (OST) programs, such as after-school programs, summer camps, and museum settings, as sites for science, technology, engineering, and mathematics (STEM) learning. The chapters illuminate the types of STEM that are learned by young people participating in these programs, and also the types of insights on learning that educators and researchers can gain when working in or studying OST programs.

The organized OST program has been described by Mahoney, Larson, Eccles, and Lord (2005) as:

> *characterized by structure, adult-supervision, and an emphasis on skill-building. These activities are generally voluntary, have regular and scheduled meetings, maintain developmentally based expectations and rules for participants in the activity setting (and sometimes beyond it), involve several participants, offer supervision and guidance from adults, and are organized around developing particular skills and achieving goals. These activities are often characterized by challenge and complexity that increase as participants' abilities develop. (p. 4)*

Such programs are an anchoring force in the learning ecologies of many individuals and communities and have been shown to play important roles in the development of young people's trajectories of learning (see Barron et al., Chap. 8; Basu & Barton, 2007; Fusco, 2001). But because they are organized (with allocated settings, resources, and staffing), and therefore by necessity they are funded, they are subject to forms of accountability that are inextricably linked to larger societal views on what counts as learning in science and mathematics. As some have noted, how systems of schooling define learning is beginning to frame "what counts" in OST (Hull & Greeno, 2006; Nocon & Cole, 2006). As such, better understanding the nature and potential of these sites, with a view to documenting their contributions

B. Bevan (✉)
Center for Informal Learning and Schools (CILS), Exploratorium,
San Francisco, CA, USA
e-mail: bronwynb@exploratorium.edu

B. Bevan et al. (eds.), *LOST Opportunities: Learning in Out-of-School Time*,
Explorations of Educational Purpose 23, DOI 10.1007/978-94-007-4304-5_13,

Fig. 13.1 Angel Columns. © Exploratorium, www.exploratorium.edu

to participants' learning and development in STEM, becomes important not only for developing our understanding of how people learn across their life span (how they develop interest, understandings, and commitments), but also for informing the future directions of a specific educational sector, shorthanded as "out-of-school time."

This volume is just one expression of the growing interest and investment in the OST or informal sector. As a person who has worked in a science museum for over two decades, it is always with a sense of chagrin that I write about our institutions, organizations, and activities in terms that are framed in the negative: *in*formal learning, *out*-of-school time, *after*-school, etc. We have several exhibits at the Exploratorium that address the phenomenon of "negative space." This is the in-betweenness, the other, the stuff that maybe is the background, or maybe is the foreground, or maybe is something unto itself. *Angel Columns* is one of these exhibits, as is *Talking in Circles*, where spinning a slightly asymmetrical three-dimensional urn allows you to see the profiles of two people chatting with one another. What are we looking at? What do we care about? The spinning urn or the chatting profiles? (Figs. 13.1 and 13.2)

Science-rich cultural institutions is a useful term that has been proposed for science centers and museums, but this term excludes other structured settings (such as after-schools or summer camps) that provide opportunities for young people to engage in STEM learning. Perhaps our difficulty with finding the right term is evidence that we are actually misconceptualizing the phenomenon – maybe there is more than one thing, each to be investigated as a type, and not one thing that is defined as *not* or *after* school.

Fig. 13.2 Talking in Circles. © Exploratorium, www.exploratorium.edu

As is explored in several of the chapters in this volume, stepping outside of the physical place that people typically think of when they think of learning, immediately leads us to more complex understandings about the nature of knowledge, STEM, or learning. This complexity is important not only for understanding what happens outside of school but also, as many have described in this volume, for rethinking schools themselves. This possibility may be more salient than ever today as the future of schools, in the context of the new electronic technologies now available, is a matter of discussion extending beyond the boundaries of the usual suspects.

Structured OST Programs in the Larger Learning Ecology

As a part of mapping the complexity of the learning ecologies and trajectories that are created by, and also made available to, learners in our communities, we need to understand how structured OST STEM programs and activities provide support to youths' life-long, life-wide, and life-deep (Banks et al., 2007) trajectories of learning and development. The chapters in this section serve to complicate our understanding of these spaces. They focus on what we are calling (perhaps still in the negative frame?) the "hybrid" nature of the settings and of the practices within the settings – where localized situated knowledge is leveraged alongside more formalized approaches or conceptual tools. The OST spaces are presented as possibly fertile meeting grounds for supporting deeper engagement with and commitments to STEM. They are also presented as complicated and sometimes contradictory spaces – perhaps reflecting the convergence of different sets of expectations, social arrangements, goals, participation structures, and resources; perhaps reflecting the negative framing.

Bevan and Michalchik's analysis of the structured OST program, based on a study of 16 National Science Foundation-funded STEM OST programs, reveals the complex and variable ways in which science is positioned and experienced in the OST setting. In particular, they argue that there are multiple and often quite indirect links to school STEM, and that this is one reason to eschew an "additive" model that conceptualizes STEM OST as a supplement/complement to school science. There is a need, they argue, for better and more nuanced analyses and assessments of learning in OST than are commonly used today, as many current studies and initiatives rely on school or school-like assessments, or omit the role of STEM as a disciplinary context for documenting developmental outcomes.

Baker, Remillard, and Lim's study of an after-school robotics club details the hybrid nature of after-school programs set in school buildings and staffed by school personnel. This study investigates what formalized mathematics looks like in an informal setting. Their study shows how the informal structures of the after-school setting generate what they call hybrid practices, in which students develop and implement localized, context-specific mathematical practices while also appropriating, to different degrees, formal mathematical practices to achieve meaningful goals. They note the critical role of adults in the introduction of formalized mathematics to advance localized purposes. This finding raises many interesting questions about the goals and purposes of these programs. How much formalized knowledge is important in these settings, and how critical is it that the knowledge is held by a student versus a teacher?

Khisty and Willey's chapter reports the results of a study of an after-school program that they designed explicitly as a place where they could study how children use language resources in their engagement with mathematics. *Los Rayos de CEMELA* is an adaptation of the *Fifth Dimension* program (Cole & The Distributed Literacy Consortium, 2006). In this bilingual and supportive setting, they document the ways in which Latina/o children's language resources can both advance engagement and learning and can also be subjected to micro-aggressions, primarily among the children themselves, that reflect deep-seated devaluations of children's linguistic and cultural resources. They also note the ways in which the after-school program constitutes an important social network support, in which children are inducted into mathematical identities and practices, which may be missing for many children from nondominant communities.

The chapter by Vomvoridi-Ivanović, Varley, Viego, Simić-Muller, and Khisty, which also studies *Los Rayos*, shows how the low-stakes nature of the after-school setting can provide classroom teachers, both novice and experienced, opportunities to develop new stances with respect to science and mathematics. Their study finds that bilingual teachers, most of whom were raised in Spanish-speaking households, nevertheless struggle with teaching mathematics to English language learner (ELL) students without resorting to English for technical mathematical terms. They show how the low-stakes and more fluid social arrangements of the after-school setting provide teachers with new views on their students, particularly in terms of their use of linguistic resources. This, they show, raises awareness among the teachers of how the classroom setting constrains their students in many ways. The authors of this chapter note that many of the classroom teachers who are brought into the after-school setting bring school-like expectations with them for how the learning activities will be managed and will unfold. This finding, coupled with the new types of awareness that they show teachers to develop in this setting, raises important considerations for thinking about the staffing of after-school programs. The opportunities for reflection that the *Los Rayos* teachers had seem to be critical for their awareness and articulation of the way in which the structures and social arrangements of the setting leverage the linguistic and cultural resources of students.

Crain, Loomis, and Ogawa's chapter describes the results of two studies conducted at the Exploratorium. They explore the ways in which the cultural script of "hands-on" as defining science animates much of our understanding of what "real" science looks like. They describe the way that hands-on as a cultural script has become institutionalized, is used to legitimize specific forms of activity as science, tends to be oversimplified, and also is resistant to change. The dominance of this cultural script, they argue, has made it more difficult for other forms of science – forms that may be as equally relevant to "authentic science" such as observation, collaboration, and argumentation – to be included and "counted as" science by both the educational institutions and the general public.

Michael Cole closes this section with a commentary on the potential and the problems associated with OST as a learning space in today's high-stakes assessment society. As he notes, many who work closely with these sites, including the authors of the chapters in this section, are committed to the powerful developmental effects the programs under study are seen to provide. Yet, how to understand, document, and relate these experiences to a broader inquiry into children's learning and development across time and setting as well as to the policymakers whose decisions set the course of many of these programs remains a significant challenge to educators committed to the design of these learning spaces.

In sum, the chapters in this section are contributions to better defining the organized OST space and potential, and they lay the groundwork for future studies and research that can more adequately position and expand the power of OST to support children's engagement with STEM. It is no small irony that just as researchers in the learning sciences are beginning to seriously examine the organized OST setting as a place to complexify and expand our understanding of what counts as science and what counts as learning, the education policy community, in the context of high-stakes accountability in schools, appears poised to weigh in and perhaps appropriate the OST space as a site for extending and expanding the school day. The chapters in this section make it clear that the assumptions, structures, and possibilities of the OST setting are fundamentally different from school and classroom settings. Further research is needed to inform educational policies that can more appropriately support organized OST settings as alternative, not expanded, educational spaces in ways that leverage the learning affordances of their hybrid social, structural, and developmental features.

References

Banks, J. A., Au, K. H., Ball, A. F., Bell, P., Gordon, E. W., Gutiérrez, K. D., et al. (2007). *Learning in and out of school in diverse environments*. Seattle, WA: The LIFE Center and the Center for Multicultural Education, University of Washington.

Basu, S. J., & Barton, A. C. (2007). Developing a sustained interest in science among urban minority youth. *Journal of Research in Science Teaching, 44*(3), 466–489.

Cole, M., & The Distributed Literacy Consortium. (2006). *The fifth dimension: An after-school program built on diversity*. New York: Russell Sage.

Fusco, D. (2001). Creating relevant science through urban planning and gardening. *Journal of Research in Science Teaching, 38*(8), 860–877.

Hull, G., & Greeno, J. G. (2006). Identity and agency in nonschool and school worlds. In Z. Bekerman, N. C. Burbules, & D. Silberman-Keller (Eds.), *Learning in places: The informal education reader* (pp. 77–97). New York: Peter Lang Publishing.

Mahoney, J. L., Larson, R. W., Eccles, J. S., & Lord, H. (Eds.). (2005). *Organized activities as developmental contexts for children and adolescents*. Mahwah, NJ: Lawrence Erlbaum Associates.

Nocon, H., & Cole, M. (2006). School's invasion of "after-school": Colonization, rationalization, or expansion of access? In Z. Bekerman, N. Burbules, & D. Silberman-Keller (Eds.), *Learning in places: The informal education reader*. New York: Peter Lang Publishing.

Chapter 14
Out-of-School Time STEM: It's Not What You Think

Bronwyn Bevan and Vera Michalchik

Introduction

This chapter reports findings from a national study of 16 different multiyear out-of-school time (OST) science programs. As several authors included in this volume have noted, the designed after-school/out-of-school time space constitutes a sort of hybrid space – it is not school and it is not home. There are both structured and unstructured dimensions to it, and patterns of interaction in the programs vary widely. These variations occur across different programs and also within a given program according to the time of year or the time of day. For example, some programs may offer unstructured playtime as well as structured homework supervision or activity time. Some programs may be primarily worksheet driven for part of the year and primarily outdoors during other times.

Our study has found that the ways in which programs are structured have significant implications for the design and assessment of student learning activities in organized OST settings. Because after-school programming, as well as science, technology, engineering, and mathematics (STEM) in after-school programs, has been expanding over the past decade (Afterschool Alliance, 2005), and because of increased policy interest in extending the school day into OST time, we argue that there is a pressing need to better specify and understand the opportunities and

B. Bevan (✉)
Center for Informal Learning and Schools (CILS), Exploratorium,
San Francisco, CA, USA
e-mail: bronwynb@exploratorium.edu

V. Michalchik
Center for Technology in Learning, SRI International, Menlo Park, CA, USA

B. Bevan et al. (eds.), *LOST Opportunities: Learning in Out-of-School Time*,
Explorations of Educational Purpose 23, DOI 10.1007/978-94-007-4304-5_14,

constraints of the settings. Such an understanding might guide program leaders and policymakers in their decisions about how much it is possible, and how much it is desirable, to orient the after-school setting to address the needs of school STEM.

Learning in OST Settings

Our study focuses specifically on how the designed after-school/OST setting supports children's interest and learning in STEM.

Research has found that learning of and commitment to a field of study develops over multiple timeframes and settings (Bransford et al., 2006; Nasir, Rosebery, Warren, & Lee, 2006; National Research Council, 2000, 2009; Sawyer, 2006). As Bell and colleagues point out in Chap. 9 in this volume, children navigate a wide variety of settings and resources throughout their days and years. Structured informal (out-of-school) learning experiences have been shown to be important in supporting STEM engagement and learning (National Research Council, 2009), including developing children's commitments to school STEM and career pathways (Fadigan & Hammrich, 2004). As a consequence, there is growing interest on the part of STEM educators, scientists, and policymakers to introduce STEM learning activities into the expanding field of organized OST activities or programs, which includes after-school programs, summer camps, and weekend classes for children and youth. Federal agencies such as the US Department of Education, NASA, NOAA, and the National Science Foundation have all, over the past several years, increased funding to support STEM education in OST sites. Furthermore, a recent study found that almost 90% of after-school program leaders would like to expand science in their programs (Chi, Freeman, & Lee, 2008). Yet, the nature, possibilities, and constraints of the OST STEM setting remain underspecified in the research literature. STEM educators who bring expectations about learning designs and outcomes based on their experiences in schools, or even in research labs, are sometimes confounded by the structural constraints in OST that directly impact the possibilities for learning, and they sometimes do not capitalize on the developmental affordances that after-school settings provide for a type of engagement with STEM that may be different than what is typically found in schools.

By better specifying the nature of the after-school STEM setting, this study complicates the question about how structured OST STEM programs contribute to children's interest in STEM. The results of this work argue against the adoption of an "additive" model that assumes that if children participate in after-school STEM programs by x amount, their overall interest, capacity, and engagement in STEM, and particularly school STEM, should rise by an amount equivalent to x. The variation we find among programs, combined with research detailing the complex and highly contingent learning ecologies that children traverse, suggests that such an assumption would be problematic at best.

Learning and Development in OST STEM Programs

The opportunities and challenges within the OST setting as a *developmental* environment (i.e., addressing the social, emotional, intellectual, and physical well-being of participants and building children's capacities to act in and on the world) have been well documented (Halpern, 1999, 2002; Honig & McDonald, 2005; Lerner, Dowling, & Anderson, 2003; Mahoney, Larson, Eccles, & Lord, 2005; Vandell et al., 2006). Among the features that provide developmental opportunities are flexible uses of time, socially supportive adult-child relationships, positive peer groups and role models, and low-stakes contexts that are intellectually safe and encourage children to take on new roles and stances vis-à-vis specific domains. Research has shown that *developmental* features of the OST setting can support an openness, readiness, and willingness to try new things and work with peers (see Mahoney et al., 2005).

At the same time, high-quality programming in the OST context is challenged by fluctuating attendance, a lack of dedicated space for planning and implementation, unstable funding, and high staff turnover resulting in part from lack of adequate training and compensation. These are fundamentally *structural* features related to the ways in which OST programs are positioned (funded, valued, leveraged) in most communities today.

In comparison to findings regarding the developmental supports and structural constraints in OST programs, there is little known – and translated into policy – about the potential of the OST setting as a *learning* environment and even less about it as a *STEM learning* environment. In part, this is because until recently the literature has tended to dichotomize learning and development. Many studies of OST STEM programs provide important insights on the developmental outcomes for children engaged in STEM activities (outcomes include gains in self-efficacy, collaboration skills, and community-mindedness); however, often, these studies do not include analysis of the kinds of conceptual knowledge, including epistemological understandings and procedural skills in STEM, that relate to, and perhaps have led to, these outcomes (e.g., see Furman & Barton, 2006; Fusco, 2001; Rahm, 2007). Others studying OST settings look mostly at instrumental knowledge, and even test scores, to measure learning, without analyzing the developmental conditions and affordances which might effect changes in instrumental knowledge (e.g., see the studies of the impacts of the 21st Century Community Learning Programs by Bodilly & Becket, 2005; James-Burdumy et al., 2005). This apparent dichotomy between development and learning in OST may be due to methodologies used in studies that attempt to aggregate outcomes across disparate programs (i.e., cost concerns may demand easily acquirable and comparable "common denominators," and all children have test scores that can be examined). It also may be due to the way in which learning is conceptualized by the researchers (e.g., defined as "knowing," as made manifest in verbal tests, and not conceptualized as a broader and integrated process of knowing, doing, and being [e.g., see Herrenkohl & Mertl, 2010]).

The lack of a robust literature linking developmental and academic learning outcomes across programs has led to increasing policy pressure on OST programs to focus on academic outcomes at the expense of developmental ones (for discussion of this growing pressure, see Hull & Greeno, 2006; Nocon & Cole, 2006).

Our research conceptualizes learning as the taking up, use, and mastery of cultural tools – in this case, the tools of STEM, which include physical, conceptual, procedural, epistemological, and others. Examples of cultural tools range from common formulae, to discipline-based uses of evidence, to scientific practices and procedures, etc. *Learning* is made evident in a child's (or adult's) growing and increasingly purposeful familiarity, use, and, ultimately, transformation of cultural tools (Stetsenko, 2009). *Development* is conceptualized as a person's evolving (and also continuous) capacity to act in and upon the world. We understand learning to be the central driver of development (Vygotsky, 2004), since it is through the appropriation, use, and transformation of cultural tools (such as language, instruments, or concepts) that children increase their capacity to act in and upon the world. At the same time, a child's development leads to the creation of opportunities to learn. For instance, as a child develops the capacity to walk or to converse with adults or to seek information on the Internet, he or she generates and accesses new opportunities to learn. Thus, we view learning as driving development and development as expanding opportunities for learning. In the context of OST STEM programs, we seek to identify the learning that is driving the development and we seek to identify the developmental activities that afford opportunities for learning.

Studying Learning and Development in OST STEM Programs

The dichotomy in the literature notwithstanding, programs that attend to children's development generally use learning activities as the drivers of development (Cole, 2006; Fusco, 2001; Gutiérrez, Baquedano-Lopez, & Tejeda, 2000; Honig & McDonald, 2005; Rahm, 2007). That is, they capitalize on the hybrid nature of the OST setting to both provide supportive social contexts and to introduce concepts, instruments, terminology, and other conceptual tools that students can use to engage in disciplinary subject matter (see Chap. 15 by Baker et al. in this volume, for further discussion of hybridity in the OST setting). However, attempts to document the impact of these programs tend to split along the lines of child/youth development outcomes (which is of increasingly less interest to program funders) and academic learning outcomes, using standardized tests as the assessment tools (which rely on narrow conceptions of learning and which are often not sensitive to the contributions of the OST programs).

We do not propose a solution to this problem in this study or in this chapter. However, we believe that before this issue can be thoroughly addressed, researchers must better describe OST learning environments, so that the various OST program designs, levels of participation, and actual experiences can be accounted for in ways that can begin to be studied across disparate contexts (Honig & McDonald, 2005).

Moreover, the methods used must produce descriptions that have valence with various stakeholders concerned with development, on the one hand, and academic knowledge, on the other. That is, student learning in STEM must be analyzed and explicated in relation to developmental activities, and at the same time, developmental conditions and features of learning activities must be considered in relation to their power for engaging students in, and sustaining, STEM learning activities. Through such efforts, it may be possible to conceptualize, at scale, how development and learning afford one another in the context of STEM. This is a relationship we seem to know well as parents and practitioners, but it is a relationship that is sometimes lost when programs and systems of programs are subjected to external evaluation.

In 2006, the authors of this chapter received a grant to study a federally funded initiative supporting 16 different STEM programs in 15 urban and rural communities. The goal of the study was to characterize and specify the nature and benefits of the OST setting as a site for STEM learning, particularly with respect to how the programs advanced children's interest in and commitment to STEM. Our study sought to identify the ways in which the cultural tools of STEM were positioned and introduced in the OST learning activities and how the developmental and structural features of OST settings operated to support children's participation and taking up of such tools.

In this chapter, to contribute to the development of more complete ways of analyzing the potential of OST programs to support STEM learning, we identify several dimensions that reflect interactions between structural and developmental features of the OST environment and that we have found to vary across programs. Our goal is to complexify investigations and understanding of the nature of the OST STEM setting: it is neither the idealized developmental space for deep inquiry-based learning that some suggest (e.g., Coalition for Science After School, 2007) nor is it constrained in ways that preclude learning, as others might have it (e.g., Halpern, 2006); at least, we posit that it is neither of these things consistently and it is both of these things at times.

We find that developmental features of OST settings tend to create environments where all children are encouraged to participate and build on their prior knowledge and interests to engage in group activities. Thus, OST programs tend to be receptive to, seek, and even build children's demand for more STEM. We also find that structural features of OST settings, that challenge the stability of staffing, resources, and time use, can lead to STEM being positioned as a set of instrumental artifacts to be mastered or memorized in order to complete a given task or project.

These structural features make it more difficult to position STEM as a means of inquiry into questions of real interest and salience to the learners. We also note, however, that even in structurally constrained environments, children can experience hands-on activities in highly positive ways, which may lay the groundwork and build interest for more STEM. They also often support children's appreciation of the cultural and social relevance of scientific activity by taking them out into the world to see STEM in action.

Science in OST Settings: Dimensions for Analysis

Before we describe the dimensions, we provide three short program examples, which we will later reference in the discussion (extended transcripts and detailed field notes are the source for these brief vignettes).

Example #1: Hurricanes

In a midsized mid-Atlantic city, a group of seven lower elementary grade students are seated around a table with their adult facilitator (a preservice teacher) discussing hurricanes, which they had been learning about for several weeks. The program meets each day in a community center, with children attending on different days depending on their family's needs. The STEM program we are here to observe is the STEM enrichment activity for the program, which occurs once a week. When we walk in, the children have just completed watching a short video about hurricanes. Now, each child is asked to share with each other what they know about hurricanes. After the first two children speak, sharing observations of "They have strong winds" and "They can blow your house down," a young boy stands up and began to spin around the table like a hurricane as he describes how hurricanes move. He spins around and around the table while his peers and the program facilitator sit patiently, watching his progress and waiting for him to make it around the table and back to his chair. When the young boy makes it back, he drops into his chair and begins to recount his plans, for when he grows up, to build his house on stilts so that the house can't be flooded. He discusses this for several minutes before switching to his thoughts on how dogs fare in hurricanes. The facilitator, who has been smiling, nodding, and responding to his comments with "Really?" and "Uh-huhs," lets him finish his thoughts and then suggests that they ask the next child at the table what she knows about hurricanes. A young girl then takes her turn to say what she knows about hurricanes.

Example #2: Motors

In a middle school program in a midsized city in the South, 12 students are going to make battery-operated motors using copper wires, paper clips, magnets, tape, and batteries. Before starting the activity, the two classroom teachers pass out ice cream cones to the children, who sit in two rows of desks facing the front of the class. One of the teachers leans back in her chair with her own ice cream and banters for several minutes before asking the children what they know about motors. For some time, the students and teacher have a casual conversation about their experiences at home with their parents' cars, boats, or lawn mowers, leading to the identification of several components of car motors (air, oil, gasoline, batteries), the teachers pass out the battery-operated motor activity materials and worksheet instructions. Children work individually at their desks to construct their motors, and teachers roam through the room to assist them. After some 30 additional minutes, it becomes apparent that something is not working. The copper wire will not wrap around and cling to the battery and therefore nobody can create an electrical connection. Children continue to good-naturedly work at the activity, and, although there are some groans of frustration, there is no recrimination. But after another 20 minutes of wrapping and rewrapping the wires around the batteries, there have been no successful motors built. The clock nears 5 pm and parents start to drift in to pick up their children. The children start packing up their bags and begin to leave one by one. Afterwards, the teachers state that they intend to repeat the experiment at a future session.

Example #3: Racing Rovers

There are 15 middle school boys and girls in this Saturday morning OST program that meets throughout the academic year in this large economically depressed Midwestern city. The goals of the program, which has a strong track record of success over many years, are to support children through high school and into STEM careers. The Saturday classes, held in university classrooms, are taught by middle school math and science teachers from local school districts.

During the class, the adults are positioned at the front of the classroom, where they are using overhead projectors, screens, and chalkboards to structure the classroom discourse. The activity this semester is for children to build "steel can rovers" using coffee cans, rubber bands, plastic wheels, and washers in film canisters. Pairs of children will experiment with the different variables to design rovers and test how the variables affect speed and distance. The winning rovers from each of the six cohorts of students will race each other at the end of the semester. On the chalkboard is written:

Objectives:

To use graphing calculator to observe and discuss performance of the rover.
To explore the relationship between the ballast and the maximum number of windup turns to give the rover.

At the head of the class, the teacher talks through the worksheet with the children. He reviews the purpose of the worksheet, "This is probably the most critical part of the lesson because now we start making observations of what we see in our rover. And from our observations we can start making predictions on how the rover is going to do [in the trials]." He asks the children to identify the variables that they have to work with. One child says, "Weight. The number of washers." Another child, "The number of wind up turns." The teacher repeats the comments in affirming ways. He then asks, "How about question number one: What effect do the rubber bands have on the performance?" A child responds, "The rubber bands propel the ballast." The teacher repeats what the child says and then asks him to put his statement "in layman's terms." The child quickly replies, "The rubber bands add energy." The teacher affirms the child's statement and continues. As the children respond to his questions, he records their observations on the projected overhead slide:

Changing weights change the speed and distance
Few washers → more distance
More washers → slower

Our larger study mixes quantitative and qualitative methods to document three main facets of the 16 OST programs: (1) who is participating (demographics and participation rates), (2) what they are doing (nature of the STEM learning activities and the OST context in which they take place), and (3) how the programs afford participation that can develop children's interest, dispositions, and STEM learning. In this chapter, we report on findings related to the second facet.

To provide a framework for our initial set of observations, we adapted an instrument created by Vandell and colleagues (2006) to document the developmental and structural features of the OST programs, adding a set of items focused on the nature of the STEM learning activities in the settings. We also collected field notes and interview data that we analyzed to identify emergent themes and constructs using naturalistic inquiry methods (Lincoln & Guba, 1985). Through these means, we distinguished four fundamental dimensions that varied across programs and appeared to directly affect, and be affected by, the nexus between the OST setting's developmental and structural features and the potential for learning outcomes. These features are (1) "what counts" as STEM, (2) goals for student learning, (3) activity structures, and (4) connections with school STEM.

(1) What Counts as STEM?

The subject-matter focus of the 16 programs we studied varied across projects, including, for example, astronomy, robotics, environmental sciences, physics, mathematics, earth sciences, and chemistry. We found that STEM itself was conceptualized in three different ways: (1) as a set of concepts, facts, and procedures; (2) as a set of inquiry processes that characterize investigation of phenomena; and (3) as a set of professional practices, careers, and fields. As such, the goals of the programs could primarily be characterized in one of three ways:

- Engaging children with STEM phenomena and developing their STEM content understanding ($n=10$)
- Engaging children in sustained inquiries and developing their understanding of the nature of science through investigations of content areas ($n=4$)
- Engaging children in a process and commitment to pursuing STEM in academic and career contexts ($n=2$)

In the first case, as illustrated in Example #2 and implied in Example #1, programs tended to be organized around engaging students in a range of loosely grouped hands-on, table-top activities that generally related to school STEM curricula, such as building a battery-operated motor or creating musical instruments using different length straws. Programs might shift topics, for example, moving in a single day from the topic of sink and float to that of force and motion. Adult facilitators provided instruction and assistance and led discussions about "what's going on."

In the second case, programs engaged students in semester-length, or longer, investigations of particular phenomena, typically local ecologies. In many cases, children worked in small groups to research (and later publicly present) particular topics or activities of interest to them, such as experimental testing of water quality or documenting arsenic levels in local playgrounds. Facilitators were assigned to one or more small groups and acted as coaches. Programs were augmented by table-top activities and, often, field trips.

In the third case, as illustrated in Example #3, programs tended to cover school curricular topics, with more direct instruction, often in front of chalkboards, augmented by hands-on experiments and field trips to or visits from local STEM-based industry professionals. These programs emphasized developing identities of achievement and readiness for academic success in STEM fields; they typically measured their effects through participants' graduation and college enrollment rates.

Although these three ways in which STEM was conceptualized were not mutually exclusive (as our examples make clear), and indeed must operate together to address what most policy documents argue constitutes scientific literacy (e.g., American Association for the Advancement of Science, 1993; National Academies of Sciences Committee on Science Learning K-8, 2007; National Research Council, 1996, 2009), we found that the time constraints and attendance patterns in OST programs usually forced one conceptualization over the others. Programs with the most stable attendance were able to engage children in sustained inquiries and investigations. Programs, like Example #3, Racing Rovers, that committed to develop (and measure) children's pursuit of STEM into college also generally had strict attendance requirements. Less stable attendance patterns, as in Examples #1 and #2, often were associated with programs that were organized around loosely grouped sets of activities. This approach made allowances for children coming and going, but we also found that some of these programs tended to have less well-prepared adult facilitators and the combination of fluctuating attendance, staff preparation,[1] and curriculum design posed challenges to STEM content learning. However, despite these challenges, our observation data show that students were provided multiple experiences with STEM terms, tools, and phenomena, and generally these experiences were undergirded by developmental conditions that enabled positive, peer-supported engagement that sustained students' participation. For many stu-

[1] Our definition of staff preparation includes content background, familiarity with the activities, teaching experience, as well as time to plan and transition from the school day to the after-school day.

dents, hands-on engagement with STEM was a limited option in their K-8 classrooms, and these programs offered them both novelty and the opportunity to bring themselves, physically, into engagement with STEM.

In summary, structural features appeared to shape the way in which STEM was conceptualized and positioned. Developmental features encouraged participation in the activities, however they were positioned. Thus, the types of STEM engagement and learning – with STEM positioned as primarily knowledge, processes of inquiry, or academic commitments – that were afforded in these programs varied as the structural and developmental conditions shaped the nature of STEM.

(2) Learning Goals for Children

Our study is guided by theory that posits that interest in a subject-matter domain develops dynamically through experiences that allow children to exercise and learn to use the cultural tools of STEM, developing skills and capacities that enable deepening engagement with the subject matter (Barron, 2006; Bronfenbrenner, 2001; Hidi & Renninger, 2006). Most programs we studied provided students opportunities to develop several of a range of capacities needed to productively engage in STEM. These[2] include children's understanding and use of:

- STEM concepts and phenomena
- STEM procedures and practices
- Scientific terms and other symbolic representations
- Scientific instrumentation
- Scientific ways of knowing, rules of evidence (epistemologies)
- STEM implications and applications, or meaning and relevance
- STEM questions, ideas, and curiosities about the natural world
- STEM directions, roles, careers, and pathways for future pursuit

While all of these capacities are implicated with one another, and most learning activities encompassed several of them, we found that the structural constraints in programs often made it difficult to move beyond initial stages of noticing and wondering, which is, of course, an essential starting point for further exploration and understanding. Sometimes programs enabled children to explore or build on their

[2] These capacities are based on the NRC 2007 "strands" of disciplinary expertise (entailing understanding of scientific concepts, practices, epistemologies, and endeavors). We call out four others that our studies found were specific contributions OST programs make, such as the ways in which OST learning opportunities sometimes operate simply to stimulate curiosity and wonder, how they can elucidate the relevance of science to young people's lives, and how they can provide access to scientific instruments and also to symbolic representations that might not be included in the standard curriculum. The NRC 2009 "strands" for informal environments added interest and identity to the 2007 set; however, we hold that interest and identity are not strands of science but rather dynamic and situated properties of the developing person who meets and engages in science.

questions; other times, they did not. Whether children had opportunities at school or at home to further explore these questions was not a part of this study.

In fact, we found that the developmental features of low-stakes (nonjudgmental) and peer-supportive environments proved particularly powerful for encouraging the generation of children's questions and ideas, as well as for providing time for digressions and personal stories (as in Example #1) that allowed children to make connections between their prior home, school, and family experiences and the STEM phenomena. In Example #3, the goals of the program, achieved with the commitment of parents, to support students' academic pursuits radically changed the nature of the experience, allowing for an extended multi-week curriculum, the introduction of formalized language and tools, while at the same time building on a nonjudgmental climate that positioned all children as capable in math and science.

Fluctuating attendance, as well as variable staff capacity, seemed to challenge the potential of the programs to support students' epistemological understandings of the nature of scientific evidence and argumentation or to firmly develop conceptual understanding and inquiry skills. For example, the same program as described in Example #1 was also implemented with upper elementary school-aged children at another location; in this second setting, in a church basement, students built houses made of paper and cardboard and then used leaf blowers to blast them with "hurricane forces." Because of fluctuating attendance, different children's projects were at different stages of completion. As they finished their constructions, children subjected the houses to the blasts of the leaf blowers. The children were excited about the ways in which their houses skidded across the floor and withstood, or did not withstand, the blasting air. But the different stages of completion made it difficult for children to observe the outcomes for each of the houses (some were still focused on assembling their houses, while others were blasting them with air, and still others were dancing around with the remains of their air-blasted houses). Classroom management and timing issues (the adult was wielding the leaf blower) prevented the facilitators from keeping records that would allow the sorts of comparisons about what materials or constructions were more durable, for how long, or why. Thus, children experienced the general phenomenon of forces acting on their engineering designs, but the ways in which one would systematically test for durability, how valid experiments would be conducted, and what constitutes evidence or a fair test were not included in the activity, even though all of the elements were in place for such an approach. Indeed, it is quite possible that on another day or in another setting, the program might well have included these elements. This is a typical example of the types of instruction we observed in many programs and again exemplifies how the developmental features encourage participation, but the structural features impose constraints on what is typically possible. We emphasize that we do not deem these experiences to be inferior in any way; indeed, research in the learning sciences shows that they are productive and necessary to developing a commitment to pursue STEM. We also believe that it is essential that such features of OST STEM programs are acknowledged (perhaps celebrated) and also taken into account in any efforts to develop systematic assessments or evaluation methods.

(3) Activity Structures

By activity structures, we refer to the design and implementation of STEM learning activities, including the materials, pedagogies, and relationships among teachers and learners. STEM in the OST setting is often idealized in the literature to be hands-on, inquiry-oriented, and project-based (Coalition for Science After School, 2007). Our study found that activity structures were widely varied and included stand-alone hands-on activities, worksheets, field-based research, design and engineering activities, lectures and guest speakers, field trips, and parent and community events. Students exercised varying levels of agency in these activities, from following prescriptive worksheets to designing and developing their own investigations.

Almost all of the 16 projects we studied involved hands-on table-top activities (using "recipes," activity sheets, worksheets, or kits common to school science). Although many table-top activities were quite school-like, the developmental features that included more flexible uses of time, the relative freedom of choice, and the low-stakes environment altered the nature of these activities. We observed less pressure on children than one typically finds in classrooms to follow exact directions, work at the same pace, or reach the same conclusion. In many cases, a more important goal than mastery of the content or completion of the instructions was positive engagement with the activities, which sustained students' role and standing in the group and encouraged them to persist in undertaking further STEM activities. In some cases, participants were directly engaged in field work and even contributing to a current body of scientific research.

In this sense, the developmental features allowed for children to engage positively with the teachers and with each other in the context of STEM. But in Example #2, Motors, the structural features, which had precluded advance planning and testing, foiled the goals that teachers had for the activity that day. The learning that resulted from this nexus of features certainly included developing senses of self in relationship to participating in a community of STEM engagement; it also helped to relate "school" STEM to everyday experiences. It did not allow children to successfully build a motor. We do not take the position that therefore the activity was a failure; on the contrary, as we stated, the children experienced the social and inclusive nature of engaging in STEM. However, typical assessments that would seek to test specific conceptual knowledge might not detect the kinds of things that were experienced and learned that day. The potential described here is that the developmental conditions which support ways of being create the conditions for further development and expansion of children's knowing, doing, and being in the context of STEM.

(4) Connections with School STEM

Many policymakers assume that if there are connections between OST and school STEM, then what is learned in OST programs should be detectable in school assessments. However, our observations suggest that connections between OST and school

STEM may be less direct and less apparent than would be detectable by school measures. For example, an OST program that focuses on explorations of local wetland ecosystems might involve conceptual understanding, encompassing epistemologies of science and inquiry practices, that do not appear on school tests. Example #2 with the battery-operated motor activity we cited above might well directly connect to the state tests, but what was learned in the day we witnessed had more to do with ways of being and knowing that would not show up on a test. What also might have been learned (if made explicit) was not so much the content of the activity but the fact that things do not always work as planned, that different variables affect the experiment or project, and that science is, at least part of the time, a process of trial and error. The camaraderie in the room, as well as a lengthy discussion about children's experiences with motors, in which children talked about their parent's boats, while the teacher described her father's lessons on car engines, communicated ideas about the social nature of scientific activity and phenomena. These are starting points for deeper explorations of STEM, and they may be essential for engagement with school STEM, but they may not be tested or appear on standardized scorecards.

One prevalent notion regarding OST STEM is that children's interest and capacities are developed in OST settings and that these enhanced degrees of interest and competency will directly translate to more actively engaged classroom STEM learners, which in turn will affect children's scores on standardized tests. This may well be true in certain circumstances, and programs like the one described in Example #3, Racing Rovers, are predicated on such an understanding. However, whether the opportunities presented in the OST setting (e.g., the opportunity for developing interested, confident, and engaged STEM learners) are capitalized on by schools, families, or even the after-school programs themselves is somewhat left to chance. Moreover, if school STEM differs substantially from OST STEM (e.g., being more text-based), it is not at all clear that children will see "science" as appealing in both cases – or even see "science," as it is practiced in each setting, as an essentially and generally unified domain. There are a large set of assumptions, reflecting a mechanistic or additive worldview, that presumes that if children are interested in one setting in what adults call "science," they will be interested and engaged with it in another, despite inconsistencies between the way that STEM is positioned and experienced.

Our study has identified the following ways in which OST programs connect with school STEM:

- Providing children with firsthand experience of participating productively and successfully in STEM learning activities
- Enhancing children's awareness of purpose, relevance, and future trajectories related to STEM that require persistence in school STEM
- Expanding children's interest and capacities (see above) to engage in STEM learning activities
- Increasing children's experience with particular tools, concepts, or processes used in school classrooms
- Supporting teacher practices to make STEM more meaningful and engaging to students in their classrooms

All of these points of connections are built on developmental factors that can be particularly potent in the OST setting, including positive relationships; powerful role models; purposeful engagement; intellectual comfort and safety; the social, creative, and relevant nature of STEM and STEM learning; and the ability to contribute to processes that are valued at both individual and group levels (Fadigan & Hammrich, 2004; Honig & McDonald, 2005; Larson, 2000; Mahoney et al., 2005; McLaughlin, 2000). They also depend on the nature of structural features present in the programs. For example, children's experiences with concepts or tools will be developed to different degrees depending on consistency of participation.

Each of the brief program examples we have provided indicates in some way that they are making these connections between STEM learning in formal and informal setting. Almost none of the connections listed above, however, would show up on a standardized test or even in grades. Whether they translate to new forms of participation depends significantly on many factors, beyond the scope of the OST program or the control or influence of the child, including the quality and quantity of STEM learning opportunities in school.

We would argue that only if schools and OST programs work together to build on the experiences in each setting, would one ever expect to detect the outcomes of OST experiences in the school setting, especially on standardized tests. There are many examples of such intentional bridging between formal and informal settings, although these efforts have not yet been subjected to methods of documentation and assessment that capture the developmental features, account for the structural features, and reveal learning outcomes that are salient to both school and OST settings (Bevan et al., 2010). Moreover, we believe that there is an essential question that needs to be asked about whether collaborations between school and OST programs should be oriented toward supporting school outcomes and goals or whether they should be oriented toward expanding children's experiences by giving them more real-world-based, inquiry-oriented, and personally compelling experiences with STEM. As some of the chapters in this volume make clear, it is debatable whether or not these two views of the goals of OST STEM are, should be, or could be mutually exclusive.

Conclusion

The goal of this chapter was to outline four different program dimensions that are shaped by the nexus of developmental and structural features and which affect, in broad strokes, the opportunities for children's STEM learning in OST settings. We argue that greater specification of the particular conditions of OST programs as a type, and as particular cases, is necessary before the field can appropriately consider and "measure" the contributions and outcomes of OST STEM. STEM in OST is many different things and needs to be implemented and assessed in ways that take these differences into account. We argue that the potential for STEM learning – which we define broadly as entailing processes of knowing, doing, being, and becoming in the context of STEM – is significantly shaped by both developmental and structural features of the OST setting.

Structural constraints, outlined above, drive many OST programs to forego long-term inquiry projects in favor of discrete module-like, table-top activities that may reinforce conceptions of STEM as a collection of disparate facts and procedures. At the same time, positive developmental features of the OST setting create more flexible conditions in which programs may promote key outcomes – such as questioning, meaning-making, and developing positive dispositions and epistemological stances toward STEM – that are often not addressed in school STEM. These "life-deep" dispositions and understandings may be important ingredients for "lifelong" commitment to and success in STEM (Banks et al., 2007). More research is needed to understand how different learning opportunities, across settings and time, come together to support sustained interest, capacity, and persistence in STEM.

A Suggestion

We close with an argument that OST STEM activities should be designed, selected, and evaluated in ways that leverage the developmental power of the space, while taking into account its potential structural constraints. Fluctuating attendance is a given, not a problem: children and their parents schedule events such as doctor appointments, soccer practice, and other family obligations for the hours after school. One can unfortunately expect that other constraints, such as low-paid staff and undedicated spaces, will also continue as a given.

For this reason, we suggest that activities that are designed to provide children with opportunities to explore and test scientific materials, phenomena, or ideas – activities that are not predicated on "one right way" of accomplishing them – may be especially well suited to the OST setting in that they are less dependent on adequate time and staff support for children to experience "success." Thus, they account for the structural constraints that confound activities that are designed to ensure that all students arrive at the same precise understanding (such as how to complete a circuit) by not prescribing an endpoint, and they expand on developmental conditions that encourage active participation and engagement of prior knowledge and interests. (Of course, some OST programs have great stability and therefore may be rich developmental spaces for more sustained and epistemological investigations over time; these are vital parts of the learning ecology and should be supported and expanded.)

To recapitulate our main point, scientists, STEM educators, and policymakers interested in expanding STEM programs to the OST setting need to be aware of the many structural and development features that characterize OST programs and shape the opportunities and possibilities for children's learning in STEM. These conditions are fundamentally different from school conditions. STEM educators and policymakers need to take these differences into account and design and assess OST STEM programs accordingly.

Acknowledgments Work on this project and the analysis for this chapter come from the entire research team, which includes the authors, Ruchi Bhanot, Julie Remold, and Patrick Shields from SRI International and Noah Rauch and Robert Semper from the Exploratorium. This chapter describes work conducted as part of the Center for Informal Learning and Schools and the Academies for Young Scientists, supported in part by the National Science Foundation (Grants ESI-0119787 and DRL-0639656). Any opinions, findings, and conclusions or recommendations expressed in this material are those of the authors and do not necessarily reflect the views of the National Science Foundation.

References

Afterschool Alliance. (2005). *Afterschool issue overview*. Retrieved December 20, 2005, from http://www.afterschoolalliance.org/after_over.cfm

American Association for the Advancement of Science. (1993). *Benchmarks for science literacy*. New York: Oxford University Press.

Banks, J. A., Au, K. H., Ball, A. F., Bell, P., Gordon, E. W., Gutiérrez, K. D., et al. (2007). *Learning in and out of school in diverse environments*. Seattle, WA: The LIFE Center and the Center for Multicultural Education, University of Washington.

Barron, B. (2006). Interest and self-sustained learning as catalysts of development: A learning ecology perspective. *Human Development, 49*, 193–224.

Bevan, B., Dillon, J., Hein, G. E., MacDonald, M., Michalchik, V., Miller, D., et al. (2010). *Making science matter: Collaborations between informal science education organizations and schools*. Washington DC: Center for Advancement of Informal Science Education.

Bodilly, S., & Beckett, M. K. (2005). *Making out-of-school time matter*. Santa Monica, California: RAND Corporation.

Bransford, J. D., Barron, B., Pea, R. D., Meltzoff, A., Kuhl, P., Bell, P., et al. (2006). Foundations and opportunities for an interdisciplinary science of learning. In R. K. Sawyer (Ed.), *The Cambridge handbook of the learning sciences* (pp. 19–34). New York: Cambridge University Press.

Bronfenbrenner, U. (2001). The bioecological theory of human development. In N. J. Smelser & P. B. Baltes (Eds.), *encyclopedia of the social and behavioral sciences* (Vol. 10, pp. 6963–6970). New York: Elsevier.

Chi, B., Freeman, J., & Lee, S. (2008). *Science in afterschool market research study*. Berkeley, CA: Lawrence Hall of Science, University of California.

Coalition for Science After School. (2007). *Science in after-school: A blueprint for action*. New York: Lawrence Hall of Science and Exploratorium.

Cole, M. (2006). *The fifth dimension: An after-school program built on diversity*. New York: Russell Sage.

Fadigan, K. A., & Hammrich, P. L. (2004). A longitudinal study of the educational and career trajectories of female participants of an urban informal science education program. *Journal of Research in Science Teaching, 41*(8), 835–860.

Furman, M., & Barton, A. C. (2006). Capturing urban student voices in the creation of a science mini-documentary. *Journal of Research in Science Teaching, 43*(7), 667–694.

Fusco, D. (2001). Creating relevant science through urban planning and gardening. *Journal of Research in Science Teaching, 38*(8), 860–877.

Gutiérrez, K. D., Baquedano-Lopez, P., & Tejeda, C. (2000). Rethinking diversity: Hybridity and hybrid language practices in the third space. *Mind, culture, and activity, 6*(4), 286–303.

Halpern, R. (1999). After-school programs for low-income children: Promise and challenges. *The Future of Children, 9*(2), 81–95.

Halpern, R. (2002). A different kind of child development institution: The history of after-school programs for low-income children. *Teachers College Record, 104*(2), 178–211.

Halpern, R. (2006). *Confronting the big lie: The need to reframe expectations of afterschool programs*. New York: Partnership for After School Education.

Herrenkohl, L. R., & Mertl, V. (2010). *How students come to be, know, and do: A case for a broad view of learning*. New York: Cambridge University Press.
Hidi, S., & Renninger, K. A. (2006). The four-phase model of interest development. *Educational Psychologist, 41*(2), 111–127.
Honig, M., & McDonald, M. (2005). *From promise to participation: Afterschool programs through the lens of socio-cultural learning theory*. New York: The Robert Bowne Foundation.
Hull, G., & Greeno, J. G. (2006). Identity and agency in nonschool and school worlds. In Z. Bekerman, N. C. Burbules, & D. Silberman-Keller (Eds.), *Learning in places: The informal education reader* (pp. 77–97). New York: Peter Lang Publishing.
James-Burdumy, S., Dynarski, M., Moore, M., Deke, J., Mansfield, W., & Pistorino, C. (2005). *When schools stay open late: The national evaluation of the 21st century Community Learning Centers program: Final report*. Washington, DC: US Department of Education, National Center for Education Evaluation and Regional Assistance.
Larson, R. W. (2000). Toward a psychology of positive youth development. *The American Psychologist, 55*(1), 170–183.
Lerner, R. M., Dowling, E. M., & Anderson, P. M. (2003). Positive youth development: Thriving as a basis of personhood and civil society. *Applied Developmental Science, 7*(3), 172–180.
Lincoln, Y. S., & Guba, E. G. (1985). *Naturalistic inquiry*. Newbury Park, CA: Sage Publications.
Mahoney, J. L., Larson, R. W., Eccles, J. S., & Lord, H. (Eds.). (2005). *Organized activities as developmental contexts for children and adolescents*. Mahwah, NJ: Lawrence Erlbaum Associates.
McLaughlin, M. W. (2000). *Community counts: How youth organizations matter for youth development*. Washington, DC: Public Education Network.
Nasir, N. S., Rosebery, A. S., Warren, B., & Lee, C. D. (2006). Learning as a cultural process: Achieving equity through diversity. In R. K. Sawyer (Ed.), *The Cambridge handbook of the learning sciences* (pp. 567–580). New York: Cambridge University Press.
National Academies of Sciences Committee on Science Learning K-8. (2007). *Taking science to school: Learning and teaching science in grades K-8* (Prepublication Copy: Uncorrected Proofs ed.). Washington, DC: National Academies Press.
National Research Council. (1996). *National science education standards*. Washington, DC: National Academy Press.
National Research Council. (2000). *How people learn: Brain, mind, experience, and school*. Washington, DC: National Academy Press.
National Research Council. (2009). *Learning science in informal environments: People, places, and pursuits*. Washington, DC: The National Academy Press.
Nocon, H., & Cole, M. (2006). School's invasion of "after-school": Colonization, rationalization, or expansion of access? In Z. Bekerman, N. Burbules, & D. Silberman-Keller (Eds.), *Learning in places: The informal education reader*. New York: Peter Lang Publishing.
Rahm, J. (2007). Learning and becoming across time and space: A look at learning trajectories within and across two inner-city youth community science programs. In W.-M. Roth & K. Tobin (Eds.), *Science, learning, identity: Sociocultural and cultural-historical perspectives* (pp. 63–80). Rotterdam, the Netherlands: Sense Publishers.
Sawyer, R. K. (2006). *The Cambridge handbook of the learning sciences*. New York: Cambridge University Press.
Stetsenko, A. (2009). Personhood: An activist project of historical becoming through collaborative pursuits of social transformation. *New ideas in psychology.*
Vandell, D. L., Reisner, E. R., Pierce, K. M., Brown, B. B., Lee, D., Bolt, D., et al. (2006). *The study of promising after-school programs: Examination of longer term outcomes after two years of program experiences*. Madison, WI: Wisconsin Center for Education Research, University of Wisconsin-Madison.
Vygotsky, L. S. (2004). *The essential Vygotsky*. New York: Kluwer Academic/Plenum Publishers.

Chapter 15
Examining Youth's Mathematics Practices in an After-School Robotics Team

John Y. Baker, Janine T. Remillard, and Vivian Lim

Ms. Miller, we need geometry!

Sixteen-year-old Joseph and his teammate, Kim, were in the process of determining a diagonal length on the robot competition field. After consulting the competition manual and measuring several side lengths, Joseph realized that he might need to use some geometric principles to calculate the missing diagonal length. It was at this point that he called across the room to Ms. Miller, the team advisor, who was also a math teacher in the high school, indicating that, "We need geometry!"

In our ethnographic study of youths' mathematics practices in three out-of-school-time programs, moments like this are of particular interest to us. Our aim is to examine the knowledge and practices the youth employ in undertaking the activities of the program and how these relate to formal mathematics – the mathematics primarily taught in high schools. In this instance, Joseph identified the need for mathematics learned in school, mathematics that was not immediately at his fingertips.

Joseph and Kim (all names used in this chapter are pseudonyms) were members of the Beech Hill High School robotics team. The team was part of an initiative in a large urban school district that aimed to increase youth interest in STEM careers through active engagement in engineering practices. Each team designed and constructed a robot and prepared supporting documentation. The robotics team was located at an intersection of school and out of school; it was housed in school buildings and run by school faculty but was structured so as to support youth autonomy and decision-making.

J.Y. Baker (✉) • J.T. Remillard • V. Lim
Graduate School of Education, University of Pennsylvania, Philadelphia, PA, USA
e-mail: jybaker@hotmail.com

B. Bevan et al. (eds.), *LOST Opportunities: Learning in Out-of-School Time*, Explorations of Educational Purpose 23, DOI 10.1007/978-94-007-4304-5_15,

The team, though affiliated with Beech Hill High School, operated very much like the out-of-school or informal learning contexts studied by Heath (2001) and McLaughlin (2000). Specifically, the youth were engaged in creating an authentic product that would be judged by practitioners using real criteria from the field. In this way, the team involved youth in "sustained, long-term comprehensive projects" that required participants to plan, commit to seeing through, and evaluate. Team members had responsibility for decision-making about every aspect of the robot. McLaughlin characterizes these types of settings as "intentional learning environments" that are "youth-centered, knowledge-centered, and assessment-centered" (2000, p.10). Indeed, we were drawn to the robotics team because of its potential to support these components: the program was responsive to the youth it served; it was designed around a particular area of expertise, engineering; and it included meaningful measures of youth accomplishment through the competitions.

In selecting research sites, we were interested in settings where youth might be authentically engaged in using formal, school mathematics. This interest grows out of two trends noted in the literature on out-of-school mathematics practices. First, there is evidence that when engaged in authentic, meaningful activity, youth and adults develop and use informal mathematics practices that are deeply intuitive, meaningful to its users, and tied to particular features of the problem's environment (Lave, 1988; Nunes, Schliemann, & Carraher, 1993; Scribner, 1984). The practices and their contexts are inherently tied together, and the users of the practices display flexible and robust knowledge within the particular domain of the practice. Second, there is little evidence in the research that those engaged in such intuitive and contextually grounded practices naturally make connections between their knowledge and formal mathematics, nor develop abstracted or generalized knowledge related to the particular mathematics ideas (Baker, 2007; Nunes et al., 1993). It is this disconnect between formal and informal mathematics practices that we set out to understand through our study of mathematically oriented, youth development programs.

In this chapter, we analyze some of the mathematics practices of the youth on the Beech Hill robotics team. We were curious about whether engaging in meaningful activity that uses ideas taught in school mathematics could help youth make connections between formal and informal mathematics. Furthermore, we were interested in whether and how the practices of the youth were mediated by the youth-centric nature of the program. We found that the hybrid nature of the setting fostered the development of practices that integrated aspects of formal and informal mathematics. We identify them as *hybrid practices,* drawing on the work of Gutiérrez, Rymes, and Larson (1995), Moje et al. (2004), and Barton, Tan, and Rivet (2008). However, as we elaborate below, the process by which these hybrid practices emerged involved ongoing negotiations among multiple scripts, or expected patterns of behavior, which were fundamentally tied to the nature of the setting.

Theoretical and Empirical Background

International studies in out-of-school mathematics have focused on unschooled youth and adults in their work practices. A number of studies have examined the informal practices of youth who sell fruit and candy on city streets in Brazil (Nunes et al., 1993; Saxe, 1988). These researchers found that the mathematics practices of the sellers were oral in nature and were conducted in ways in which the solution strategies remained close to the details of the problem situation. The youth broke down difficult calculations into simpler combinations of calculations that they were familiar with and were able to accurately solve. Other research with bookmakers (Schliemann & Acioly, 1989), fishermen (Nunes et al., 1993), and carpenters (Nunes et al., 1993) had similar findings.

In the United States, there have been studies of everyday practices of adults while shopping (Lave, 1988; Murtaugh, 1985), dieting (de la Rocha, 1985; Lave, 1988), and making financial decisions (Lave, 1988; Chap. 2 by Esmonde et al., this volume). These authors found, among other things, that people tend to use both quantitative and qualitative considerations while making numerical decisions. They argue that problem creation is an important part of problem-solving in people's everyday lives and that the problem contexts are meaningful to their solvers. Scribner (1984) looked at arithmetic and spatial reasoning of dairy workers as it applied to the packing and delivery of dairy products. She found, like others (cf. Masingila, 1994), that people used informal practices that were flexible given the many parameters with which both trades had to work. This flexibility was a function of experience, with the more experienced workers often finding solutions that use the least amount of effort to complete.

Nasir (2000) looked at the statistical reasoning of teenage youth who were part of a basketball team. She found that youth who played on the high school basketball teams began to internalize methods of accounting for different player statistics kept by the school and media. Their methods were informal and quite accurate. Beyond this study, there is very little work done with middle- and high-school-aged youth. Such work, however, is critical to understanding where and how connections can be fostered between the informal mathematics that adolescents use out of school and the formalized mathematics that they are taught in school.

Our work follows in the sociocultural tradition (Engeström, 1993, 1999; Vygotsky, 1978), focusing on the cultural practices of the youth during activity (see Chap. 16 by Khisty & Willey, this volume for elaboration). To understand the role of out-of-school and in-school aspects of the robotics team, we draw on hybridity theory (Gutiérrez et al., 1995). Looking at school classroom language practices, Gutiérrez et al. use *first space,* or the monologic script, to describe normative teaching practices that privilege adult knowledge and classroom management. They use *second space,* or the counterscript, to describe the disruptive behaviors students typically employ against the dominant script. The authors argue that authentic learning occurs in the *third space*, where the first two scripts intersect. They believe that "the potential for a third space salvages the classroom as a locus for social change...the teacher and the students are the loci of 'internal dialogic meaning'" (p. 452). Gutiérrez et al. speculate that the creation of a third space in classrooms can cause changes in the types of knowledge and representations that are privileged. Barton et al. (2008) and Moje et al. (2004) found similar shifts while investigating youth-adult interactions and youth learning in classrooms where teachers explicitly tried to create third spaces.

Using the notion of hybridity, other researchers have extended the concept of third space to examine youth-adult encounters in informal learning settings (cf. Chap. 16 by Khisty and Willey, this volume). Gutiérrez, Baquedano-López, Alverez, and Chui (1999) examined an after-school activity system to further investigate how hybrid language practices play out. These researchers observed powerful ways in which youth "draw from their own as well as each other's linguistic and sociocultural resources to collaborate in problem-solving activities, creating rich zones of development" (p. 88). Informal settings, especially those described by McLaughlin (2000) that foster authentic engagement, can thus create important places where we can learn about how youth draw on various resources.

The robotics program fits somewhere between the formal classroom and the informal out-of-school program. We call settings that draw on the practices and structures of both formal and informal settings *hybrid spaces*. These spaces include the monologic and counterscripts as the youth and adults come to the space with relationships from the formal setting (i.e., teacher-student). However, other scripts exist and compete for legitimacy in the setting where youth voluntarily choose to participate, where they share some decision-making authority, and where their own practices are valued. There are also scripts attached to the knowledge and assessment-centered components (McLaughlin, 2000) of an informal setting. For the robotics team, these were introduced through the purpose of the activity – competing successfully in the robotics competition. Our research seeks to understand what hybrid practices look like in this hybrid setting.

Methods

We use qualitative methods designed to create *rich descriptions* (Maxwell, 2005) of the mathematical behavior of the youth in the study. Our work is both exploratory and descriptive. Our analysis draws on data collected through participant observation, interviews, and focus groups over three years. For the following discussion, we use data from Beech Hill robotics team in the fall of 2007 to illustrate a theme observed across all three of our sites.

Beech Hill Robotics

Beech Hill robotics team was located in a comprehensive high school in a large, postindustrial city in the Northeast. The school served around 1,500 students, 99% of whom were African-American and 70% of whom qualified for free or reduced lunch. The robotics team is one of a half dozen after-school activities offered by the school.

In the fall of 2007, 22 youth attended the meetings, with a cadre of 12 youth contributing the majority of the work toward all parts of the competitions. Of these 12, four were female and eight were male. There were participants from all grades in the school, and half of the youth had participated the previous year. The demographics of the team matched those of the school. The advisor, Susan Miller, an African-American woman in her fifties, was a mathematics teacher in the school. She had come to teaching six years prior, after many years in the marketing industry. The team was also supported, once a week, by college students from two local universities.

Team meetings took place in Beech Hill High School. Each fall, the team took part in a regional contest in which they were given six weeks to prepare: (1) a robot, which competed against others in a timed task; (2) an engineering notebook that contained a record of the design and construction process and decision-making; (3) an oral presentation, through which the team described the engineering process; (4) a table display, which marketed their "product" and illustrated the work of the entire team; and finally, (5) demonstration of spirit and sportsmanship. Teams who placed at the local competition were given four more weeks to prepare for the national competition. The team met after school nearly every day during the period before the regional competition, at first staying for a couple of hours each day and then later staying until 8 and 9 at night.

Typical meetings started off with the youth recapping their work from the previous session, followed by the advisor doling out tasks for the day that the youth worked on in small groups. The youth then broke out into groups – decided by Susan and the youth themselves based on level of participation, seniority, and knowledge of robotics – and worked on tasks related to the different parts of the competition. The main activity was always around building, practicing with, and refining the robot. Susan tended to let the team work on their own, taking part only when asked for help.

The theme of the competition in the fall of 2007 was *Extraterrestrial Exploration*. Teams were to build robots that could recover four different types of supplies from a cargo bay and return them to a depot in three minutes. The starting area and depot were on a lower level; the cargo bay was raised 18 inches and was accessible by a 30% grade ramp. Four teams competed concurrently for shared resources in the competition field, though each team was assigned their own starting square and depot. The task was designed to be complex, with a level of difficulty similar to the tasks in an undergraduate engineering design course. The competition field is shown in Fig. 15.1.

Data Collection

We collected our data primarily through participant observation, artifact analysis, interviews, and focus groups. A group of three graduate assistants participated in 80% of team meetings for the six weeks prior to the local competition and the four

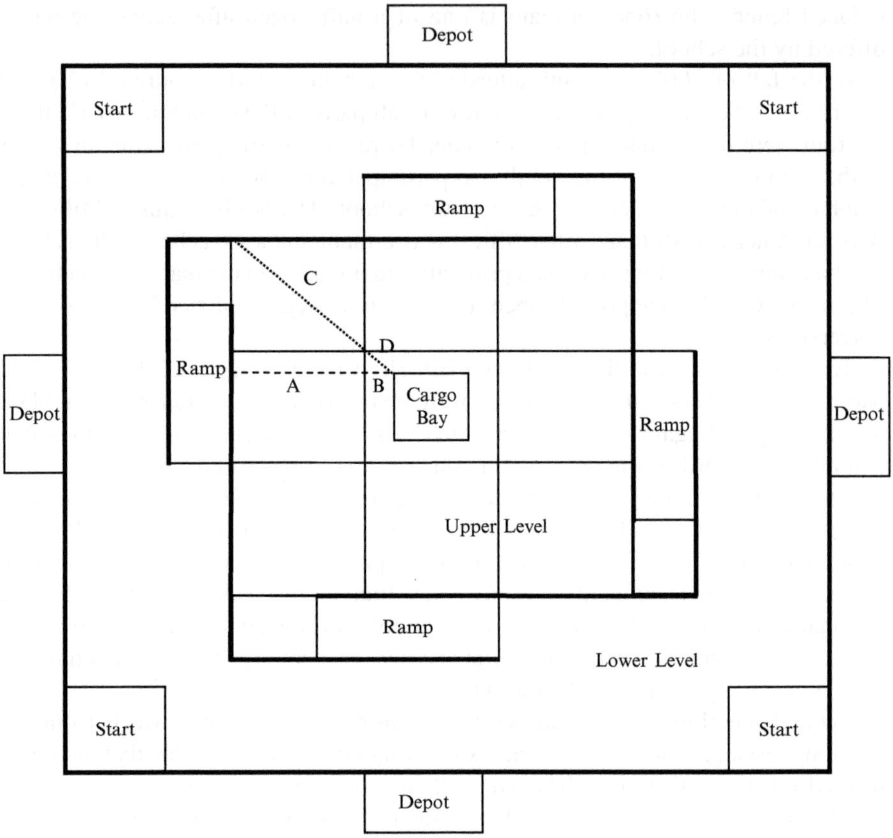

Fig. 15.1 Competition field

weeks leading up to the national one. Participant observation involved the researchers attending team meetings, taking part in team activities, and talking to the youth about their practices. Field notes were written directly after observations. Most of the youth took part in a focus group that was used to help create a protocol for an experience interview. The youth were then interviewed with this protocol to gather information about each participant's perceived role in the team, as well as their experiences in school and with mathematics.

Analysis

During the fall of 2007, we gathered 30 sets of field notes. Although some mathematical problem-solving occurred each day, we identified ten events during which significant mathematics or problem-solving took place. While much of the team's mathematics practices occurred quickly and often unnoticed in the course of team activity, significant mathematics events included explicit and prolonged mathematical problem-solving. Each event was coded and analyzed using the following coding categories: (a) how the use of mathematics was initiated; (b) the role of adults and youth in structuring the approaches taken; (c) the nature and role of mathematical representations, conventions, and tools and their connections to formal or informal mathematics; (d) the kinds of reasoning and strategies that emerged; and (e) the particular mathematics content. Our coding strategy was based on previous theoretical and empirical work (Engeström, 1993; Lave, 1988; Nunes et al., 1993), as well as being grounded in the experience and voices of the youth. This analysis led to the key finding presented here: that the youth adapted the formal, process-oriented strategies presented by adults in the program to the concrete task of creating a product, thus creating what we call *hybrid practices*.

Hybrid Practices

Our analysis of the mathematical events occurring in the robotics team revealed two important patterns. First, similar to findings in the literature on out-of-school mathematics practices, when left on their own to solve mathematical or engineering problems, the youth relied on *local and concrete* problem-solving strategies. That is, they tended to use approaches that maintained a close tie to the particulars of the context and did not draw on abstract tools or strategies to assist them. Second, in a number of instances, various adults with some form of mathematical or engineering expertise worked with the youth to assist them in problem-solving. In these cases, the adults tended to encourage the youth to rely on *generalized and abstracted* approaches to solving problems. These approaches were needed to address the complexity of the task in a way to make a team's robot competitive. Despite this guidance, the youth tended to develop practices in which they adapted strategies offered by the adults to reflect key elements of their local practices. In this way, formal mathematics and engineering practices became embedded in the competition as additional scripts to the aforementioned set of formal and informal scripts. It was in the arena of these multiple, overlapping, and competing scripts that *hybrid practices* were produced.

The following vignette, written from field notes taken by two of the authors during the October 4, 2007 team meeting, provides an example of the mathematics practices we observed. It illustrates the kind of concrete and local practice we observed when youth had sole decision-making authority over their strategies, as well as the nature of adult intervention on their practices and the hybrids that resulted. We follow with an analytical discussion of the hybrid nature of the practices common to the team and then consider the way hybridity is a critical component of the program.

Finding a Diagonal Length

Susan started off by telling the five participants present that the meeting would last only one hour because she had another commitment. By this date, the team had already designed a prototype for its robot and made a plan for how they wanted to accomplish the task. Rather than having the youth continue yesterday's activity of building parts of the robot, Susan gave them a task that they could finish in a short amount of time. She proposed that they find the diagonal distance from the corner of the *upper level* of the field to the *cargo bay* in the center (this distance is labeled as *C* and *D* on Fig. 15.1). The *upper level* was composed of nine square boards set 18 inches above the *lower level*, with a raised *cargo bay* in the center. She explained that the purpose of this particular task was to ascertain whether their robot could reach the center of the *upper level* from the *lower level* without ascending the ramp. The youth had already discussed this option and had decided that the robot would need to climb the ramp to collect the supplies. When the youth asked Susan for clarification about this task, she said that having the robot pick up pieces from the lower level was a strategy that the team had discussed and rejected. She reminded them that while they had already decided to climb the ramp, they needed to record their decision-making process in the engineering notebook.

Two youth engaged in the task while the other three in attendance quietly looked on and talked among themselves. Joseph was in 11th grade. He had been on the team for three years and was enrolled in Algebra II. He was central to the design and creation of the robot, and Susan described him as a "tinkerer." Kim was a senior not enrolled in mathematics at the time. She was on track to be the first member of her family to graduate from high school. She had been on the team for three years and was also one of the primary contributors to the robot's design and construction.

Joseph and Kim sat on opposite sides of a long table with an enlarged printout (36 inches by 36 inches, orthographic view) of the competition field between them. Fig. 15.1 is a drawing of the competition field from a top view.

Joseph began the task by estimating the orthogonal distance, the dashed line in Fig. 15.1 (lengths *A* and *B*), from the side of the upper level to the cargo bay. Kim found a diagram that provided the dimensions of the large squares as 48 inches by 48 inches and the cargo bay as 24 inches by 24 inches. Joseph reasoned that the orthogonal distance was 60 inches; 48 inches to transverse one square, length *A*, and 12 inches into the second square to reach the cargo bay, length *B*. He described how he figured out the size of length *B* by explaining that the cargo bay took up 24 of the second square's 48 inches. Using his hands, he "moved" the cargo bay to one side of the middle square, saying that it took up 24 of the 48 inches across this square. He then reasoned that since the cargo bay was actually in the middle of this square, the remaining 24 inches were split evenly between the two sides.

Joseph then began to figure out the length of the diagonal from the corner of the upper level to the cargo bay, represented by the dotted line in Fig. 15.1 (lengths *C* and *D*). He broke the task down into parts again, looking for the distance across the first square, length *C*. He said it was 48 inches. When Kim asked him how he knew this, Joseph responded, "This is 48 and this is 48, and this is 48 also," pointing to the two sides and then the diagonal of the first square. Vivian, the researcher observing the group, asked them if all of the dimensions of the square were the same. Kim grabbed a wooden dowel and placed it along the base of a drawing of one of the squares, using her finger to mark the length. She then moved the dowel to the other side and noted that the lengths were the same. She then shifted the dowel so that it went along the diagonal of the square. She told Joseph that all the distances were not the same and that the diagonal was longer than the sides.

To find the length, Joseph began looking for a tape measure that was longer than 48 inches while Kim looked through a binder of diagrams of the competition field for help. She came upon a drawing of the under-support of the square, which included a diagonal piece that traversed most of the entire diagonal of the square. The measurement given for this diagonal was 65 5/8 inches. The diagonal was not exactly the correct length because of some trim, but it was close. Looking at the large diagram together, they agreed on a number and wrote the distance "65 in."

Joseph and Kim then began to figure out the distance on the second square from the upper corner to the cargo bay, length *D*. After several minutes, Joseph said the distance was 7 inches but quickly recanted. Kim went through the diagrams again looking for some clue but did not find anything. At this point, Vivian intervened. She pointed to the second square and asked them if they knew the orthogonal distance from the side to the cargo bay; they responded that it was 12 inches. She then completed a right triangle with her finger. Both team members thought for a minute. The researcher suggested that they think about the larger triangle they had been working from and make an estimate. Kim subtracted 65 from 48 with the calculator on her phone but was not sure how to use that number. Joseph turned around and called across the room to Susan, "We need some geometry." Susan did not respond. Vivian then mentioned the Pythagorean theorem and guided the youth to use it to solve both the large and small triangle problems. The youth were able to name the theorem and remembered part of the formula, but they needed assistance utilizing it. Through a joint effort, the youth calculated the long distance at 67 inches and the short distance at 17 inches.

After they finished the task, the youth sat and talked to each other until Susan told the group it was time to go. She asked Joseph and Kim what they had found, and they said that the diagonal from the corner of the upper level to the corner of the cargo bay (lengths *C* and *D*) was 65 inches. Susan said that that was far too long, commenting on last year's robot not being able to reach 3 ft. She told them to write down what they had done and the number they got. Kim wrote "65 in." on the large printout before the team left.

Analysis: The Making of Hybrid Practices

Several characteristics of this exchange, which was typical of many observed in this site, have particular implications for our analysis of the hybrid nature of the youths' mathematical practices. The first is the relative concreteness of their self-initiated argumentation. Joseph, for instance, used a physical and spatial explanation for how he knew length *B* was 12 inches. He described how he mentally moved the 24-inch cargo bay to the side of the 48-inch second square and then argued that the other 24 inches of the square was split in half. Kim used an ad hoc and concrete method – comparing lengths using a wooden dowel – to disprove Joseph's assertion that the diagonal was the same length as the side of the first square. Similarly, Joseph sought a measuring tool longer than 48 inches, once his assertion was disproved. Their use of a mathematical theorem, one commonly taught in high school geometry and familiar to both youth, occurred as a result of Vivian's intervention and under her guidance. The use of available resources to make informal mathematical assessments aligns with Scribner's (1984) findings with dairy workers, where the workers used the physical array of products to assemble orders.

Second, the two youth completed this task together, problem-solving jointly to come to a consensus about the distance. Joseph was able to reason about the orthogonal distance from the side to the cargo bay and explain his reasoning to Kim, and Kim was able to estimate the diagonal distance across the first square. They used each other's respective and joint knowledge and processes to move toward a solution. Importantly, they did not necessarily take one another's solutions as fact; rather, they assessed their credibility through empirical tests. The locus of authority for knowing was shared by the youth, although the last part of the vignette shows that this joint authority was secondary to the authority of Susan and the researchers. During the last segment, they recognized the need for school mathematics but were unable to use it without Vivian's help. To come to a final solution, they integrated the information the school mathematics yielded with the distances they had already determined using concrete local practices. This type of collaborative problem-solving is a goal of youth-centered programs like this and greatly adds to youth ownership of the process (McLaughlin, 2000).

Finally, the vignette illustrates the typical hybrid nature of the robotics team's practices in several ways. As we just described, the youth were inclined to rely on concrete approaches that were grounded in the physical material they had before them. Yet, they were willing to incorporate abstract strategies into their practices when encouraged and guided to do so. Furthermore, the task itself was a hybrid. The team worked on their engineering notebook but in a post hoc way that had little authentic meaning given its temporal separation from the larger team's earlier decision. Because participants were required to record their decision-making in their notebooks (an authentic engineering task), Susan instructed the youth to devise and record formal processes for a decision that had already been made informally. Joseph and Kim understood the significance of the notebook and undertook the measurement and recording activity not from an engineering perspective but from a need to follow the rules of the competition. As a result, while the youth's engagement in the problem-solving task was authentic on one level, their imprecise treatment of the findings suggested that they had little stake in them. After determining the length to be the sum of 67 inches and 17 inches, they reported it to Susan as 65 inches. Satisfied that they had enough data to formally reject the approach, Susan instructed them to record the quantity 65 inches.

Connections to Hybridity Theory

In their research on the relationship between adult and student discourse in learning environments, Gutiérrez et al. (1995) call the tendency for students to rebel against adult control "underlife," and they use it to describe the subversive acts that youth take up to distance themselves from school. Both Moje et al. (2004) and Barton et al. (2008) provide powerful examples of ways teachers create hybrid spaces in classrooms by bringing together different knowledge, discourses, and relationships in ways that break down oppositional binaries and allow them to work together to generate new practices and identities.

Our analysis of the robotics team, as well as the other out-of-school programs we examined, suggests that the hybridity of practices that we observed had different roots than those found in schools; it emerged out of the hybrid nature of the program, at the intersection of formal and informal contexts. The monologic script and the counterscript existed to some extent but were complicated by shifted power relations and competed with other scripts. The robotics program was located at the margins between in and out of school. It was run by a teacher, in the school building, whose role was not to teach. The program engaged the youth it served in ways very different from school. Youth decision-making, including whether or not to follow adult guidance, was a central value of the team's practices. Moreover, the youth attended voluntarily and had significant responsibility for the group's work. Their commitment to building a competitive robot was authentic and not imposed by an external authority. Yet, through taking up this commitment, they also were subject to the mathematical and engineering scripts introduced by the robotics competition.

The practices described in the vignette taken up by Joseph and Kim brought together the informal, joint problem-solving strategies of the youth, the formal mathematical convention offered by Vivian, and the procedural expectations of engineering (that the notebook was a place to document design decisions) prompted by Susan. It illustrates the way the hybrid practices in this hybrid space crossed multiple boundaries.

Conclusion

Our analysis of the hybrid practices and the hybrid nature of Beech Hill Robotics Program highlights the importance of understanding the significance and complexity of settings that are hybrid by design. Just as analyses of student and teacher engagement in classrooms have revealed the ways that hybrid spaces break down oppositional binaries and allow them to work together to generate new knowledge, we contend that examination of hybrid spaces in out-of-school settings can shed light on the particular potential of these settings to place new demands on youths' performance including providing considerable degrees of authority over their performances. More importantly, the hybrid setting thus represents an opportunity for learning theorists to explore how youth draw on idiosyncratic, communal, and formal knowledge/practices in a space where the youth have autonomy and their ways of knowing and doing are valued.

Two key questions have arisen from our analysis that merit further study. First is the question of the relationships between the hybrid mathematics and engineering and formal mathematics practices in schools. Although the youth engaged in meaningful activity that used and adapted ideas taught in school mathematics, it is unclear whether the demands of the setting helped the youth make significant connections between formal and informal mathematics, connections that might position them as powerful users of formal mathematics. As noted above, instances that involved abstract and formal mathematics practices occurred when suggested and supported by adults, but we did not observe instances of youth using these practices independently. So the question remains, to what extent can the mathematics practices that emerge in hybrid settings serve as bridges to the learning of formal mathematics and engineering practices that will grant youth access to new educational opportunities?

The second, related, question is about the role of adults in hybrid, youth-centered spaces. It is evident that the adult role in these settings is complex and involves achieving a delicate balance between imposing the expectations associated with particular scripts or practices, in this case the engineering and competition scripts, and allowing youth to rely solely on the scripts that are most familiar to them. Although hybridity theory helps us examine these spaces in new ways, it does not diminish the challenge of creating these youth-centered spaces. The field would benefit from further research that focuses specifically on the role of adults and other authorities in youth-centered, out-of-school programs.

Acknowledgments Work on this project and the analysis for this chapter come from the entire research team, which includes the authors, Elaine Simon, Christine Massey, Lynda Ginsburg, and Stefan Grudza. This chapter describes work conducted as part of MetroMath, supported in part by the National Science Foundation (Grant ESI-0333753). Any opinions, findings, and conclusions or recommendations expressed in this material are those of the author and do not necessarily reflect the views of the National Science Foundation.

References

Baker, J. Y. (2007). *Out-of-school mathematics: Synthesis and analysis of research concerning the use of mathematics in non-school settings*. Unpublished master's thesis, University of Pennsylvania, Philadelphia, PA.

Barton, A. C., Tan, E., & Rivet, A. (2008). Creating hybrid spaces for engaging school science among urban middle school girls. *American Educational Research Journal, 45*(1), 68–103.

De la Rocha, O. (1985). The reorganization of arithmetic practices in the kitchen. *Anthropology and Education Quarterly, 16*(3), 193–198.

Engeström, Y. (1993). Developmental studies of work as a testbench of activity theory: The case of primary care medical practice. In S. Chaiklin & J. Lave (Eds.), *Understanding practice: Perspectives on activity and context* (pp. 64–103). Cambridge, UK: Cambridge University Press.

Engeström, Y. (1999). Activity theory and individual and social transformation. In Y. Engeström, R. Miettinen, & R.-L. Punanmäki (Eds.), *Perspectives on activity theory* (pp. 19–38). New York: Cambridge University Press.

Gutiérrez, K. D., Baquedano-López, P., Alverez, H. H., & Chui, M. M. (1999). Building a culture of collaboration through hybrid language practices. *Theory into Practice, 38*(2), 87–93.

Gutiérrez, K., Rymes, B., & Larson, J. (1995). Script, counterscript, and underlife in the classroom: James Brown versus Brown v Board of Education. *Harvard Educational Review, 65*(3), 445–471.

Heath, S. B. (2001). Three's not a crowd: Plans, roles, and focus in the arts. *Educational Researcher, 30*(7), 10–17.

Lave, J. (1988). *Cognition in practice*. Cambridge, UK: Cambridge University Press.

Masingila, J. O. (1994). Mathematics practice in carpet laying. *Anthropology & Education Quarterly, 25*(4), 430–462.

Maxwell, J. A. (2005). *Qualitative research design: An interactive approach* (2nd ed.). Thousand Oaks, CA: Sage.

McLaughlin, M. W. (2000). *Community counts: How youth organizations matter for youth development*. Washington, DC: Public Education Network.

Moje, E. B., Ciechanowski, K. M., Kramer, K., Ellis, L., Carrillo, R., & Collazo, T. (2004). Working toward third space in content area literacy: An examination of everyday funds of knowledge and discourse. *Reading Research Quarterly, 39*, 38–72.

Murtaugh, M. (1985). The practice of arithmetic by American grocery shoppers. *Anthropology & Education Quarterly, 16*, 186–192.

Nasir, N. S. (2000). "Points ain't everything": Emergent goals and average and percent understanding in the play of basketball among African American students. *Anthropology & Education Quarterly, 31*(3), 283–305.

Nunes, T., Schliemann, A. D., & Carraher, W. (1993). *Street mathematics and school mathematics*. New York: Cambridge University Press.

Saxe, G. B. (1988). Candy selling and math learning. *Educational Researcher, 17*(6), 14–21.

Schliemann, A. D., & Acioly, N. M. (1989). Mathematical knowledge developed at work: The contribution of practice versus the contribution of schooling. *Cognition and Instruction, 6*(3), 185–221.

Scribner, S. (1984). Studying work intelligence. In J. Lave & B. Rogoff (Eds.), *Everyday cognition: Its development in social context* (pp. 9–40). Cambridge, MA: Harvard University Press.

Vygotsky, L.S. (1978). *Mind in society: The development of higher psychological process*. Cambridge, MA: Harvard University Press.

Chapter 16
After-School: An Innovative Model to Better Understand the Mathematics Learning of Latinas/os

Lena Licón Khisty and Craig Joseph Willey

Introduction

At the heart of our discussion is a concern to better understand how to transform the inequitable schooling conditions of Latinas/os, especially in mathematics. People experience mathematics differently based on their gender, race, ethnicity, language, and socioeconomic status (Gutstein, 2005; Gutstein, Lipman, Hernández & de los Reyes, 1997; Khisty & Chval, 2002; Martin, 2006). Mathematics learning, like other educational components, is a racialized form of experience (Martin, 2006) and is not, as some perceive it, a neutral discipline, free of social and political biases and consequences. Latinas/os' disproportionate patterns of mathematics failure influence their continued underrepresentation in higher education, professional careers, and higher socioeconomic levels. Even in light of efforts since the 1960s to address the educational and social conditions surrounding Latinas/os, radical transformation of hegemonic views of Latinas/os has fallen short, and perhaps no other place is this more evident than in the classroom and in mathematics.

A number of probable causes for the miseducation of Latinas/os have been offered, but these frequently focus on the individual student, the student's home and community, and/or the student's language. The problem continues to be defined in terms of a pathology of the student (Khisty, 1995). Since we have not improved the general educational and mathematical status of Latinas/os, new perspectives and research methods are clearly warranted (Mercado & Santamaría, 2005) along with concomitant practice.

L.L. Khisty (✉)
Curriculum and Instruction, College of Education,
University of Illinois at Chicago, Chicago, IL, USA
e-mail: llkhisty@uic.edu

C.J. Willey
School of Education, Indiana University-Purdue University, Indianapolis, IN, USA
e-mail: cjwilley@iupui.edu;

B. Bevan et al. (eds.), *LOST Opportunities: Learning in Out-of-School Time*,
Explorations of Educational Purpose 23, DOI 10.1007/978-94-007-4304-5_16,

In this chapter, we describe an after-school project (i.e., *Los Rayos de* [the thunderbolts of] *CEMELA*) and lessons we have learned from our three-year ethnography of working with elementary grade Latinas/os in mathematics. *Los Rayos* (for short) was not designed to be a tutorial activity like many other after-school projects, nor was it our intention to study "an after-school" project itself. Instead, *Los Rayos* was intended to be a radically different learning environment for Latinas/os, one that would inherently define their characteristics as valuable learning capital and that would be a space between out-of-school (i.e., home and community) and in-school contexts which potentially could offer us needed, new perspectives and understandings about Latinas/os' mathematics learning and schooling. Furthermore, *Los Rayos* was not designed such that we would measure what children learned. Instead, the goal was for *Los Rayos* to teach us, as researchers, by offering new contexts for new insights (Khisty, 2004). Ultimately, the after-school represents a challenge to prevailing assumptions about learning contexts and patterns and hopefully contributes to new directions for research and practice in mathematics education.

We begin our discussion by briefly describing *Los Rayos*, its characteristics, and its intellectual framework. We then discuss critical lessons about schooling and mathematical development that we have gained from the after-school experience. These lessons represent unexpected insights that emerged because of the design features of the after-school; we present three of these lessons. We conclude by considering the implications of what we have learned for creating more effective school (mathematics) learning environments for Latinas/os (practice) and for research of informal learning contexts such as *Los Rayos.*

The After-School and Its Intellectual Framework

A General Description of Los Rayos

Originally, *Los Rayos* was designed so that we could intimately investigate the unrecognized and undocumented language and social practices (learning capital) and knowledge Latinas/os have for doing nonremedial mathematics, that is, mathematics that is just a bit beyond what they usually do in their classes (Khisty, 2004). We assumed that to adequately and positively transform Latinas/os' current educational status, we needed to have a better understanding *in situ* of these learning resources and the processes in which they are used in mathematics. In general, *Los Rayos* represents our attempt to create an alternative context, or an alternative learning culture to that of the typical school, in order to help students develop a repertoire of tools and experiences to transform their own schooling.

Los Rayos met in a technology room in a school building right after classes ended. The school had a student population that was 100% Mexican or Mexican descent and was situated in an inner-city, working-class neighborhood that had a similar demographic composition. In this relatively large neighborhood, Spanish was frequently heard in all social and business contexts. As with many other schools in the neighborhood and in the district itself, there was concern about students' performance in the high-stakes areas of reading and mathematics. For several years, the school had worked to implement a school-wide curriculum that would develop biliteracy in Spanish and English. However, mathematics had not effectively been included in this process.

Los Rayos was a general adaptation of the *Fifth Dimension* after-school project (e.g., Cole & The Distributed Literacy Consortium, 2006) and related projects such as *La Clase Mágica* (e.g., Vásquez, 1994, 2003) and *Las Redes* (Gutiérrez, Baquedano-López, & Alvarez, 2001). However, *Los Rayos* specifically focused on mathematics. As in *La Clase Mágica* and *Las Redes*, bilingualism and biliteracy were at the center of all activities and thinking in *Los Rayos*. Our goal was to also use this activity system to better understand how best to support linguistically diverse learners in mathematics. Furthermore, like the *Fifth Dimension* and the other after-school projects, *Los Rayos* had a playful atmosphere and orientation; for example, the mathematics was presented as whimsical or humorous short stories and games, and the pressures or anxieties typically found in mathematics classrooms were minimized. Lastly, *Los Rayos* was an activity system (e.g., Engeström, 1999) – discussed in a later section – with multiple layers of complementary interacting elements: the organization of students and other participants, the mathematics activities, the roles of all participants, and the norms and expectations that set direction for the after-school club.

For three years, *Los Rayos* met twice a week, serving approximately 14–20 Latina/o students on any given day. In *Los Rayos,* students were encouraged and supported to be self-directed, creative, interactive, and self-responsible in everything they did. Students made decisions of various kinds such as what particular sets of problems they wanted to work on (e.g., probability, patterns, etc.). They chose whether to work in groups that included other students, university representatives, and/or sometimes parents or they chose sometimes to work one-on-one with a favorite partner; however, they were discouraged from choosing to work alone. Students could even change the direction of activities as they did when they asked to use some of the after-school time to draw and connect this to mathematics.

Through voluntarily responding to an invitation to participate in *Los Rayos*, students implicitly self-identified themselves as "someone interested in mathematics." The program began when the students were in third grade; this report was written two years later when the children were in the 5th grade. Most students continued to participate in *Los Rayos* each year, even though the school offered a variety of after-school activities such as tutoring, guitar lessons, and martial arts. In essence, we came to know most of these Latina/o students for three academic years. Furthermore, a majority of the students were female. Given the current status of mathematics among Latinas/os, we were amazed by all of the foregoing.

One objective of our after-school project was to give students experiences in mathematics that they would not normally get in classrooms for many reasons. Our assumption was that Latina/o students were able to do much more in mathematics than was often expected and that with various mediational tools, especially dialogue with more experienced others (Vygotsky, 1978), they could accomplish a good deal, including traditional skill development. In general, the after-school activities focused on problem-solving and nonremedial mathematics. Many of the mathematics problems were adapted from reform-oriented, upper-level, or high school curricula, such as the *Interactive Mathematics Program* (Fendel, Resek, Alper, & Fraser, 2000). Students solved open-ended problems in probability and algebraic thinking (patterns) and did projects that emphasized rational numbers (i.e., creating recipes) and that explored mathematics in workplaces. Students also created digital stories to "retell" their work on these projects and put forth related, self-devised mathematics problem situations for the audience to solve.

As in the other *Fifth Dimension*-inspired projects, students communicated electronically with a whimsical mathematics wizard, *El Maga* (who lived in cyberspace). At the end of each after-school session, students wrote to *El Maga* about their mathematical experiences for that day, asked any questions they might have about mathematics or *El Maga*, and/or posed their own problems. Such electronic writing was a key mediational tool and was part of creating a natural purpose and context for communication (Cole & The Distributed Literacy Consortium, 2006; Gutiérrez et al., 2001; Vásquez, 2003). However, our purpose for fostering electronic communication between *El Maga* and the students extended to engaging students in writing mathematically, a part of mathematics that is frequently neglected in the classroom (Chval & Khisty, 2009).

In *Los Rayos*, Spanish was privileged in that steps were taken to reinforce its social and cognitive value and to dispel its association with deficits. All materials were in both Spanish and English, which in itself was radically different from most classrooms in the school, district, and country. We also considered it critical for young Latinas/os to interact with "role models" (e.g., Latina/o undergraduate students, faculty, and graduate students) who proudly and capably used two languages for doing mathematics and not for just social communications. Therefore, all undergraduate students (facilitators) were native speakers of Spanish and self-identified as having proficiency in the language. Since most of the other after-school "personnel" spoke Spanish, everyone was encouraged to speak Spanish during the sessions even when students might speak in English. However, the use of two languages was still a common linguistic tool for meaning-making for everyone.

Lastly, the after-school involved multiple kinds of interacting participants: grade school students, Latina/o undergraduate students (many of whom were preservice teachers), graduate student researchers, postdoctoral researchers, university faculty, and frequently parents of the students. Everyone acted as a facilitator for assisting students with comprehending the tasks to be accomplished and doing the mathematics. However, emphasis was placed on encouraging students to be active problem solvers and minimizing "telling" students how to do the mathematics. Also, everyone was involved in some aspect of research; for example, the undergraduate students (who were also referred to as research helpers) and parents took field notes on various aspects of students' involvement in the activities and engaged with other researchers offering their own interpretations of phenomena. Graduate students and faculty were active participants in the activities and with the students and were not detached observers. In some cases, they actually became the object of study.

In essence, all after-school participants had many roles – and roles that were different from their usual ones.

The Intellectual Framework of Los Rayos: The After-School as an Activity System

In this section, we place our work with an after-school program in a broader intellectual framework, one that is both our rationale for its design and implementation and also the lens we use for understanding it. Our work draws from sociocultural-historical theory (e.g., Engeström, 1999; Moll, 1990; Vygotsky, 1978) and, in particular, activity theory. Conceptually, our work rests on the premise that development is a social activity, one where development stems from increased familiarity with cultural-historical mediating tools and artifacts, especially language (Vygotsky, 1978). Here language emphasizes dialogue as a key element in development (Bakhtin, 1981). This perspective shifts us from the view that learning is an individual internal phenomenon to a view that learning is rooted in social interaction. As Holt (2008) explains:

> concepts such as identity exist and persist because of encounters with, and orientations to, the language, manners, and material arrangements of the social world.... We control who we are and create a new identity...through the external control of mediating artifacts.... Our experience of the world is shaped by our existing competence in using objects, itself influenced by the experience of our peers and the accumulated wisdom of previous generations. (p. 55)

From this perspective, all activity is goal-directed (Leont'ev, 1978) and conscious learning emerges from activity. Therefore, to fully understand development, we "...must start by analyzing the development of the child's activity, as this activity is built up in the concrete conditions of its life" (Leont'ev, 1978, p. 395) or the societal nature of an individual's life (Tolman, 1999). Within this framework, activity and activity systems become the focus of attention.

An activity system (see Fig. 16.1) represents the ongoing interplay between various social elements: the relations among subjects (who we refer to as actors), objects of activity (competing or complementary motives or purposes of actions), and the communities they are part of or from. These relations are mediated by norms that are enacted or created, artifacts and tool use, and division of labor (organizational structure) (Engeström, 1999). An activity system is not fixed. While the object remains stable, the activity system can be thought of as an open design that is continuously shaped by shifting collective desires, intentions, and norms (Holt, 2008). Activity is, in essence, an evolving complex structure of mediated and collective human agency (Engeström, 1999). It is here that activity systems recognize the transformative effects of actions, meaning that human activity is creative and able to exceed or transcend given constraints and instructions (Engeström, 1999).

The significance of this is that individuals, each with different knowledge developed through various social contexts, interact through action-oriented relationships

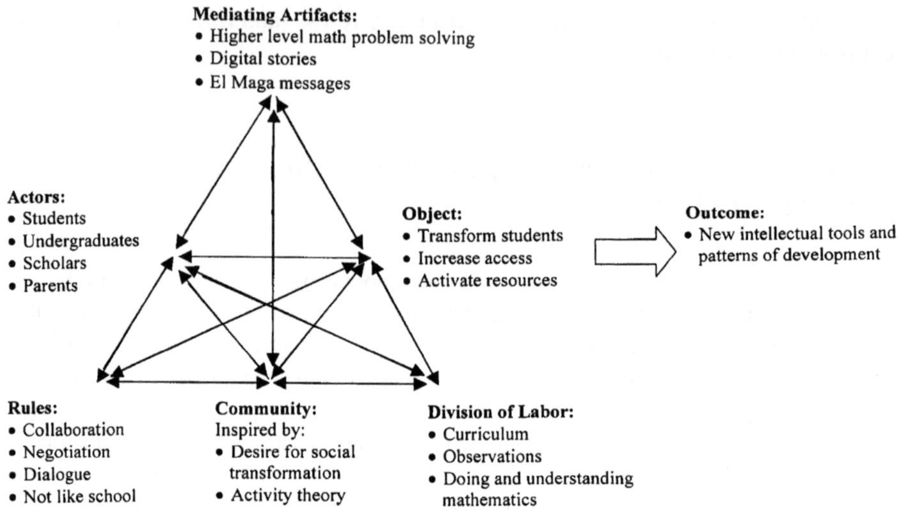

Fig. 16.1 Activity system (Adapted from Engeström, 1999)

of familiarity, challenge, and even anxiety (Becker, 2001). These interactions can be collisions between conflicting or simply differing histories, norms, etc., and as a result can create what Gutiérrez and colleagues (1999) call a hybrid "third space" (see Chap. 15 by Baker et al., this volume). These third spaces can potentially transform competing and differing points of view into rich zones of collaboration and learning. In after-school activities, this becomes highly evident as children bring their own classroom-based perspectives, norms, tools, manners, and histories (i.e., knowledge) into a context that is very different from what they are used to. Hybrid third spaces were evident in that the after-school was playful but not play, was concerned with doing a school subject (mathematics) but not for school purposes, and was in a school but was not school. But, just as the after-school is a third space for children, it is similarly a third space, or rich zone for learning, for all other participants as well.

Lessons Learned from the After-School Club

We now turn to three lessons we have learned from our work in the after-school context. These lessons are significant because they represent pressing insights about Latinas/os' schooling that are often hidden, overlooked, or ignored but which we have come to believe are more critical than is often assumed.

Our discussion is based on extensive ethnographic work carried out during three years of conducting the after-school. Qualitative data were gathered as part of our work to investigate and document the language and cultural resources Latinas/os bring to doing mathematics. Since the after-school club itself served as a generative research methodology, new and varied research questions naturally emerged, and existing data could be used to pursue these questions. The lessons we discuss are a result of reflecting on the patterns or themes that surfaced as we conducted analysis on some of these questions and from our own participant observations.

In addition to gathering observation field notes (by both undergraduate facilitators and university representatives), the various working groups of student participants were videotaped during each after-school session, or twice per week, for approximately 18 weeks of each school year. Each session yielded approximately five sets of video data, one set per working group, along with electronic writing by each student. Additionally, students were periodically interviewed both individually and in focus groups. Data were analyzed using an iterative process drawn from methods of grounded theory (Glaser & Strauss, 1967), whereby patterns or themes emerged from initial holistic viewing of the data, and then analytic categories and themes were refined by revisiting the data. No less than six people collaborated to identify relevant themes (including those presented in this chapter), survey the extensive data set for additional instances, and collectively analyze and interpret the situation to assure accuracy and validity.

Informal Mentors, Role Models, and the Establishment and Nurturing of Networks

Stanton-Salazar and Spina (2003) argue that there are significant differences in the social networks of urban adolescents as compared to those of their middle-class counterparts. As a result of these network differences, the educational, social, and economic outcomes are grossly disparate. We can think of a network as a system of relationships and interactions that convey meaning to youth about their physical, social, or educational realities (Stanton-Salazar & Spina, 2003).

While there likely are plenty of community members who can individually serve as informal mentors or role models, there is, more importantly, the lack of "mechanisms and institutional resources necessary for systematically generating [social] connections for large numbers of youth" in urban communities (Stanton-Salazar & Spina, 2003, p. 250). The examples that Stanton-Salazar and Spina (2003) share illustrate a way in which university students and local community members can create, or model, relationships and interactions that help youth construct meaning of their world. Through this process, a new culture is developed that demonstrates and values a "communitarian" form of coping and goal striving (p. 249). That is, emotions – such as confusion, frustration, failure, and isolation on the one hand, or ambition, passion, and inspiration on the other – do not need to be dealt with individually or internalized. To some extent, one's struggles likely are or have been experienced by another person, and effective networks make this concept clear.

While this sort of modeling is less visible in working-class environments, middle- and upper-class children are surrounded by such networks:

> Middle- and upper-class children do not acquire the attributes and social competencies necessary for success merely through modeling the behavior of parents and other key, yet atomized, figures. Much of what they learn is through participation in a culture of multiple networks, social circles, and meeting places organized on the basis of a particular shared identity, social structures, and ethos. This shared identity, structure, and culture together govern social relations that foster exchange, interdependence, and solidarity oriented toward the maintenance of social privileges, political and economic power, and a particular (legitimating) worldview. (Stanton-Salazar & Spina, 2003, p. 249)

One of the most valuable effects of the after-school program was the establishment of a sustainable social network. This network served many different purposes for the various participants. Most prominently, *Los Rayos* became an emotional and academic net, or system, that supported the holistic development of the elementary students. As we have mentioned, *Los Rayos* was enacted in an urban environment where students often faced harsh social conditions. Therefore, students did not attend each session of *Los Rayos* free from distraction or emotional issues that might weigh on their minds. In fact, there were numerous situations in which students arrived clearly bothered by one issue or another. Each time this occurred, other participants of *Los Rayos* took time to deal with the issue seriously, dedicating themselves to counsel the student and, as a result, noticeably altering the child's disposition from one that was not prepared to engage in mathematical tasks to one that appeared relieved and ready to collaborate with peers.

The most important message from this generalized anecdote is the fact that students increasingly experienced *Los Rayos* as a group of individuals upon which they could rely. They could count on the after-school club for academic development, emotional support, life-options coaching, and fun and engaging activities, among many other things. The best evidence to illustrate this phenomenon is to take notice of the retention rate of *Los Rayos* participants: 14 participants from Year 1 continued to come in Year 3, despite many other after-school sports and activities offered simultaneously. Moreover, there were at least six new students who joined in later years. Undoubtedly, this speaks to the energy and contentment students found within the program activities, but, at least as important, it speaks to the connections and relationships children developed with the older participants (e.g., university representatives and undergraduate students). For example, over two years, a group of four students (who have become friends) grew very fond of their undergraduate facilitator (UG, for short). In a concluding interview, one student made it clear that she thought Maria (the UG) "is the best helper ever." The respect the child developed for Maria over the years was unmistakable, as she finished her final evaluation of the after-school program with these words: "Lov[e] you Maria."

The young students began to inquire about and express their interest in becoming an UG in the future, in essence, becoming a person with education. Interestingly, Maria, the same undergraduate student admired by the younger student, paralleled this interaction by inquiring about and expressing interest in earning a Ph.D. in the future, or following in the "footsteps" of the doctoral students who were part of *Los Rayos*. While we do not suggest that these ambitions would never have occurred without the structure of the after-school club, it does provoke us to wonder about the lasting effects created and sustained as a result of the emergent social network, reinforced by the program participants.

Most likely, the elementary students were the primary beneficiaries of this social network. However, as the last example shows, inevitably all other *Los Rayos* participants were affected by the regular interactions. From our observations of the undergraduate preservice teachers, we noticed that they were experientially discovering the invalidity of the deficit-model view of minority families that dominates popular opinion. Simultaneously, they were acquiring a critical anthropological/ethnographical perspective with respect to the cultural resources available to these learners through the multigenerational interactions with the children's parents. This is precisely what has been advocated for by countless researchers over the past two decades (e.g., Moll, Amanti, Neff, & González, 1992; Stanton-Salazar & Spina, 2003). Conversely, the parents were witnessing firsthand their children's interactions with future, present, and past school personnel as they engaged in rich, meaningful mathematical activities, many of which were rooted in the social context of the surrounding community.

In summary, if Stanton-Salazar and Spina (2003) assert that a lack of established or sustained social networks in segregated urban neighborhoods is one contributor to class and racial exclusion in our society, we suggest that, on a less macrolevel, a lack of social networks is similarly a mediating factor in the underachievement and misguided learning of mathematics for Latina/o students. While we agree with other researchers who have provided evidence that working-class populations surely *do* possess multiple funds of knowledge (e.g., González, Andrade, Civil, & Moll, 2001; Moll et al., 1992) and their children certainly *do* come to school with everyday mathematical knowledge (Ginsburg, 2006; Saxe, 1988), we also note that there are social structures that function as a "sorting machine," allowing some groups advancement through an assortment of opportunities and systematically neglecting to provide other groups opportunities that would position them for upward mobility.

Intuitively, it makes sense that children's academic and personal successes correlate with the support they find in their homes, schools, and communities. Indeed social networks can be found in any neighborhood. However, it is critically important that we concentrate our efforts into transforming existing networks in Latina/o communities into ones that embolden youth to persist in their personal struggles, selecting and executing the life course that aligns with their ambitions. In *Los Rayos,* this message was implicitly reiterated in our activities and actions: what has historically been inaccessible can become accessible.

Language and Issues of Racism

During the three years of working with the children of *Los Rayos*, we observed, through field notes and videotapes, several instances where students expressed, including through unspoken signals, embarrassment that they spoke Spanish, were reluctant to use the language even with older Latinas/os who spoke Spanish with them, and even ridiculed other students because they spoke Spanish. Indeed the sociopolitical context of the country defines Spanish as an "outlawed" language (e.g., antibilingual education legislation in Arizona and California) and historical memories persist regarding corporal punishment of and blatant discrimination against Spanish speakers; therefore, the students had reasons for some of these behaviors and attitudes. Yet the extent to which the third-grade Latinas/os displayed such actions was unanticipated and disturbing. Given the school context that outwardly worked very hard to reinforce Spanish as a valuable social and cognitive tool and that had many Latina/o teachers who taught in Spanish or bilingually, we had assumed students would have much more positive dispositions toward Spanish. We assumed that they would feel linguistic pride in an environment where so many participants spoke Spanish. That some of the students, by their actions, expressed disregard for Spanish, suggested a wider and deeper internalization of denigrating language ideologies that characterize our country. Moreover, since all students and undergraduate students were Latina/o, such negative definition of Spanish pointed to an equally negative self-definition.

The most startling aspect of this internalization of society's racism was evident in the microaggressions (Solórzano, 1998) along language lines that children inflicted on one another. Microaggressions can be characterized as subtle, sometimes seemingly unsubstantial, but continuous verbal or physical signals that denigrate and subjugate a person based on race, language, and other demographic indicators. Microaggressions, because they are not obvious, are difficult to address, yet their impact is clearly felt. In the case of the after-school club, microaggressions took two forms. First, there were many instances when a student would aggressively and righteously "demand" that materials be read and/or discussions be conducted in English, even though all the other group members read or spoke more proficiently in Spanish. In these cases, English took precedence because of its social status, and the needs of Spanish speakers were ignored and made insignificant. For example, López Leiva (2009) documents the interactions and exclusionary practices among four third-grade boys and their undergraduate facilitator, José. In the episode below, the boys are solving a critical thinking puzzle. Rodrigo begins by reading the problem in Spanish:

1. Rodrigo: [reading] que son comunes al círculo y al pentágono pero no en el triángulo o el rectángulo (*that are common to the circle and the pentagon, but not in the triangle or the rectangle*)
2. Alfonso: Do it [read] in English!
3. José: He can do it however he wants. He can do it in English or Spanish.
4. Alfonso: I can't understand him!
5. José: [to Rodrigo] Can you tell him what you mean? Ok, let's take a look at it. Que son comunes al círculo y al…, ¿qué es eso? (That are common to the circle and to the…, what is it?)
6. Alfonso: Read it in English!
7. José: [to Alfonso] I will.

Both Rodrigo and Alfonso were bilingual, but Rodrigo had a higher level of proficiency in Spanish and preferred to speak in Spanish. Alfonso's Spanish proficiency was not as developed as Rodrigo's but he did speak Spanish as evidenced by later observations of him with his parents. However, Alfonso strongly objected to Rodrigo reading in Spanish (lines 2 and 6). Jose attempted to "stand up" for bilingualism (line 3) but eventually gave in (line 7) to Alfonso's demand for English, and the rest of the discussion was done in English. Interestingly, Rodrigo solved the puzzle very easily and quickly, and Alfonso struggled with it – never using Rodrigo as a resource because of his preference for Spanish.

Second, some undergraduate facilitators at times ignored requests from Spanish-speaking students for clarifications of statements made in English or simply spoke in English, again, knowing full well that not all students were proficient in English. In these cases, the UGs did not use their bilingualism to ensure everyone's participation. The impact of such actions is to make invisible and exclude students on the basis of their language. The actions of some of the undergraduate students in this regard is highly disturbing given that all undergraduate students in *Los Rayos* self-identified themselves as bilingual and expressed support for Spanish.

In the after-school context, where students and undergraduates had freedom to decide for themselves how to communicate, or how to use Spanish and English, it became apparent through their actions how much they had internalized anti-Spanish or antibilingual attitudes. In the case of the UGs, it did not even occur to them (initially) that they had a tool to reach all students or that they excluded and marginalized – just as they themselves more than likely had been on different occasions – students who they purportedly wanted to serve.

Unnaturalness of Schooling

A third important lesson that we gained through *Los Rayos* is a deeper recognition that schools truly are unnatural contexts for learning. While this may be an old idea, the reality and depth of it is easily taken for granted. The unnaturalness of schools becomes especially evident through tensions among students that resulted from having to transition from an environment dominated by rules and consequences to an environment marked by choices and freedoms. For third-, fourth-, and fifth-grade students, the concept of self-management is not automatically acquired. And, as we all know, schools tend to operate autocratically, where expectations are outlined, and student behavior is punitively altered as needed by appropriate authorities.

Understandably, schools have enormous responsibilities to function at a level where there is little space for inefficiency or error. Additionally, they are often stretched thin in terms of personnel and resources to optimally educate their students. Therefore, it becomes necessary to create a master schedule and corresponding routines and procedures in order to move the masses through the educational process with as few bumps as possible. However, as a result of our work, we have grown increasingly more aware of the unnatural conditions surrounding students as they are processed through the system and the tremendous negative effects they have on learning. These conditions are not inconsequential; they impact Latinas/os' cognition, identity, and sense of agency (Varley, Willey, & Khisty, 2009).

For example, it is rather outlandish to expect a person, be it adult or child, to sit quietly for hours on end, intensely engaged in developing meaning around particular subject matter. Oftentimes, the opportunities to move around or interact socially with peers are severely limited in the name of proper order and educational achievement (i.e., test scores). While some teachers and administrators are cognizant of this phenomenon and interject regular physical and mental breaks throughout the day, Pelligrini and Homes (2006) have documented how recess and time for play, which was at one time normalized, is increasingly being brushed aside and replaced by more time-on-task, or academic time, in efforts to reach the elusive goals of higher standardized test scores as promoted by the federal government in this age of accountability. As expected, this movement does not come without implication. While few empirical data are available to show the negative effects of reduced "free" time, Pelligrini and Homes (2006) offer a thorough overview of the importance of breaks and unstructured time for social, emotional, and cognitive development. They even go as far as to suggest that regular peer interactions interjected throughout the school day can serve as a predictor for academic achievement, especially among young children. This point takes on even greater significance when considered in light of research on effective instructional practices for bilingual learners (of whom the largest group is Latinas/os) (e.g., Dalton, 1998; Garcia, 1993) that highlights the need for extensive opportunities for students to engage in extended and functional talk around content (Mohan & Slater, 2005). Unfortunately, Latinas/os too often do not have these opportunities in mathematics (e.g., Brenner, 1998; Lipman, 2004).

With this trend in mind, it became clear to the *Los Rayos* design team that an environment must be created that would not be like school because natural instincts within the child could be oppressed or compromised. We also recognized that we could not be completely free of school influence, as the setting would be within the school walls with familiar classmates. However, the rules could be changed drastically. As we alluded to earlier, the transition from the norms of the traditional classroom to the new structures of *Los Rayos* was far from seamless. It was evident that the children had no previous parameters for handling these freedoms and affordances. Students' energy erupted in outlets other than mathematical activity: they socialized, chased one another, effectively avoided anything that resembled school-like tasks, and predictably engaged in whichever other activities presented themselves. However these energetic behaviors did not last long. Meaningful projects and mathematical discussion soon replaced the conversational chatter, and after some time, students began to accept their new roles, and one by one, they became accustomed to the "new" expectations.

More importantly, students began to remind us about what is natural for learning. As one student recalled:

> Instead of – you can't get up on your feet in the normal class, like you have to stay with the person you are working with. You can't go around and check what they're doing to see if you or your answer… to see if whoever you are working with… to see if you got the answer right with another pair. But when you're at the after-school, you can move around and ask them, "Oh, what did you get? Because I got this." And then we look at each other's work, and we see if one of us got it wrong. And, it's kind of better than in class.

The students in the after-school program were experiencing two starkly different mathematics learning environments. This student's comparison of the after-school learning environment to that of her mathematics class is telling. She pointed to the reality that peer resources were often stifled in the name of orderliness, and at the expense of the students. Students, indeed, have a natural tendency to socialize and utilize social interactions to increase understanding or obtain clarification of particular topics – even in mathematics.

Another little girl noted very confidently to us: "What do you expect us [students] to do when we have been sitting all day?" Students' reactions to our sometimes less than "not-like-school" curriculum attempts also challenged us to rethink the role and nature of project-based learning or activities that were more owned by the students. We found that typical "control" of students had the effect of silencing them in ways that ultimately excluded them from acquiring dialogic tools for learning. Also, it was very difficult during the first two years of our project to engage students in communication (verbal, drawing, modeling) about their mathematical thinking. They had become accustomed to sitting silently in classrooms, and they had come to assume that offering extended talk was a way for "adults to show how a person is dumb." Another little girl angrily responded to a university representative's question about her mathematical process with this statement: "All you do is ask me questions. You just think I'm dumb...."

The point here is that current reforms in mathematics practices emphasize active engagement, communication of reasoning (NCTM, 2000), and ultimately, proficiency in mathematical talk. What we learned was that, in an environment like *Los Rayos* where students have the freedom to talk, they demonstrate that they cannot do it. Their actions suggest that they have been socialized to view talking mathematically negatively and to resist it. Altogether, this does not bode well for their classroom participation and learning – unless students have alternative contexts such as *Los Rayos* to develop these skills.

Moreover, we consider shifts in their active engagement and communication as evidence that students are developing mathematical agency and that a more natural, autonomously oriented environment leads to stronger student agency and affiliation with mathematics. Hull and Greeno's work (2006; as cited in Cobb, Gresalfi, & Hodge, 2009) points out that the normative identity established in a particular setting is the result of the distribution of authority and the ways that students exercise agency. Through both deliberate and evolving design principles, *Los Rayos* adult participants created an environment that established these norms: students selected and even created the mathematical tasks they were to complete, they were the authority with respect to the problem situations they developed, and they solved problems with whichever method that made sense to them, exercising what Pickering (1995) refers to as conceptual agency.

A new community of practice (Wenger, 1998) emerged, and students gradually adopted the new practices and ways of being. Informal observations of the after-school participants in their regular classrooms suggested that they had appropriated a newfound identity and corresponding mathematical agency: they took risks, they raised their hands more often than others to answer questions, they offered longer responses, and they used more tools (e.g., verbal communication, representations, peer resources, etc.). We are increasingly convinced that an acute focus on activity, social interactions, and agency – features often absent from classroom environments serving working-class students – can empower Latina/o students and challenge their persistently low academic status.

We acknowledge that relinquishing some of the control of the children's mathematics learning is not easy to do. As we pointed out, the implementation of a radically different learning environment comes at a temporary expense of chaos. We are well aware that with the first taste of this disorder, it is tempting to revert to the familiar practices that have "worked" in the past: textbook-like problems, large-group learning, direct instruction, a teacher-centered orientation, and rules of silent behavior. However, our point has nothing to do with advocating anything close to instructional anarchy. Instead, observations from *Los Rayos* point to how typical "unnatural" classroom environments – with their direct or implicit restrictions and inhibitions of activity, talk, and agency – have detrimental social and cognitive consequences for Latinas/os. We view these consequences as part of their subjugation.

Implications for Practice and Concluding Remarks

The lessons we have highlighted from the alternative after-school learning environment of *Los Rayos* have significant implications for teaching practices. In considering preservice teachers, the learning environment of *Los Rayos* is significantly different from what they will likely observe in their student-teaching classroom. There is much to be learned in this alternative learning context: the importance of networks, the critical relevance of bilingualism and biliteracy and insidiousness of inflicting microaggressions on students on the basis of language, and different ways of conceptualizing and implementing learning contexts. However, the design issues we have highlighted here need to be made transparent to preservice and in-service teachers alike, in part because they have likely experienced many of the oppressive learning conditions we described, and in part because it is innovative instruction and dynamic learning environments that will change the educational outcomes of minority students currently stuck in a perpetual state of underachievement.

Our purpose has been to highlight the potential of an after-school environment as a context for better understanding how to improve the educational status of Latinas/os. We described the efforts that went in to creating a context that challenged conventional wisdom about language, social practices, and schooling practices with Latinas/os. It was not our purpose to demonstrate student learning, although there is evidence of student transformation. Our emphasis remains on the idea that context matters and profoundly matters for subjugated "minority" students.

Based on our data and the lessons we have learned, we cannot subscribe to the notion that minority families insufficiently provide their children the resources they need to succeed in school or that the problem of learning resides anywhere with the general population of minority and/or poor children. *Los Rayos*, and other programs like it, provides evidence that Latina/o children can productively and successfully engage with sophisticated mathematics activities and learning given different instructional contexts (see Díaz, Moll, & Mehan, 1986). On the contrary, we maintain that schools do not have the mechanisms in place to capitalize on the social, linguistic, and cultural capital that non-White students bring to the classrooms. This is not suggesting that schools are malintended institutions working conspiratorially to neglect to provide Latinas/os and African Americans high-quality educational opportunities. Rather, we concur with many in the research community that schools and teachers are not equipped to counter dominant ideologies that position minorities as culturally deficient, nor are they equipped to utilize the funds of knowledge (Moll et al., 1992) that children bring to class from their lives, their homes, and their communities. An after-school context is one vehicle for reducing the possibility that the learning resources and assets of Latinas/os continue to go unrecognized, underutilized, and unaccepted and that the conditions of their schooling go unnoticed and unchallenged.

Acknowledgments This chapter describes work conducted as part of CEMELA, supported in part by the National Science Foundation (Grant ESI-0424983). Any opinions, findings, and conclusions or recommendations expressed in this material are those of the author and do not necessarily reflect the views of the National Science Foundation.

References

Bakhtin, M. (1981). *The dialogic imagination*. Austin, TX: University of Texas Press.

Becker, M. (2001). Managing dispersed knowledge: Organizational problems, managerial strategies, and their effectiveness. *Journal of Management Studies, 38*(7), 1037–1051.

Brenner, M. E. (1998). Development of mathematical communication in algebra problem solving groups: Focus on language minority students. *Bilingual Research Journal, 22*(3–4), 149–174.

Chval, K. B., & Khisty, L. L. (2009). Bilingual Latino students, writing and mathematics: A case study of successful teaching and learning. In R. Barwell (Ed.), *Multilingualism in mathematics classrooms: Global perspectives* (pp. 128–144). Bristol, UK: Multilingual Matters.

Cobb, P., Gresalfi, M., & Hodge, L. L. (2009). An interpretive scheme for analyzing the identities that students develop in mathematics classrooms. *Journal for Research in Mathematics Education, 40*(1), 40–68.

Cole, M. & The Distributed Literacy Consortium. (2006). *The fifth dimension: An after-school program built on diversity*. New York: Russell Sage.

Dalton, S. S. (1998). *Pedagogy matters: Standards for effective teaching practice*. Santa Cruz, CA: Center for Research on Education, Diversity & Excellence.

Díaz, S., Moll, L. C., & Mehan, H. (1986). Sociocultural resources in instruction: A context-specific approach. In California State Department of Education (Ed.), *Beyond language: Social and cultural factors in schooling language minority students* (pp. 187–230). Los Angeles: California State University, Los Angeles.

Engeström, Y. (1999). Activity theory and individual and social transformation. In Y. Engeström, R. Miettinen, & R.-L. Punanmäki (Eds.), *Perspectives on activity theory* (pp. 19–38). New York: Cambridge University Press.

Fendel, D., Resek, D., Alper, L., & Fraser, S. (2000). *Interactive mathematics program*. Emeryville, CA: Key Curriculum Press.

Garcia, E. E. (1993). Language, culture, and education. In L. Darling-Hammond (Ed.), *Review of research in education* (pp. 51–98). Washington, DC: American Education Research Association.

Ginsburg, H. P. (2006). Mathematical play and playful mathematics: A guide for early education. In D. G. Singer, R. Michnick Golinkoff, & K. Hirsh-Pasek (Eds.), *Play = Learning: How play motivates and enhances children's cognitive and social-emotional growth*. Oxford, UK/New York: Oxford University Press.

Glaser, B. G., & Strauss, A. L. (1967). *The discovery of grounded theory: Strategies for qualitative research*. Hawthorne, NY: Aldine de Gruyter.

González, N., Andrade, R., Civil, M., & Moll, L. (2001). Bridging funds of distributed knowledge: Creating zones of practices in mathematics. *Journal of Education for Students Placed at Risk, 6*(1&2), 115–132.

Gutiérrez, K. D., Baquedano-López, P., & Alvarez, H. (2001). Literacy as hybridity: Moving beyond bilingualism in urban classrooms. In M. de la Luz Reyes & J. Halcón (Eds.), *The best for our children: Critical perspectives on literacy for Latino students* (pp. 122–141). New York: Teachers College Press.

Gutiérrez, K. D., Baquedano-López, P., & Tejeda, C. (1999). Rethinking diversity: Hybridity and hybrid language practices in the third space. *Mind, Culture, and Activity, 6*(4), 286–303.

Gutstein, E. (2005). *Reading and writing the world with mathematics: Toward a pedagogy for social justice*. New York: Routledge.

Gutstein, E., Lipman, P., Hernández, P., & de los Reyes, R. (1997). Culturally relevant mathematics teaching in a Mexican American context. *Journal for Research in Mathematics Education, 28*(6), 709–737.

Holt, R. (2008). Using activity theory to understand entrepreneurial opportunity. *Mind, Culture, and Activity, 15*, 52–70.

Hull, G., & Greeno, J. G. (2006). Identity and agency in non-school and school worlds. In Z. Beckerman, N. C. Burbules, & D. S. Keller (Eds.), *Learning in places: The informal education reader* (pp. 77–97). New York: Peter Lang.

Khisty, L. L., & Chval, K. (2002). Pedagogic discourse: Its role in equity and mathematics learning. *Mathematics Education Research Journal, 14*(3), 4–18.

Khisty, L. L. (2004). "Los Rayos de CEMELA" after-school project: The UIC CEMELA Activity. Unpublished manuscript. University of Illinois at Chicago.

Khisty, L. L. (1995). Making inequality: Issues of language and meanings in mathematics teaching with Hispanic students. In W. Secada, E. Fennema, & L. Adajian (Eds.), *New Directions for Equity in Mathematics Education* (pp. 279–297). Cambridge University Press.

Leont'ev, A. N. (1978). *Activity, consciousness, and personality*. Englewood Cliffs, NJ: Prentice Hall.

Lipman, P. (2004). *High stakes education: inequality, globalization, and urban school reform*. New York: Routledge Falmer.

López Leiva, C. (2009). *Mathematical and bilingual identities in after-school settings*. Unpublished manuscript, University of Illinois at Chicago.

Martin, D. (2006). Mathematics learning and participation as racialized forms of experience: African American parents speak on the struggle for mathematics literacy. *Mathematical Thinking and Learning, 8*(3), 197–229.

Mercado, C. I., & Santamaría, L. J. (2005). A new vision for Latino/a education: A comparative perspective on research agendas. In P. Pedraza & M. Rivera (Eds.), *Latino education: An agenda for community action research* (pp. 11–43). Mahwah, NJ: Lawrence Erlbaum Associates.

Mohan, B., & Slater, T. (2005). A functional perspective on the critical "theory/practice" relation in teaching language and science. *Linguistics and Education, 16*, 151–172.

Moll, L. (1990). *Vygotsky and education: Instructional implications and applications of sociohistorical psychology*. New York: Cambridge University Press.

Moll, L., Amanti, C., Neff, D., & González, N. (1992). Funds of knowledge for teaching: A qualitative approach to developing strategic connections between homes and schools. *Theory into Practice, 31*, 132–141.

National Council of Teachers of Mathematics (NCTM). (2000). *Principles and standards for school mathematics*. Reston, VA: NCTM.

Pelligrini, A. D., & Homes, R. M. (2006). The role of recess in primary school. In D. G. Singer, R. Michnick Golinkoff, & K. Hirsh-Pasek (Eds.), *Play = Learning: How play motivates and enhances children's cognitive and social-emotional growth*. Oxford, UK/New York, NY: Oxford University Press.

Pickering, A. (1995). *The mangle of practice*. Chicago: University of Chicago Press.

Saxe, G. B. (1988). Candy selling and math learning. *Educational Researcher, 17*(6), 14–21.

Sol6rzano, D. (1998). Critical race theory, racial and gender microaggressions, and the experiences of Chicana and Chicano scholars. *International Journal of Qualitative Studies in Education, 11*, 121–136.

Stanton-Salazar, R. D., & Spina, S. U. (2003). Informal mentors and role models in the lives of urban Mexican-origin adolescents. *Anthropology and Education Quarterly, 34*(3), 231–254.

Tolman, C. W. (1999). Society versus context in individual development: Does theory make a difference? In Y. Engeström, R. Miettinen, & R.-L. Punanmäki (Eds.), *Perspectives on activity theory* (pp. 70–86). New York: Cambridge University Press.

Varley, M., Willey, C., & Khisty, L. L. (2009, April). *Latina/o student beliefs about language, mathematics, and agency*. Paper presented at the annual meeting of the American Educational Research Association, San Diego, CA.

Vásquez, O. A. (1994). The magic of *La Clase Mágica*. *Australian Journal of Language and Literacy, 17*(2), 120–128.

Vásquez, O. A. (2003). *La clase mágica: Imagining optimal possibilities in a bilingual community*. Mahwah, NJ: Lawrence Erlbaum Associates.

Vygotsky, L. S. (1978). *Mind in society: The development of higher psychological processes*. Cambridge, MA: Harvard University Press.

Wenger, E. (1998). *Communities of practice: Learning, meaning, and identity*. New York: Cambridge University Press.

Chapter 17
Teacher Development in After-School Mathematics Contexts: Insights from Projects that Capitalize on Latinas/os' Linguistic and Cultural Resources

Eugenia Vomvoridi-Ivanović, Maura Varley Gutiérrez, Gabriel Viego, Ksenija Simić-Muller, and Lena Licón Khisty

Introduction

The preparation of teachers for diverse populations has been the subject of a growing body of research and discussion over the last two decades (e.g., Cochran-Smith, Fieman-Nemser, McIntyre, & Demers, 2008). However, as it stands now, there is not a great number of teachers who are adequately and appropriately prepared with the skills and knowledge to teach diverse students (Darling-Hammond, 2006). Since achievement in mathematics is highly dependent on teachers' capabilities (e.g. Khisty, 2002; NCES, 2001), the underpreparedness of teachers does not bode well for Latinas/os and other nondominant students receiving the support they need to perform well in mathematics (Gutiérrez, 2002). While research has pointed to the importance of linguistically responsive learning environments for Latinos/as in mathematics (e.g. Khisty, 1995; Moschkovich, 1999a, 1999b) and to practices teachers can use to facilitate bilingual students' content learning (Khisty & Viego, 1999), the question remains about how to prepare and support teachers in creating such learning environments.

E. Vomvoridi-Ivanović (✉)
Department of Secondary Education, University of South Florida, Tampa, FL, USA
e-mail: eugeniav@usf.edu

M.V. Gutiérrez
Elsie Whitlow Stokes Community Freedom Public Charter School, Tucson, AZ, USA

G. Viego • L.L. Khisty
Department of Curriculum and Instruction, College of Education,
University of Illinois at Chicago, Chicago, IL, USA

K. Simić-Muller
Department of Mathematics, Pacific Lutheran University, Tacoma, WA, USA

B. Bevan et al. (eds.), *LOST Opportunities: Learning in Out-of-School Time*,
Explorations of Educational Purpose 23, DOI 10.1007/978-94-007-4304-5_17,

Critics have suggested that teacher education programs have not done an adequate job in preparing teachers to teach diverse populations (Ladson-Billings, 1999; Zeichner & Hoeft, 1996). In response to these challenges, greater emphasis has been given to redesigning field experiences. While these experiences have positively impacted teacher candidates' awareness and acceptance of students from culturally diverse backgrounds (Valdés, Bunch, Snow, Lee, & Matos, 2005), what is often missing in these experiences is the explicit focus on the nature and role of language in teaching subject matter. In addition, one aspect of diversity in teacher education that has received little attention is the preparation of teachers to teach English language learners (ELLs) (Zeichner, 2005). This is particularly evident in the domain of mathematics teacher education (Khisty, 2002). The issue, therefore, becomes how to develop mathematics teachers who can create effective learning environments for Latinas/os who are also ELLs. Moreover, what kinds of experiences do teachers need in order to help them develop appropriate understandings and skills for integrating home language (Spanish) and mathematics and to bridge the theory they learn in their courses to practice in order to create effective learning environments for Latinas/os?

This chapter seeks to address these issues. Specifically, the purpose of this chapter is to describe insights about the development of teachers that were gained from implementing after-school projects for elementary grade Latinas/os in mathematics. We begin our discussion by describing two after-school mathematics projects that served as nontraditional field experiences for teachers. We then highlight three patterns that emerged that are relevant to the development of teachers for Latinas/os in mathematics. We close our discussion with some concluding thoughts related to the implications of these insights.

Background

The work discussed comes from two after-school projects created by the Center for the Mathematics Education of Latinas/os (CEMELA) which is concerned with issues, research, and practice related to the teaching and learning of mathematics with Latinas/os. CEMELA has developed two after-school mathematics projects, which are located at two different geographical sites in the US: *Los Rayos de CEMELA* is located in a large Midwestern district and *After-school Math Club* is located in a large Southwestern district, near the Mexican border. Both after-school projects are housed in elementary schools that predominantly serve working-class students of Mexican descent (see Chap. 16 by Khisty & Willey, this volume, for more details).

The after-school projects involved teachers in two ways: first, as facilitators who worked with groups of third to fifth grade students twice a week for a school year, and second, as "coresearchers" with the university after-school research team (faculty, postdocs, and graduate students). The facilitators were primarily undergraduate students, many of whom were recruited from teacher education courses, whom we will refer to as *novice teachers*. The coresearchers were a group of in-service teachers who worked with us during the summer in an experimental form of professional development in mathematics.

The CEMELA after-school mathematics projects are loosely modeled on the work of the *Fifth Dimension* (e.g., Cole and the Distributed Literacy Consortium, 2006) and other similar after-school projects (e.g. Gutiérrez, Baquedano-López, Alvarez, & Chiu, 1999; Vásquez, 2003). They are designed to engage novice teachers as facilitators of mathematical activities with Latina/o elementary students and to meet weekly with graduate fellows and postdocs for a debriefing and planning session. Together, these activities make up our "experimental practicum."

This work assumes that learning at any age occurs in a social context (Vygotsky, 1978) that emphasizes active dialogue among participants (Wells, 2001). Furthermore, we assume that what is known by an individual is the outcome of continuing co-construction processes that depend on multiple opportunities to encounter and make sense of challenging new experiences (Wells, 2001). The after-school projects serve as nontraditional field experiences for preservice teachers as they participate in a unique context where they must form interpersonal relationships with students, negotiate mathematical ideas, and engage in dialogue across two national languages.

This field experience goes beyond preservice teachers' traditional or formal university training that emphasizes what it means to implement culturally relevant pedagogy. Many of the after-school mathematical activities are developed around students' experiences and interests. Moreover, the novice teachers are not expected to teach or tutor in the traditional sense. Instead they are expected to interact with the children and engage in various activities with them. They are afforded the opportunity to experiment and use a multiplicity of resources with the students while doing mathematical activities, without the constraints of a student teaching experience where there is a set curriculum to be taught and where large groups of students are supposed to meet certain benchmarks.

Instead, novice or preservice teachers interact with small groups of Latina/o students in a relaxed environment, where they can focus on how students learn mathematics through interacting with them. Involvement in the after-school gives facilitators the opportunity to examine children working on nonroutine mathematics in a setting that capitalizes on their language and cultural resources. In addition to this, the weekly debriefing meetings, where the novice teachers reflect on their experiences in the after-school project, provide spaces for them to make sense of their own and their students' mathematical behaviors as related to issues of language, culture (defined here as social practices), and identity, among others.

We also made teacher development an integral component of our work at one of the after-school projects that was held after hours in a school that was striving for biliteracy in Spanish and English (Khisty & Viego, 2007). Our approach to this activity is rooted in the early work of Díaz, Moll, and Mehan (1986) where they had a teacher view videos of her Latina/o students in a different context with the result of unsettling the teacher's perceptions of her students and ultimately fostering positive redefinitions of students' abilities. Consistent with this work, we invited the teachers from the after-school site to work with us in a two-week teacher research seminar as coresearchers analyzing various kinds of data we had gathered in the after-school. Through the careful examination of activities conducted on the neutral ground of the after-school setting, and using a theory-based, dynamic system of cooperative alternatives, teachers had the opportunity to consider (a) how students did mathematics, (b) what language and social practices they used as resources to do the mathematics, (c) how dialogue and problem solving contributed to students' participation in the subject, and (d) how their insights into student activities connected to their own practices.

Methods

Our discussion is based on extensive ethnographic data gathered over the three years of implementing the after-school projects. During this time, approximately 32 novice teachers participated in both sites. Most of the facilitators were Spanish-English bilinguals and many were Latinas/os whose first language was Spanish. The majority of the novice teachers were female and about a third of all novice teachers were preservice teachers at the time they were involved in the after-school. Finally, most of the novice teachers were involved in the after-school for a semester, while a few remained for an entire academic year or longer. We collected video data of the facilitators interacting with the students as they did mathematics and other activities during every session of the after-school. We also videotaped debriefing sessions where facilitators discussed their insights and concerns about students, mathematics, teaching, and the valuable tensions brought up by a nontraditional learning environment. Along with this, we collected the facilitators' field notes. Similarly, we videotaped the full-time two-week professional development sessions held for 12 teachers who viewed, analyzed, and discussed selected after-school video clips. It is from this data – and our own observations – that our insights emerged. We used qualitative methods (Lincoln & Guba, 1985) and discourse analysis to examine dialogic and other relevant patterns regarding teachers and preservice teachers that emerged and related to how they developed through their involvement in the after-school projects.

Insights

Data analysis reveals three patterns of teachers' insights on the teaching and learning of mathematics to Latinas/os:

1. Novice teachers gained insights regarding the role of language in mathematics.
2. Novice teachers' beliefs about the teaching of mathematics shifted toward centering learning in students' lives and experiences.
3. In-service teachers gained critical insight into the socialization processes of schooling.

The Role of Language in Mathematics

The novice teachers often commented and reflected on their own and the students' uses of language when doing mathematical activities. As a result, they gained several insights regarding the role of natural language in mathematics. One of the most interesting insights they gained relates to the different aspects of proficiency in Spanish. We chose to discuss the insights of a group of novice teachers who were elementary preservice teachers and who had identified themselves as Latinas/os and as having native-like proficiency in Spanish. In fact, all of them noted that their first language was Spanish, that they were born in Mexico, and that they were raised in Spanish language-dominant households. Some of them described themselves as having been in a bilingual school program and/or preparing to receive a bilingual teaching endorsement. However, doing mathematics in Spanish, facilitating mathematical discussions in Spanish, and assisting students in mathematical activities in Spanish proved to be a great challenge for most of them.

During one debriefing meeting, the novice teachers were asked to do the mathematical tasks they were going to use with the children in the after-school. All activities were provided in both Spanish and English, in the same format that students would receive them. Julie[1] chose to use the Spanish version of the activities and informally teamed up with two other novice teachers, Diana and Sonia, who also chose the Spanish version. The following describes Julie's experience with the Spanish materials and gives a sense of some of the issues that came to the fore.

By observing Julie's gestures, facial expressions, and remarks, it became apparent that Julie had difficulties in comprehending some aspects of the Spanish version of the activity. Julie looked puzzled every time she read the Spanish version and kept going back to the English version of the activities. Even though she discussed the context of the activities in Spanish she would resort to English when talking about mathematics. Later on, she reflected on her difficulty in understanding mathematical Spanish:

> say your first language is Spanish but you know English, if they give you a word problem and it's in English you would be able to do it but if they give it to you in Spanish you won't be able to do it…I learned English math so if they give me a word problem like they did like today and I read the Spanish one I had to reverse and I had – I know how to read Spanish and I understand it but I comprehend math a lot better if I read it in English

Julie was seeking bilingual endorsement at the time of the study and had read in her bilingual education course that conversational fluency in one cultural language does not translate to academic fluency in the same cultural language (Cummins, 1981). However, it was not until she experienced trying to comprehend, discuss, and solve a mathematical problem in Spanish that she fully realized this. Julie had a hard time comprehending the mathematics in the problem she was given and had to resort to the English version of the problem because, as she explained, she had learned

[1] All names have been changed.

mathematics in English. She started to realize that the reason she was not able to think about mathematics in Spanish was because she had never been taught mathematics in that language. In other words, Julie was not familiar with Spanish mathematical discourse and this resulted in her difficulty comprehending the Spanish mathematical text, even though the problem involved elementary mathematics in which she was competent.

Julie expressed many times during the debriefing meetings that it was necessary for her to read the English version of the activities and then solve the activities in English before attempting to assist students in Spanish. Deborah and Rex shared the same experience with Julie.

Julie: *I feel like if I could read an English worksheet then I could translate it in Spanish but if I get a Spanish worksheet and I read it then I'll take a lot longer to understand it…cause I could translate it and I could teach it to them in – well not teach – in Spanish.*
Deborah: *Me too. Exactly.*
Rex: *Yeah, the same way.*

Most novice teachers who participated in the after-school during another semester faced similar challenges as the ones discussed previously in doing mathematics, facilitating mathematical discussions, and assisting students in mathematical activities in Spanish. Like Julie, Deborah, and Rex, other novice teachers expressed that when trying to solve a mathematical problem they would have to think about how to solve it in English and then translate their solution into Spanish. They attributed their inability to think about mathematics in Spanish to the fact that they had never been taught mathematics in Spanish. All novice teachers said that not knowing mathematical terminology in Spanish prevented them from using Spanish when doing mathematics with the students in the after-school.

This example shows the difficulty that teachers like the ones described above, who are Spanish bilingual but not biliterate, face in thinking and talking about mathematics in Spanish. It is often taken for granted that since a teacher is Latina/o and is fluent in both Spanish and English, that teacher will be able to facilitate discussions about mathematics in both languages with ease. However, the data suggests that this is not the case. The Latina/o bilingual novice teachers who participated in the after-school projects faced challenges in doing so. What is significant here is that the context in which they came to the realization that conversational fluency in Spanish does not translate to mathematical fluency in Spanish was not their teacher preparation program but the after-school program where for the first time they attempted to communicate mathematically in Spanish. Even though the preservice teachers had been taught in their bilingual education courses that conversational fluency and academic fluency are different kinds of fluencies, they did not expect to run into any difficulties when trying to do mathematics in Spanish. This may be due to the fact that they did not have the opportunity to explore these different kinds of fluencies in their home language through their teacher preparation program. As they attempted, however, to hold conversations about mathematics in Spanish in the context of the after-school, they realized that even though they were fluent in conversational Spanish, they were not fluent in mathematical Spanish.

A Shift in Teachers' Approaches to Pedagogy

As noted earlier, a few of the novice teachers were involved in the after-school for an extended period of time. Due to these novice teachers' increased participation in the after-school, we were able to observe changes in their approaches to pedagogy. Specifically, two novice teachers who were involved in the after-school for an entire academic year gained great insights regarding mathematics pedagogy and the importance of relationships with students. Gloria, a mathematics major from Mexico who was Spanish dominant, and Jill, an English-dominant Spanish-proficient, Anglo elementary preservice teacher, both experienced a shift in their views about the teaching of mathematics from a didactic approach to a more student-centered activity. Due to the interactive and community-based nature of the mathematics learning in the after-school setting, both Gloria and Jill formed very close relationships with the students and described the role of these relationships as being imperative to effective learning environments as well as significant in their shifts in thinking. The unique learning environment afforded these teachers the opportunity to develop relationships and engage in dialogue with the students, and they came to appreciate their students' experiences as a basis for mathematics learning.

Both Gloria and Jill's expectations in the beginning, like those of the other novice teachers, were that the after-school would be more like school. Gloria described her own mathematics schooling as traditional and commented that she had enjoyed the structure that it provided. For this reason, it was all the more impressive to observe her adopt a completely different approach in Math Club and with much success. In the beginning, she said, "I tended to give the kids the answer, like I didn't let them think about it. I tried to guide them. [Now,] I feel kind of good. I am more conscious about it and try to step back and give them clues and let them think first." This comment reflects a shift in her thinking from one of knowledge being transmitted to a more sociocultural approach to learning as a social activity.

Just like Gloria, Jill placed great emphasis on relationships with students. Yet, when talking about her initial expectations from after-school, she said that she hadn't expected to get to know the kids as much because of the traditional role of a teacher as expert: "Really, I was expecting more to have – you know like a leader, more of a teacher role and not really getting to kind of personally know the kids and so that has been something that I didn't expect to happen." She shifted her thinking in terms of the relationship between students and teacher being one that is not personal, possibly because of the "leader" or expert role of a teacher, to one in which these close relationship are central to a learning environment.

The after-school was a significant experience for Jill also, in terms of her attitude about how mathematics should be taught, which differed from the practicum experience she was having in a traditional classroom. During the time that she was a facilitator in the after-school, she was also observing a second-grade class regularly. She remarked that the teacher she was observing used very standard teaching methods – which she described as few projects and many worksheets. This frustrated her, since through her after-school experience she had started to believe that the entire mathematics curriculum could be based on projects and hands-on activities, most of which were related to students' everyday lives and experiences (such as posing and solving problems related to immigration during the immigration rallies that took place in the spring of 2006). When this second-grade teacher asked her to teach his mathematics class one day, she created a fractions lesson that included reading a book and hands-on activities related to problem solving based on students' experiences. She said that the after-school experience had influenced her greatly and that without it she never would have been motivated to create a mathematical activity like this one. This development speaks to the importance of exposure to nontraditional mathematics learning environments in supporting teachers' development of effective and engaging practices.

Jill was convinced that mathematics should be taught in connection with other subjects and with the students' lived experiences:

> *And also the connection of math to everything else and that just kind of as a general idea in teaching, keeping cross subject connections. I mean we [in the Math Club] don't get just plain arithmetic or just plain math you know, we incorporate it with what's going on around them in their community.*

Jill was particularly involved in bringing in literature-based mathematics projects as well as community-based activities such as community math hunts using photography. Students toured their community and took photographs of what they believed represented mathematics. Then they posed and solved problems related to these photographs, such as community murals. Jill commented on the importance of this project in getting to know the students' community, which she now sees as a central aspect of teaching mathematics as reflected in the above quote.

The insights articulated by both Gloria and Jill point to the power of nontraditional field experiences for novice teachers. Although Gloria was not a preservice teacher, her experience still sheds light on important issues, such as shifting perspectives on mathematics learning, that are important to consider for preservice teacher preparation. Particularly, coming from a math major that described her own schooling experience as traditional, it is interesting to see how this experience in a unique after-school learning environment that centers on student experience and fostering a community of learners has influenced her perspective on mathematics learning. Gloria described a shift in her own approach to the students' mathematical learning from one of transmission to a more collaborative and exploratory-based experience.

Jill also described her dedication to teaching mathematics through projects and hands-on experiences as a result of this experience. For Jill in particular, it became imperative to teach using projects, especially if based on students' experiences, rather than through lecturing and worksheets, as she described she saw in a classroom where she was currently doing observations.

Insights into Effects of the Socialization of Schooling

One of the insights that emerged from teachers and that was discussed repeatedly as they viewed videos of "their" Latina/o students was a concern about what the school was teaching students. The teachers assumed that students were developing as critical thinkers who could explain their thought processes and who could express themselves in two languages. However, as they analyzed students' actions and talk in the after-school where children were freer to act and talk as they wished, teachers found evidence that students actually were hesitant to offer their own points of view or solutions to problems, and too often simply just settled for "the" answer. They found that students had difficulty in expressing their mathematical thinking. Overall, they were concerned about the ways that the school socialized children into being passive learners – a characteristic that may be desirable for some teachers but not for students' learning, as most of the teachers noted. One of the teachers captured this concern in the excerpt below; she wondered what her experiences meant in terms of beliefs that teachers have about how children learn:

Annie: I was thinking a lot about teachers and seeing Maria and Sandra [two students in a video] and how if they're both in third grade...what behaviors of theirs has been praised and have been ridiculed. You know, Maria is an ideal student for many teachers. And, you know...sit in your seat, follow directions...what does that say for us and what kind of behaviors are we fostering in our school....I just think about that and...she's quiet, she follows directions.

In another session, Annie returned to this topic but connected it to her own experiences as a learner.

Annie: You know those are habits too, your learning habits, your study habits. When I went...to the writing conference, we were talking about the teachers modeling the writing, reading, writing, modeling it aloud and saying, "Oh, that doesn't make sense." ... And thinking about that and showing the kids how you can think about things and constantly rereading what you're writing, and it's the same thing [in mathematics].

Through the videos from the after-school, teachers saw students in a less constrained environment than they normally did and were confronted with the idea that in classrooms children were being socialized to be passive learners. This idea was eventually connected to whether the school really valued children's thinking or any learning that involved exploration, wondering, and asking questions. Otherwise, they felt their students would display such characteristics.

Also as the teachers watched videos of "their" students interacting either proficiently or poorly in Spanish and/or English in mathematics, they grappled – as many of the novice teachers similarly did in their discussions – with a common issue among many teachers of language minority students: How can bilingual students be appropriately assessed on their content understandings when their proficiency in their second language may confound their expression of this knowledge? One teacher, Connie, summed it up thusly:

I have a wonderment in how our children express themselves, and how it's not really laid out the difference between explanations in English versus explanations in Spanish. The ease may come in English or the ease may come in Spanish. But we've really not gotten side-by-side transcripts nor side-by-side children. We've looked at one student Carla... and we see how she's thinking in English when she has [to speak] in English, or when she's in a...group she has Spanish [with]...another student who seems to excel in Spanish but we don't [ordinarily] have all that evidence.

Connie wondered how children expressed themselves in two languages. She recognized that it was impossible to tell what a student knew unless a teacher could see the same student talking about the same thing first in one language and then in the other ("side-by-side transcripts"). Unless a teacher had such "evidence" for a student, it would be difficult to know what the student's abilities or understandings really were. More importantly Connie pointed out that what we might observe also depended heavily on who the student was interacting with and the nature of the other person's language strengths. In typical school contexts, teachers might not recognize these critical aspects of learning in two languages (Khisty, 2006; Khisty & Morales, 2004).

Conclusions

In this chapter we have aimed to describe various insights that teachers gained through their involvement in the CEMELA after-school mathematics projects. In our discussion, we identified three themes or patterns that emerged from our interactions and observations with teachers.

First, teachers came to a deeper and clearer realization of the different aspects of proficiency in Spanish or any language. Even though many of them had encountered this idea in classes, they had not fully realized the extent of the differences in proficiencies there were between social and academic aspects of language or among subject areas. Their "face-to-face" collisions with speaking mathematically and the tensions that arose to resolve these conflicts helped bring home the need for Latina/o students to have more experiences in learning mathematics in Spanish. Teachers realized that there were challenges to doing mathematics and facilitating mathematical discussions in Spanish, something they had overlooked because they had not genuinely experienced speaking mathematically in Spanish. Teachers also became aware of the importance of providing Latina/o students with opportunities to express themselves mathematically in two languages in order to gain better insights on students' mathematical reasoning. Typical university field experiences and/or courses do not offer enough of the kinds of experiences that can lead to such insights. Because mathematics in most schools is taught in English, novice teachers in field experiences or practicum do not have exposure to thinking about mathematics in Spanish. Similarly, there are seldom teacher education courses that focus on teaching content areas such as mathematics in Spanish. Many courses that focus on content instruction for a multilingual context do so from the perspective of English as a second language where the intent is to help teachers teach content in English. The nontraditional field experience that was provided through CEMELA after-school projects afforded teachers opportunities related to doing mathematics in two languages that they could not get elsewhere. Negotiating mathematical ideas with children across English and Spanish is challenging, but these points of tension helped the novice teachers become more reflective and critical about their teaching practice.

Second, the structure of the after-school with its less school-like teacher-centered control of participation and interactions contributed to teachers also strengthening their understanding of alternative forms of teaching. The structure of the after-school provided teachers with contrasts to traditional, didactic forms of instruction, and teachers were able to compare traditional pedagogical approaches with what are more student-centered instructional modes. Furthermore, their prolonged interactions with Latinas/os, the closer relationships they developed with students, and the opportunities they had to reflect on their experiences enhanced their understandings of student experiences and the role they play in learning. They came to know firsthand what experiences and knowledge students really do bring to schools. This entire process mediated the shifts they made in understanding pedagogical structures and student knowledge. Again, in typical teacher education programs or courses, where field experiences are brief and interactions with students are more controlled, limited, and contrived, teachers cannot really learn how to know their students or how to create learning environments that capitalize on what students know. Typical teacher development structures and activities do not give teachers adequate time or space to develop an appreciation of students and student-centered approaches to teaching mathematics, and yet this is a critical aspect of teacher development (Reyes, Capella-Santana, & Khisty, 1998).

Lastly, our work with teachers and defining them as coresearchers confirmed our assumption that after-school experiences can bring about important insights into schooling processes, insights that form the foundation for instructional decisions. Too often, professional development work with teachers still focuses on transmitting new methods and misses building the fundamental ways of thinking that make the instructional change viable (Bartolomé, 2003). That the teachers were startled into questioning what their students were actually learning in terms of habits of mind instead of just skills has tremendous importance and potential for the transformation of instruction for Latinas/os and their learning. Such a revelation could not easily have been achieved without the teachers thinking of themselves as researchers and viewing their own school's students from a position "outside themselves."

We have noted how the structure of the after-school mediated shifts in what teachers understood about Latina/o students and learning mathematics, especially in two languages. These shifts were in how teachers understand pedagogical structures and commonly used teaching constructs (e.g., student background knowledge) and in how they interacted with students, with each other, and with the university representatives. The after-school contexts that we have designed were deliberately created to evoke these meaning-making opportunities about the nature of teaching mathematics with Latinas/os. The issues related to improving the mathematics education of marginalized students such as Latinas/os require that we better prepare teachers to address the specific needs of such students, and this warrants consideration of spaces that go beyond traditional forms of teacher development.

Acknowledgments This chapter describes work conducted as part of CEMELA, supported in part by the National Science Foundation (Grant ESI-0424983). Any opinions, findings, and conclusions or recommendations expressed in this material are those of the author and do not necessarily reflect the views of the National Science Foundation.

References

Bartolomé, L. (2003). Beyond the methods fetish: Towards a humanizing pedagogy. In A. Darder, M. Balfodano, & R. Torres (Eds.), *The critical pedagogy reader* (pp. 408–429). New York: Routledge Falmer.

Cochran-Smith, M., Fieman-Nemser, S., McIntyre, D. J., & Demers, K. E. (Eds.). (2008). *Handbook of research on teacher education: Enduring questions in changing contexts* (3rd ed.). New York: Routledge.

Cole, M. (Ed.). (2006). *The fifth dimension: An after-school program built on diversity*. New York, NY: Russell Sage Foundation.

Cummins, J. (1981). The role of primary language development in promoting educational success for language minority students. In California Office of Bilingual Bicultural Education (Ed.), *Schooling and language minority students: A theoretical framework* (pp. 3–50). Los Angeles: Evaluation, Dissemination, and Assessment Center, California State University.

Darling-Hammond, L. (2006). *Powerful teacher education: Lessons from exemplary programs*. San Francisco: Jossey-Bass.

Díaz, S., Moll, L. C., & Mehan, H. (1986). Sociocultural resources instruction: A context-specific approach. In California Office of Bilingual Bicultural Education (Ed.), *Beyond language: Social and cultural factors in schooling language minority students* (pp. 187–230). Los Angeles: Evaluation, Dissemination, and Assessment Center, California State University.

Gutiérrez, R. (2002). Beyond essentialism: The complexity of language in teaching mathematics to Latino students. *American Educational Research Journal, 39*(4), 1047–1088.

Gutiérrez, K., Baquedano-López, P., Alvarez, H., & Chiu, M. (1999). A cultural-historical approach to collaboration: Building a culture of collaboration through hybrid language practices. *Theory into Practice, 38*(2), 87–93.

Khisty, L. L. (1995). Making inequality: Issues of language and meanings in mathematics teaching with Hispanic students. In W. G. Secada, E. Fennema, & L. B. Adajian (Eds.), *New directions for equity in mathematics education* (pp. 279–297). Cambridge: Cambridge University Press.

Khisty, L. L. (2002). Mathematics learning and the Latino student: Suggestions from research for classroom practice. *Teaching Children Mathematics, 9*(1), 32–35.

Khisty, L. L. (2006). Language and mathematics: Toward social justice for linguistically diverse students. In J. Novotna (Ed.), *Proceedings of the 30th Conference of the International Group for the Psychology of Mathematics Education* (Vol. 3, pp. 433–440). Prague, Czech Republic: Faculty of Education, Charles University in Prague.

Khisty, L. L, & Morales, H., Jr. (2004, July). Discourse matters: Equity, access, and Latinos' learning mathematics. In *Proceedings of Topic Study Group 25: Language and Communication in Mathematics Education, 10th International Congress on Mathematical Education* (pp. 1–12). Copenhagen, DK: Technical University of Denmark.

Khisty, L. L., & Viego, G. (1999). Challenging conventional wisdom: A case study. In L. Ortiz-Franco, N. G. Hernandez, & Y. De La Cruz (Eds.), *Changing the faces of mathematics: Perspectives on Latinos* (pp. 71–80). Reston, VA: National Council of Teachers of Mathematics.

Khisty, L. L., & Viego, G. (2007, April). *Using teacher research to understand Latinos' learning resources and to improve practice.* Paper presented at the Annual Meeting of the American Educational Research Association, Chicago, IL.

Ladson-Billings, G. (1999). Preparing teachers for diverse student populations: A critical race theory perspective. In A. Iran-Nejad & D. Pearson (Eds.), *Review of research in education* (Vol. 24, pp. 211–248). Washington, DC: American Educational Research Association.

Lincoln, Y. S., & Guba, E. G. (1985). *Naturalistic inquiry*. Newbury Park, CA: Sage.

Moschkovich, J. (1999a). Supporting the participation of English language learners in mathematical discussions. *For the Learning of Mathematics, 19*(1), 11–19.

Moschkovich, J. (1999b). Understanding the needs of Latino students in reform-oriented mathematics classrooms. In L. Ortiz-Franco, N. G. Hernandez, & Y. De La Cruz (Eds.), *Changing the faces of mathematics: Perspectives on Latinos* (pp. 5–12). Reston, VA: National Council of Teachers of Mathematics.

National Center for Educational Statistics (NCES). (2001). *The nation's report card mathematics 2000* (2001–517). Washington, DC: US Department of Education Office of Educational Research and Improvement.

Reyes, S., Capella-Santana, N., & Khisty, L. L. (1998). Prospective teachers constructing their own knowledge in multicultural education. In M. E. Dilworth & M. Michael-Bandele (Eds.), *Consideration of culture in teacher education: An anthology on practice* (pp. 110–125). Washington, DC: American Association of Colleges for Teacher Education.

Valdés, G., Bunch, G., Snow, C., Lee, C., & Matos, L. (2005). Enhancing the development of Students' language(s). In L. Darling-Hammond & J. Bradsford (Eds.), *Preparing teachers for a changing world: What teachers should learn and be able to do* (pp. 126–168). San Francisco: Jossey-Bass.

Vásquez, O. A. (2003). *La clase mágica: Imagining optimal possibilities in a bilingual community of learners*. Mahwah, NJ: Lawrence Erlbaum Associates.

Vygotsky, L. S. (1978). *Mind in society: The development of higher psychological processes*. Cambridge, MA: Harvard University Press.

Wells, G. (Ed.). (2001). *Action, talk, & text: Learning and teaching through inquiry*. New York: Teachers College Press.

Zeichner, K. M. (2005). A research agenda for teacher education. In M. Cochran-Smith & K. M. Zeichner (Eds.), *Studying teacher education: The report of the AERA panel on research and teacher education* (pp. 737–759). Mahawah, NJ: Lawrence Erlbaum Associates.

Zeichner, K. M., & Hoeft, K. (1996). Teacher socialization for cultural diversity. In J. Sikula, T. J. Buttery, & E. Guyton (Eds.), *Handbook of research on teacher education* (2nd ed., pp. 525–547). New York: Macmillan.

Chapter 18
How *Hands-On* Implicitly Informs "What Counts" as Science

Rhiannon Crain, Molly Loomis, and Rodney T. Ogawa

Introduction

To ask the question "what counts as science," we need to examine the norms, values, and cultural scripts that animate our answers. What sorts of history, authority, and schemata influence our assumptions about science learning and teaching? This chapter attempts to unpack seemingly commonsense answers to this question. We argue that underlying popular and academic assumptions about "what counts" as science is the cultural script of "hands-on."

Cultural scripts are widespread, but often unrecognized, cognitive resources that help to structure our ideas about the world. They inform our popular, educational, and political notions about science teaching, learning, and practice. *Hands-on*, for example, is a ubiquitous term when it comes to describing science learning and teaching, operating as a cultural script which provides a complex but widely adopted and often implicit set of concepts that informs ideas about what activities, materials, discourse patterns, and outcomes constitute science. We argue that the notion of hands-on both informs learners' ideas about what counts in their own science learning and defines many of the science education community's ideas and decisions about science learning and teaching. Because these ideas are often taken for granted ("of course science is hands-on!"), it is often difficult to recognize the manner in

R. Crain (✉)
Cornell Lab of Ornithology, Cornell University, Ithaca, NY, USA
e-mail: wren.crain@gmail.com

M. Loomis
ECHO Lake Aquarium and Science Center, Burlington, VT, USA

R.T. Ogawa
University of California, Santa Cruz, Santa Cruz, CA, USA

B. Bevan et al. (eds.), *LOST Opportunities: Learning in Out-of-School Time*, Explorations of Educational Purpose 23, DOI 10.1007/978-94-007-4304-5_18,

which such cultural scripts implicitly define science education. It is important to examine hands-on as a cultural script because such scripts ultimately lead to blind spots. Potential solutions to problems in our science education system are overlooked because of the taken-for-granted nature about what makes good science learning.

Cultural Scripts and Social Institutions of Education

In order to move toward more inclusive practices in science education, it is an important project of academic work on science education to broaden what counts as science (Barton & Yang, 2000; Erickson & Gutiérrez, 2002). It is not enough to simply initiate reform efforts (Hewson, Kahle, Scantlebury, & Davies, 2001); rather, there needs to be an explicit challenge to (or at least reflection on) the implicit values, norms, and scripts that dictate automatic understandings of what counts as science. These cultural scripts are subject to change only after being made explicit.

Cultural scripts play an important role in providing unquestioned sources of information about how to think and act for both individuals and organizations. The emergence and subsequent institutionalization of hands-on as a successful educational technique in science museums (Ogawa, Loomis, & Crain, 2009) drove other educational organizations to adopt similar language and other symbols in order to remain legitimate providers of science education. Institutions begin as social processes (Meade, 1934) (and, arguably, as cultural scripts that provide a cognitive microfoundation to social activity [Bruner, 1990]) which, when reproduced over time, result in the acceptance of a shared vision of reality that seems to be "the way things are" (Meyer & Rowan, 1977).

Individuals do not hold cohesive, ubiquitous "cultures" (Bruni, Gherardi, & Parolin, 2007; DiMaggio, 1997; Gutiérrez & Rogoff, 2003). Instead, they rely on a complex network of cultural scripts that compose a repertoire, or toolkit, of disparate knowing bits that are strategically drawn-on to take action or make sense of an activity (Bourdieu, 1990; Rogoff & Chavajay, 1995; Rogoff, Chavajay, & Matusov, 1993; Sewell, 1992). Routine, everyday cognition relies heavily on these cultural scripts – ways of knowing that construe objects of action, roles, and unscrutinized assumptions about relationships and conditions in any situation where there is incomplete information (Weick, 1995). Cultural scripts are externally embedded in symbol systems and organizational structures (Scott, 2003), making it possible for groups of people to share scripts.

At play with the seemingly innocuous, optimistic script of hands-on are at least two threats to broadening what counts as science. First, hands-on promotes the cultural conservation of a strong boundary between science and nonscience (Gieryn, 1998). The maintenance of this boundary, and its inherent traditional location in the "white coat" notion of science (Erickson & Gutiérrez, 2002), weakens efforts to broaden what counts. While there are some hands-on programs, curricula, and exhibits that illuminate the ways in which science practice is passionate and argumentative – rather than disinterested and canonically rational – and how it requires its practitioners to be profoundly selective in their attention and aesthetically oriented to evidence in order to draw conclusions, the cultural script of hands-on often functions to hide the complexity of real science practice. Second, by enacting the idea that those learning science should act like scientists, the script of hands-on pushes science education in the direction of an education for future scientists, but not necessarily an education for future nonscientist citizens (Lehr, 2006). Although there is not normally a distinction made between these two "educations," the assumption that powers this practice is that learning science is the same as learning to be a scientist. The instantiated material practices and symbols of hands-on support a version of science education where using science for formal science practice counts and using science for other purposes does not.

To investigate this "blind spot," we first travel back in time to examine the "guts" of this cultural script, questioning how it arose, spread through the science education community, evolved, and continues – even in the face of modern critique – to influence notions of "what counts" in science learning, teaching, and practice. We then explore the function of cultural scripts in thinking, acting, and composing social institutions (especially those embedded in highly institutionalized areas such as education) and look at some evidence of hands-on acting as a script for families visiting a science museum.

How Did *Hands-On* Get to Be This Way?

The cultural script of hands-on is both old and new. Active or hands-on science education methods (e.g., Haury & Rillero, 1994; Martin, 2002) have a long history, dating back to Pestalozzi's (1746–1827) "object teaching," which valued first-hand interaction with natural objects over lecturing and recitation (Rillero, 1993). However, the specific term "hands-on" did not surface until the 1960s (Haury & Rillero, 1994). Rutherford (1993) claims that the term was first used to describe an approach for learning to use a computer. Some early employees at the Exploratorium, a pioneering interactive science center founded in 1969, introduced the term to differentiate science centers from the "hands-off" approaches espoused by schools and traditional museums (Grinell, 2007, personal communication). The term hands-on was not ubiquitous in the school science literature until the late 1980s, when it quickly became a fundamental part of the lexicon of school science pedagogy (Loomis & Crain, 2007, unpublished).

The use of the term hands-on proliferated during an era of tremendous changes in the fields of science and education. In the aftermath of the Second World War, scientists and citizens were both inspired by the power and wary of the potential of science and technology. Ideas about scientific progress and technological advancement both came under fire in the aftermath of the atom bomb and accrued momentum as the

United States sought to keep up with international developments. The launch of Sputnik in 1957 provoked a competitive fervor in Americans and led to decades of rhetoric about the USA lagging behind the world in science and technology.

The field of education responded to this context by pouring resources into science education reform. The National Science Foundation, founded in 1950, invested more in science education from 1964 to 1971 than it did at any other point during its first 30 years (Tressel, 1994). These efforts combined scientists and educators in the development of a myriad of influential and experimental curricula, which focused on material-intensive science activities. Socially motivated scientists and democratically oriented educators worked together to design programs aimed at cultivating an informed citizenry and improving society for all participants.

It was these social and political values that provided the historical backdrop for the development of interactive science centers and their calling card, hands-on science education. Appealing to the calls for innovation in science education and democratic access to information, the Exploratorium "eschewed historical and industrial collections in favor of apparatus and programs designed to communicate basic science in terms readily accessible to visitors" (Grinell, 2003, p. 2). It did away with glass cases and rigid curricula and focused on involving visitors in the process of understanding the world by interacting with authentic scientific materials and phenomena. This hands-on approach was accompanied by the belief that interactive exhibits offered a better pedagogical tool than did traditional exhibits (Bitgood & Patterson, 1987; McLean, 1993).

As one of the early interactive science centers, the Exploratorium was a key player in disseminating hands-on interaction as normative practice in science centers (Ogawa et al., 2009). Hands-on diffused across the field of interactive science centers through mission statements, exhibit design protocols, and the general media. Articles in the popular media of 1970s and early 1980s were filled with references to hands-on to describe the "revolution" in science center development (Ogawa et al., 2009). Once this term institutionalized as a descriptor for innovative pedagogy, educational organizations including "traditional" museums and schools sought to align themselves with its positive connotations (Ogawa et al., 2009). In the late 1980s, for example, specific references to hands-on proliferated in school-related curricula, when mandates, such as Science for all Americans: Project 2061 (American Association for the Advancement of Science, 1989), emerged and science kits, such as the National Science Foundation FOSS curricula, proliferated (AAAS, pp. 23–24).

Hands-on diffused into the wider environment through rules for interactivity defined by governing organizations and sanctioned hands-on activities and programs outlined by funding agencies. The National Science Foundation regulated science center development by funding the development of specific kinds of hands-on programs and exhibits, particularly ones that resembled those created by the Exploratorium (Tressel, 2007, personal communication). Hands-on also diffused through normative ideas about the obligation interactive science centers had to provide an alternative to school science. Exhibit design conferences and how-to guides disseminated these normative notions about the role of interactive science centers in the field of education. Additionally, hands-on diffused through cultural scripts about the way interactive learning is done. Throughout the 1970s, articles in newspapers and popular media proliferated stories about the "new 'hands-on'" at the Exploratorium (National Science Foundation, 1979), which supported the development of the script that science centers specialized in hands-on interaction (pp. 10–11).

The diffusion of the notion of hands-on was accompanied by several underlying assumptions: the understanding of big science and professional scientists as experts on science education; the notion of active participation as democratic access to science; the idea of science as significant, special, and symbolic of the possession of knowledge; the belief in an implied link between touching and knowing; the sense of urgency and legitimacy in exploration; the enactment of science as self-contained or occurring in discreet discursive and spatial bundles (like exhibits); and the eschewing of schooled and theoretical ways of knowing the world. These ideas, and their subsequent implications for "good" science pedagogy, implicitly underlie the actions taken and thoughts had by both professionals and consumers of science education. It is the process and consequences of the institutionalization of hands-on that concerns us, for, although the idea is not inherently misguided, unexamined action always has the potential for unintended, or unthought-out, consequences, especially as public needs for science education shift both socially and technically.

Hands-On as a Cultural Script

Hands-on has a long history in the US educational system and is used to describe a variety of activities from interactive exhibits like the Bernoulli's Ball where a jet of air levitates a beach ball, to a style of science teaching in classrooms where materials and student questions are given center stage. Given the specific historical origination of hands-on, what are the consequences of its colloquial use as a cultural script?

Cultural scripts function in very particular ways. In the following paragraphs, we show how hands-on acts as a cultural script using examples from two studies carried out at the Exploratorium (Crain, 2009; Loomis, 2009). We highlight how hands-on is (a) highly institutionalized in educational organizations generally, and in science museums specifically; (b) how it is recalled quickly and accurately to make activity recognizable; (c) how it is composed of complex, rule-like structural resources; (d) how it is employed in routine, everyday activity; (e) how it is often expressed in oversimplified terms; and (f) how it resists change.

Both studies focus on families' interactions during a visit to the Exploratorium. Examples are taken from interviews with these families. Crain's study asked families to "draw the Exploratorium," and then used those drawings as artifacts around which an interview about the families' experiences in the museum was crafted. Similarly, Loomis videotaped families at three "classic" Exploratorium exhibits and then interviewed them about their activity there. Hands-on was a common expression used by most of the families interviewed as a part of these two studies. However, families did not employ a uniform meaning for the term hands-on. Concepts of interactivity, learning, participation, involvement, accessibility, easiness, control, and authentic science were represented in families' definitions of hands-on, but these were often nonspecific and were not uniform. Lack of specificity and uniformity indicates that families were not using the technical definition of the phrase and that, instead, they were using the phrase in a symbolic manner. The use of a symbol like this in an organizational context may indicate a belief about the organization operating in the wider environmental context (Scott, 2001).

Hands-On Is Highly Institutionalized

Berger and Luckman (1966) observed that symbolic language mediates a process they refer to as institutionalization where, "actions are produced, repeated, and come to evoke stable, similar meaning in self and other" (from Scott, 2001, p. 17). Several scholars have found that symbolic cultural scripts are indicative of a wider system of meaning, especially at the level of the organizational field, that impacts how activity is framed by shared beliefs and thus recreated by people acting in an organization over and over, taking on its characteristic durability (DiMaggio & Powell, 1991; Meyer & Rowan, 1977).

The script of hands-on is deeply embedded in several organizations in the United States. Most obvious are educational organizations like schools and interactive science centers and the groups of organizations associated with them (professional bodies like Association of Science and Technology Centers, funding agencies like the National Science Foundation, and resource providers like Delta Education or the Center for Hands-on Learning). In the USA, schooling is mandatory, and science is a part of the curriculum and thus part of our public discourse. As a result, ideas about science and science education are held by much of the US population – from parents to school children to scientists themselves. Hands-on is a script that has grown to be somewhat ubiquitous across populations because it is situationally supported by organizations, professionals, and the media. For instance, Lederman, a well-recognized science education professional, said before Congress, "children trained in hands-on inquiry methods not only learn science, but they experience the joy of learning, with consequences much beyond the science class" (Lederman, US House of Representatives Testimony, September 2000). Public acts like Congressional testimony serve to support and legitimize cultural scripts, as do appearances in the media (magazines aimed at teachers and parents relentlessly reference hands-on, thereby providing complexity and legitimacy to its life as a cultural script). Hands-on is also formally institutionalized through educational standards. In an institutional history of the Exploratorium, Ogawa et al. (2009) found that the Exploratorium actively limits its own use of the phrase hands-on. They also found that the museum staff struggled to define the phrase hands-on in a consistent manner as the wider science education field began to take up the phrase to describe its own activity. Loomis (2009) found that, although long-term Exploratorium staff made consistent references to the power of materials and the importance of self-directed discovery, they rarely used the term "hands-on" to characterize learning at the Exploratorium. Consequently, the science education field itself has adopted a non-uniform use of the phrase hands-on to describe its various activities. Rather than being used in a technical manner, the phrase has come to serve as a symbol that is used to instantly legitimize membership in a progressive educational community.

Hands-On Is Recalled Quickly, Accurately, and Makes Some Activity and Information More Recognizable than Others

When faced with the need to make a decision or take action in a science education situation, the script of hands-on – as an active, supported part of a cultural toolkit – is cognitively accessible. We are able to access ideas about science that conform to this cultural script faster and more cohesively than ideas not cognitively supported by a script. As a result, we also tend to recognize activity and label it as science more often if it is congruent with hands-on. For instance, people are much more likely to label a 12-year-old measuring pH levels in unidentified liquids as science than they are to so label a woman reading a journal article in a university science library because the first example matches our script of hands-on. Arguments can be made for both activities as science, and after reflection a person might label both accordingly, but our automatic response, dictated by the script of hands-on, spurs us to recognize the 12-year-old first. Crain (2009) found that, for families in her study, the script of hands-on might have influenced the exhibits they recognized as belonging in the Exploratorium. As a result, exhibits that were most easily recognized as, literally, "hands-on" were more frequently represented in the families' drawings. The effect of this may be overrepresentation of a particular kind of exhibit that most closely matches the cultural script for hands-on.

One of the consequences of such overrepresentation may be public self-exclusion from activities that do not align with the definition of science provided by hands-on. For instance, scholars are trying to broaden what counts as science education through a movement called public engagement with science (PES) (McCallie et al., 2009). PES is a science participation model in which publics engage with formal science to establish mutual learning where both "publics and scientists have expertise, valuable perspectives, and knowledge to contribute to the development of science and its application in society (Burns, O'Conner, & Stocklmayer, 2003; Kerr, Cunningham-Burley, & Tutton, 2007; Leshner, 2003)" (McCallie et al., 2009, p.23). Crain (2009) found that families rarely drew or talked about exhibits that were built on the PES participation model even though such exhibits offered them opportunities to engage with science in an intense and personal way. Instead, their understandings of science learning as hands-on made it difficult to recognize such experiences as science learning.

Hands-On Is Composed of Complex, Rule-Like, Structural Resources Put to Strategic Use by Groups and Individuals in Activity

To think of hands-on is to think of what science is (learning about the natural world and things), how it should be done (with the hands, through explorative activity), whom it should be done by (creative, dexterous, observant people), and where it should be done (in the presence of stuff). These implicit ideas about what science is and how to do it are what constitute the resources of the cultural script of hands-on. These resources are used to make sense of situations and provide access to legitimate forms of action in situations where an idea of "what science is" is necessary to carry out activity. Hands-on is used strategically by people faced with situations where they are asked to define science or appropriate science activity. Loomis (2009) found that, during interviews with 30 families visiting the Exploratorium, 25 used the term "hands-on" to describe their activity, even though the term was not used by the interviewer and did not appear on exhibit labels. Families often used hands-on as the obvious "no-duh" answer to interview questions about the Exploratorium. One visitor, for example, described the Exploratorium's theory of learning like this: "Two words: hands on. I mean, it seems like that what this place is all about. They want you to touch things here." Similarly, Crain (2009) found this effect was widespread in her 35 families' responses to questions they perceived to be about science. Often, families would use the phrase hands-on as if it were a code word (or symbol), which in and of itself indicated their membership in the museum's accepted definition for science or learning. It provided legitimacy in the absence of technical certainty.

Hands-On Is Employed in Routine, Everyday Activity

Hands-on is infused into everyday understandings: it influences perception, interpretation, planning, and action. When a cultural script is activated, there is no space, no conscious consideration of possible alternative reactions to a situation. Hands-on functions as a cultural script by providing routinized answers in everyday situations. As such, hands-on is hard to recognize because the resources it provides are taken for granted, or accepted as natural, rational ways to conduct activity. This cultural script is acted out every time we see a child using a magnifying glass and label the action science; or when we go to a store looking for a "scientific" gift and gravitate toward colorfully boxed chemistry or entomology "kits." In a study by Crain (2009), families took the idea of interactivity activated by their use of the phase hands-on for granted. Several families did offer some justification for the phrase, but no families used the phrase (or even the idea of interactivity) in its technical sense – as a learning tool developed to more closely match the actual practice of science. Failure to fully define the phrase this way means that the script has become disconnected from its original use and, as a result, is taken for granted. The nontechnical use of the phrase hands-on means that people and educational programs (including interactive science centers) can adopt hands-on in whatever ways suit them while still maintaining social legitimacy. Clearly defining the phrase and adhering to a shared technical definition is not necessary to garner the symbolic benefits of utilizing the script to define science-learning activity generally.

Hands-On Is Vulnerable to Oversimplification

Like all cultural scripts, public versions are often extreme, simplified representations of actual cognitive scripts. With hands-on, this is evident in some mass-marketed "science kits," which might contain pictures of smiling people in safety goggles, test tubes, eyedroppers, and ideas for activities. It is also evident in public discourse, where emphasizing hands-on implicitly links science and education without the technical complexity necessary to justify such a connection ("our exhibits are hands-on" necessarily implies the exhibits promote science learning). This hints at the symbolic nature of hands-on.

Often, oversimplifications yield symbols that come to stand for otherwise complex cultural scripts. With hands-on, there are physical symbols – particular kinds of science equipment – as well as iconic and representational symbols, pictorial stamps or seals on activity kits, keywords, and even the phrase hands-on used to stand-in for the cultural script. Crain (2009) found that the lack of uniformity in families' use of the phrase hands-on indicated its symbolic use. She found that families used the phrase in their interviews with six separate definitions. Most often, families used the phrase in a general sense to symbolically link science and learning to what they drew in the Exploratorium without having to actually explain what that meant. For example, one 12-year-old boy said, "And, like, everything is hands-on and that's what makes the Exploratorium," or one Dad described his families' attraction to exhibits, "Yeah that's a big part of it, things where it was really hands-on." Use of the phrase this way made it self-defining, as if the words themselves legitimized the definition.

Loomis (2009) found that when visiting families were interviewed they described the Exploratorium as a place where they were "doing science" and felt empowered by hands-on science experiences. One visitor stated, for example, "We love to go to this place because everybody can become a scientist for a day." When asked to describe their specific actions, however, families differentiated their activity at the Exploratorium from that of scientists. This contradiction indicates that families deferred to the script hands-on in general terms, but not when asked to be specific about the relationship between their hands-on activity and science activity.

Hands-On Resists Change

The institutionalized, symbolic nature of hands-on helps it to resist change. Loomis (2009), for example, found striking consistency in the analogies used by the Exploratorium's founder in 1969 and a visitor in 2008. In 1969, the museum's founder, Frank Oppenheimer, wrote: "Explaining science and technology without props can resemble an attempt to tell what it is like to swim without ever letting a person near the water" (1968, p. 1). Forty years later, a father visiting the museum echoed the sentiment with, "When you try to tell somebody in the car how [electricity] works, it's not going to work. But if you're there, showing with wires and everything like that, you can actually visually see it." This reference to the importance of materials to understand technology, which Loomis found was echoed by long-term staff and visitors in general, reflects the persistence of the social narrative of hands-on being an effective approach to learning.

Even in the face of critique, it is easier to "tinker" (Tyack & Cuban, 1995) with non-foundational aspects of a script than to try and change it outright. This is the case with the amendment to hands-on popularized in the early 1990s ("hands-on, minds-on"). The addition of the phrase minds-on was a wide-sweeping critique of the use of hands-on by public education entities to justify programs whose activities were judged too simplistic. However, rather than toss out hands-on altogether, it was tinkered with. Despite this, Gardiner and Farragher (1997, p. 39 from Hofstein & Lunetta, 2004) found that "although many biology teachers' articulated philosophies appeared to support an investigative, hands-on, minds-on approach with authentic learning experiences, the classroom practice of those teachers did not generally appear to be consistent with their stated philosophies." Instead, they found that teachers continued using the same kinds of hands-on activities that were under attack as "minds-off," indicating that even public tinkering with such a powerful cultural script does not guarantee people will enact it.

In response to the simplistic uptake of the term hands-on, the Exploratorium itself has decreased its use of the term to describe its programming. However, rather than completely discard it, they have done extensive work to illuminate the meaning of hands-on, going so far as to develop workshops for classroom teachers that specifically address the multiple uses of the phrase hands-on by educators. One program walked participants through three kinds of material-based educational activity meant to help educators interrogate the finer details of "hands-on" learning (Kluger-Bell, 1999). The first activity showcased a teacher-proscribed, material-based form where students simply follow a set of directions to manipulate materials. The second activity was completely unguided, giving students unlimited access to an array of materials and asking them to do something with them. The third activity was based on an extensively developed inquiry model where student-derived questions are the subject of small group investigations and facilitators play a distant, but guiding, role in moving the material-based inquiry forward. Teachers engage in these different hands-on activities and then collaboratively reflect on how the different forms afford different types of experiences and could operate under different constraints and conditions. The Exploratorium is tinkering with the definition of a widespread cultural script. The families visiting the museum, however, are largely unaware of efforts to more carefully define the phrase hands-on and continue to use it in its widespread, inconsistent, and nontechnical sense.

Why Should We Care About the Role Hands-on Plays in Defining What Counts as Science?

Taking the time to understand the cultural script of hands-on elucidates the specific values and assumptions about good science learning and practice that drive reasoning and decision-making about what counts. This is especially important in light of evidence indicating that institutionalized scripts are difficult to significantly change (Tyack & Cuban, 1995). As is the case with many reform efforts, the path to change is often one where it is believed that simply introducing sweeping changes to language and rhetoric will produce actual change in activity. This belief persists even in the face of research indicating that this is not the case (Tyack & Cuban, 1995; Gardiner & Farragher, 1997; Ogawa, 1994; Ogawa, Sandholtz, Martinez-Flores, & Scribner, 2003). With hands-on, especially in the past ten years, there has been a strong movement toward discontinuing use of the actual phrase hands-on. However, the degree to which this indicates a change in the use of the actual cultural script is minor. Overall, people are still using the ideas about science learning that are institutionalized by the script of hands-on. This has a dramatic effect on practice, and perhaps most importantly, on what counts as science education practice.

Our intention here is not to automatically label hands-on as bad, but to point out the unanticipated power in unexamined cultural scripts. Hands-on emerged in a time of rich social introspection on science, optimism, and a desire for openness between formal science and citizenry. However, processes of institutionalization necessarily diluted the original characteristics of the movement as a wider and wider audience adopted it. The transition of the idea into a cultural script necessitated its simplification, its popularity, its strategic reduction to symbolism, and its resistance to change.

As the wave of support for broadening what counts as science in the service of more equitable science education peaks, it is important to put our cultural scripts up for observation. The consequences of skipping this step are continued failures of science reform efforts. Hands-on carries a particular notion of science and science learning that can work against efforts at broadening which practices count as legitimate engagement with science. For instance, Crain (2009) showed that the hands-on script may be influencing families' selections of museum experiences that count as legitimate science activity when they overlook exhibits that evoke a public engagement of science model. The public engagement of science model represents a serious attempt by many informal education organizations to broaden participation structures in science education, but public understandings of their own relationships to science learning as filtered through the cultural script of hands-on may be preventing a wider adoption of such models. Of particular concern in this example was the even stronger symbolic use of hands-on by the seven families belonging to nondominant cultural communities. These families were likely to more narrowly represent their activity in the museum; drawing exhibits that clustered even more tightly to the classic hands-on model. Arguably, these families might have felt they had more to lose if they portrayed themselves as "unscientific," than did families belonging to dominant cultural communities whose success in scientific activity was already taken for granted. Unfortunately, such stiff adherence to science learning as it is defined by the cultural script of hands-on means these families were potentially not open to rich scientific activities which fell outside the domain of those prescribed by hands-on.

Additionally, Loomis's work (2009) indicates that the use of hands-on, while easily accessed by the families to describe their scientific activity, did not help to bridge the gap between the families' understandings of their own activity as fundamentally different from the official activity of practicing scientists. So, hands-on simultaneously reinforces the perceived gap between the activity of nonscientist citizens and scientists, while limiting the potential of alternative models, like PES, to be taken seriously as science learning and broaden the means of participation in science.

Limiting the means of participation necessarily limits the expansion of science beyond the communities traditionally most successful in science in the first place. Thus, serious efforts at science education reform need to understand the ways in which cultural scripts are the products of particular histories and institutions. Ignoring the power of these scripts over learners, researchers, teachers, curriculum designers, and funders renders many reform efforts moot.

Acknowledgements This chapter describes work conducted as part of the Center for Informal Learning and Schools, supported in part by the National Science Foundation (Grant ESI-ESI-0119787). Any opinions, findings, and conclusions or recommendations expressed in this material are those of the author and do not necessarily reflect the views of the National Science Foundation.

References

Association for the Advancement of Science (1993) Benchmarks for Science Literacy. 448.

Barton, A. C., & Yang, K. (2000). The culture of power and science education: Learning from Miguel. *Journal of Research in Science Teaching, 37*(8), 871–889.

Berger, P., & Luckman, T. (1966). *The social construction of reality*. New York: Double Day.

Bitgood, S., & Patterson, D. (1987). Principles of exhibit design. *Visitor Behavior, 2*(1), 4–6.

Bourdieu, P. (1990). Structures, habitus, practices. In *The logic of practice* (pp. 52–65). Stanford, CA: Stanford University Press.

Bruner, J. S. (1990). *Acts of meaning*. Cambridge, MA: Harvard University Press.

Bruni, A., Gherardi, S., & Parolin, L. L. (2007). Knowing in a system of fragmented knowledge. *Mind, Culture and Activity, 14*(1–2), 83–102.

Burns, T. W., O’Conner, D. J., and Stocklmayer, S. M. (2003). Science communication: A contemporary definition. Public Understanding of Science, 12(2), 183–202.
Crain, R. (2009). *Institutionalization in action: Science museum interactivity and materiality from the family perspective*. Doctoral dissertation, University of California, Santa Cruz.
DiMaggio, P. (1997). Culture and cognition. *Annual Review of Sociology, 23*, 263–287.
DiMaggio, P., & Powell, W. (1991). Introduction in the new institutionalism. In W. Walter & P. DiMaggio (Eds.), *Organizational analysis* (pp. 1–38). Chicago: University of Chicago Press.
Erickson, F., & Gutiérrez, K. (2002). Comment: Culture, rigor, and science in educational research. *Educational Researcher, 31*(8), 21.
Gardiner, P. G., & Farragher, P. (1997). *The quantity and quality of biology laboratory work in British Columbia high schools*. Paper presented at the National Association for Research in Science Teaching (NARST) Meeting, Oak Brook.
Gieryn, T. F. (1998). *Cultural cartography of science: Episodes of boundary work, sociologically rendered*. Chicago, IL: University of Chicago Press.
Grinell, S. (2003). *A place for learning science: Starting a science center and keeping it running*. Washington, DC: Association of Science-Technology Centers Inc.
Gutiérrez, K., & Rogoff, B. (2003). Cultural ways of learning: Individual traits or repertoires of practice. *Educational Researcher, 32*(5), 19–25.
Haury, D. L., & Rillero, P. (1994). *Perspectives of hands-on science teaching*. Columbus, OH: ERIC Clearinghouse for Science, Mathematics and Environmental Education.
Hewson, P. W., Kahle, J. B., Scantlebury, K., & Davies, D. (2001). Equitable science education in urban middle schools: Do reform efforts make a difference? *Journal of Research in Science Teaching, 38*(10), 1130–1144.
Hofstein, A., & Lunetta, V. N. (2004). The laboratory in science education: Foundations for the twenty-first century. *Science Education, 88*(1), 28–54.
Kerr, A., Cunningham-Burley, S., and Tutton, R. (2007). Shifting subject positions: Experts and lay people in public dialogue. Social Studies of Science, 37 (3), 385–411.
Kluger-Bell B (1999) Recognizing Inquiry: Comparing Three Hands-On Teaching Techniques. In: National Science Foundation (ed) Foundations: A monograph for professionals in science, mathematics, and technology education: Inquiry: Thoughts, views, and strategies for the K-5 classroom. National Science Foundation, pp 39–50
Lehr, J. (2006). *Social justice pedagogies and scientific knowledge: Remaking citizenship in the non-science classroom*. Unpublished dissertation, Virginia Technical University, Blacksburg.
Leshner, A. (2003, February 14). Public engagement with science. Science, 299, 977.
Loomis, M. (2009). *Visitor empowerment and the authority of science: Exploring institutionalized tensions in an interactive science center*. Doctoral dissertation, University of California, Santa Cruz.
Loomis, M., & Crain, R. (2007). *Origins and diffusion of hands-on as a science education ideology*. Unpublished study, University of California, Santa Cruz.
Martin, J. (2002). *Organizational culture: Mapping the terrain*. Thousand Oaks, CA: Sage Publications.
McCallie, E., Bell, L., Lohwater, T., Falk, J. H., Lehr, J. L., Lewenstein, B. V., et al. (2009). *Many experts, many audiences: Public engagement with science and informal science education*. A CAISE Inquiry Group Report. Washington, DC: Center for Advancement of Informal Science Education (CAISE). http://caise.insci.org/uploads/docs/public_engagement_with_science.pdf
McLean, K. (1993). *Planning for people in museum exhibitions*. Washington, DC: Association of Science and Technology Centers.
Meade, G. (1934). *Mind, self, & society*. Chicago: University of Chicago Press.
Meyer, J., & Rowan, B. (1977). Institutionalized organizations: Formal structure as myth and ceremony. *The American Journal of Sociology, 83*, 340–363.
National Science Foundation. (1979, July/August). Museums to teach by. Mosaic, 17–25.
Ogawa, R. T. (1994). The institutional sources of educational reform: The case of school-based management. *American Education Research Journal, 31*, 519–548.

Ogawa, R. T., Loomis, M., & Crain, R. (2009). Social construction of an interactive science center: The institutional history of the Exploratorium. *Science Education, 93*(2), 269–292.
Ogawa, R. T., Sandholtz, J., Martinez-Flores, M., & Scribner, S. (2003). The substantive consequences of a district's standards-based curriculum. *American Education Research Journal, 40*(1), 147–176.
Oppenheimer, F. (1968). A rationale for a science museum. *Curator, 11*(3), 206–209.
Rillero, P. (1993). The enlightenment revolution: A historical study of positive change through science teacher education. *Journal of Science Teacher Education, 4*(2), 37–43.
Rogoff, B., & Chavajay, P. (1995). What's become of research on the cultural basis of cognitive development. *The American Psychologist, 50*, 859–877.
Rogoff, B., Chavajay, P., & Matusov, E. (1993). Questioning assumptions about culture and individuals. *The Behavioral and Brain Sciences, 16*, 533–534.
Rutherford, F. (1993). Hands-on: A means to an end. *2061 Today, 3*(1), 5.
Scott, R. (2001). *Institutions and organizations* (2nd ed.). Thousand Oaks, CA: Sage Publications.
Scott, R. (2003). *Organizations: Rational natural and open systems* (5th ed.). Upper Saddle River, NJ: Prentice Hall.
Sewell, W. (1992). A theory of structure: Duality, agency, and transformation. *The American Journal of Sociology, 98*(1–29).
Tressel, G. (1994). Thirty years of "improvement" in precollege math and science education. *Journal of Science Education and Technology, 3*(2), 77–88.
Tyack, D., & Cuban, L. (1995). *Tinkering toward Utopia: A century of public school reform*. Cambridge, MA: Harvard University Press.
Weick, K. (1995). *Sensemaking in organizations*. London: Sage Publications.

Chapter 19
Thoughts on Science Education During the After-School Hours: A Commentry on Part III

Michael Cole

For someone who has long been interested in after-school educational activities as a promising supplement to formal, in-school education, the papers in this section provide rich opportunities to think about the promise and the problems that such programs offer to those concerned with the infusion of science into the learning and development of their participants.

The first thing that strikes one in reading about after-school activities is the incredible variety of settings in which they occur and the equally varied social arrangements that organize them. A good many occur in schools after school hours. In these cases, the organization of the activities is likely to fall to a teacher working extra hours, perhaps with family members or community volunteers giving a helping hand. Others are to be found in nationally sponsored organizations such as the YMCA or the Boys and Girls Clubs of America, or perhaps in a church or a library. This variety makes a difference, often a very important one, in the quality of programming that is possible, particularly when one's goal is to make scientific topics the centerpiece of the activities, a topic to which I return below.

Despite all of this variety, what also stands out as a common thread among these papers is their shared concern to contrast the kind of educational environment that after-school settings routinely offer with those found in standard schools. These contrasts become the starting point for specifying ways to take advantage of the fact that the activities are NOT occurring in school in order to enrich and deepen young people's engagement with, understanding of, and appreciation for science and mathematics. Although they vary greatly among themselves, the educational settings described in this section demonstrate that after-school settings are more flexible in

M. Cole (✉)
University of California, San Diego, San Diego, CA, USA
e-mail: mcole@ucsd.edu

B. Bevan et al. (eds.), *LOST Opportunities: Learning in Out-of-School Time*,
Explorations of Educational Purpose 23, DOI 10.1007/978-94-007-4304-5_19,

terms not only of their schedules, but in their social arrangements, allowing children freedom to engage in peer interaction and less hierarchically codified interactions with adults. Precisely because they take place between the home and the school, in a time of day that—since the advent of modern schooling—has given play a privileged position, they are places where having fun is an essential ingredient, where children participate voluntarily.

This shared circumstance offers both the greatest promise and the greatest challenges to realization of the hopes placed on out-of-school time (OST) activities for infusing science into the lives of children. On the positive side, the organization of after-school settings is overwhelmingly centered on activities that are designed to be engaging so that children engage voluntarily; if you build it and it is unattractive, the kids will not come! One attraction is that after-school programs generally allow children to work in groups with their peers, and to choose the roles that they play in the various projects that are offered. Indeed, children and youth often have a voice in the projects that are offered, or allowed to walk away from those they find boring. In this kind of social environment, everyone gets to talk a lot, often using languages they feel most comfortable in, and the staff (who vary in age from their late teens to middle age) are freed of the obligation to know all the right answers, so that they can position themselves in the role of coaches and more experienced peers.

These same shared potentials of the social situation of OST help to account for the fact that virtually all of the activities that occupy center stage in science-oriented (or science inclusive) after-school programs are characterized as "hands-on" activities (just the characteristic which, according to Crain and colleagues, "implicitly inform[s] what counts as science"). Given the unrelenting call for "more time on task" for all educational activities over the past 30 years, it should come as no surprise then that policy makers look favorably on after-school as the place to make more "time on science learning" happen. It seems a sure bet.

Unfortunately, many of the characteristics that offer the greatest potential for OST to promote science learning are the same characteristics that offer the greatest challenges for using OST settings to provide the kind of educational supplement for which they appear to be perfectly suited. To begin with, because OSTs are not part of the formal school system, and it is not legally binding on children to attend; the funding sources to support such settings are scarce and uncertain, as is the ubiquity and consistency of attendance by the school-aged population such settings are designed to serve. For the same reason, restriction of a living wage to one or at most a few of the staff is the norm, leading to a high rate of staff turnover and low level of education of staff with whom children and youth are in close contact. These "structural" factors (as Bevan and Michalchik refer to them) go hand in glove with the promising social-organizational characteristic of OST activities summarized above. If a project goes unfinished, it may well not be taken up again in a later session. If a child leaves for soccer practice, no problem for the OST program, but big trouble for the goal of improving science learning. And projects do go unfinished, perhaps even un-started, as children, restless after a long day of sitting in enforced quiet in school, are freed up to "have fun" in ways that have nothing to do with planned science projects.

Another common concern that the informal, voluntary, peer-oriented, group-organized nature of after-school programs faces is that they are a nightmare to evaluate in terms of their impact on school-based academic achievement. First of all, they are problematic simply because they are voluntary. "Evidence-based" educational assessment as a synonym for "randomized group" research designs has been thoroughly absorbed into the federal institutions that use such settings as a means to promote academic learning. While some highly competent means of assessing quality of engagement at OST sites have been developed and deployed to apparently good effect (i.e., there is general agreement on whether a given program engages children in the target activities in a manner that educational practice and common sense agree are "good for the kids"), such evidence is trivially easy to dismiss on grounds of lack of appropriate control groups.

But to make matters even tougher for advocates of OST for academic learning of all kinds, including science and mathematics learning, it seems that, with relatively few exceptions, children's engagement in, and apparent success in OST rarely turns up as increases in academic performance (there are exceptions to this generalization, but they are relatively few). Moreover, measures such as school attendance often are used as a substitute for school performance, leaving open the impact of OST on such difficult-to-specify matters as scientific understanding or appreciation of science as a mode of reasoning. Perhaps as a consequence of these difficulties (and the fact that funders may support running of programs or evaluation of programs, but almost never both), some advocates of after-school educational activities, including several of the authors who describe OST programs in this volume, eschew issues of academic evaluation altogether.

There are many reasons to expect a disconnect between learning in OST activities and learning as reflected in standardized tests in school. At just the simplest level, school-based assessments are not designed to test for acquisition of the knowledge required to engage in the after-school activities. The learning experiences maybe "hands-on" and group oriented and engaging, but the standardized tasks most certainly are not. To expect such "far transfer" is folly according to contemporary educational research where broad transfer of conceptual knowledge is rare and difficult to organize even under tightly controlled experimental conditions. Unfortunately, an academician's cautions against expecting generalized transfer from specific forms of experience run counter to a taxpayer's common sense expectation, and this discrepancy is not restricted to the issue of the effectiveness of OST!

In this context, I found particularly telling Baker et al.'s case of the after-school robotics program conducted in a school by a math teacher who appeared to be able to deal comfortably with the difference between in-school and out-of-school programs. This program seemed to have a lot of promise as a setting for successful after-school science education. In addition to having a highly trained leader, it was part of a national program with graded levels of recognition. The tasks were hands-on, cutting edge, fun, and led inexorably to the kids discovering that they needed to go beyond locally devised methods for reaching their goal (in the case given, finding the hypotenuse of a triangle). What I found most sobering was the fact that, while the kids recognized that they "needed geometry," the knowledge of what that geometry was, even after the idea of the Pythagorean theorem was introduced by an older and experienced group member, came, as the authors point out, from outside the group of children. It was the explicit importation of specialized knowledge from the math teacher that helped the children progress in their work, and the authors note that this example proved the rule. School was being imported directly into after-school, which is just fine. But it was not being imported by those who might be expected to show that their science activities in after-school would also reveal themselves on a standardized test in school.

I entirely sympathize with the current authors' (often implicit) focus on the benefits to the participants, as evidenced by active engagement in pro-social, intellectually content-rich activities, the positive benefits to their self-esteem and appreciation of the dignity and power of their home language and culture, and the other important developmental challenges that such participation makes possible for participants to meet. I have been party to many such programs and I have witnessed the benefits to both child participants and (in those rare cases where college students are part of a partnership to ensure increased quality of programming on an organized basis) to college students as well. I have even been witness to many cases where participants' lives are manifestly improved by such participation: rapidly occurring decreases in antisocial behavior, and high school and even college graduation for students once headed for a job cleaning up at McDonalds, at best. But not every child's life is transformed by participation in after-school activities, even by high-quality, engaging, science- or mathematics-based activities. I have also witnessed kids who turned their backs on these same programs and studied instead what seemed to me to be the surest route to incarceration and unemployment. Just as schools cannot solve the ills of society, after-school programs cannot be expected to solve all the problems of schooling.

I find it depressing that after-school programs came into being about 100 years ago as a way of getting troublesome kids off the streets and are currently being justified by many as a means of reducing crime between 3 and 6 p.m. It reminds me that mass schooling itself was undertaken in large part to keep children who spent 6 days a week and 12 or more hours a day in factories from raising hell when they were not at work. There are deep problems with historically enduring roots at work in the cultural reproduction processes that have long been visible in formal schooling and that now appear in after-school education and policies.

Education is willy-nilly spilling outside the boundaries of the brick schoolhouse into many social institutions of contemporary life for people of all ages. There is *enormous* potential for *successfully* leveraging after-school settings in the service of improved educational attainment for the young. But in my view, until and unless the representatives of American taxpayers agree about the benefits of well-funded programs that have significant and meaningful input from their local communities, until they are convinced of the economic value of such activities, and until we as a society agree that they deserve broad economic and civil support, we will be restricted to repeated demonstrations of a potential that is realized by a few, and marginalized by the many. The hard work of the authors of these papers, and the many whose labor they write about, deserves better.

About the Authors

John Y. Baker is a Ph.D. candidate, finishing up a degree in mathematics education at the University of Pennsylvania. In his dissertation research, he investigated how youth used and learned to use mathematics as participants in an after-school robotics program. In addition to investigating the mathematics practices of the youth in this space, his work included an explicit focus on the context of a program located at the intersection of in-school and out-of-school time. John was a research fellow with the MetroMath Center.

Brigid Barron is an Associate Professor at the School of Education at Stanford, is a faculty co-lead of the LIFE center, and directs the YouthLab research group (youthlab@stanford.edu). A developmental and clinical psychologist by training, she studies processes of learning in and out of school. In a 5-year NSF supported CAREER award she documented adolescents' learning ecologies (e.g., learning opportunities across home, school, libraries, virtual communities, clubs, camps) for technological fluency development across diverse communities in the Silicon Valley region. Barron is PI on a 3-year grant funded by the John D. and Catherine T. MacArthur Foundation that follows students longitudinally as they participate in programs designed to develop their technological fluency through activities such as game design, robotics, and digital movie making.

Philip Bell pursues a cognitive and cultural program of research across diverse environments focused on how people learn about science and technology in ways that are personally consequential to them. He is the Geda & Phil Condit Professor of Science & Mathematics Education and Professor of the Learning Sciences at the University of Washington where he directs the ethnographic and design-based research of the Institute for Science and Math Education (http://sciencemathpartnerships.org/). He is a co-director of the Learning in Informal and Formal Environments (LIFE) center (http://life-slc.org/). He has a background in human cognition and development, science education, computer science, and electrical engineering.

Bronwyn Bevan directs the Center for Informal Learning and Schools (CILS), an NSF-funded center based at the Exploratorium in San Francisco. CILS conducts

B. Bevan et al. (eds.), *LOST Opportunities: Learning in Out-of-School Time*, Explorations of Educational Purpose 23, DOI 10.1007/978-94-007-4304-5,

research and provides professional development related to forging stronger links between in- and out-of-school learning. Bevan previously worked for EDC's Center for Children and Technology in New York. She holds a Ph.D. from the Graduate Center of the City University of New York and has worked at the Exploratorium since 1991.

Kristen Pilner Blair is a Postdoctoral Research Associate at Stanford University. Her research focuses on children's mathematical learning and cognition. She approaches this topic from a variety of perspectives, including designing educational technologies to help students learn mathematical concepts, studying children's informal use of mathematics in family settings, and trying to understand the neural mechanisms underlying mathematical cognition using fMRI.

Leah Bricker is an Assistant Professor of Science at the University of Michigan with a focus on language, literacy, and text. She holds a Ph.D. from the University of Washington in the Learning Sciences. Bricker pursues research in two broad areas: (a) science education (e.g., people's [youth, teachers, scientists] science learning in formal and informal settings; culturally responsive science learning and teaching), and (b) learning sciences (e.g., socio-cultural-historical theories of learning as applied to how people learn across settings and over time and as applied to how people develop expertise and identities; social disparities and learning). Bricker is also interested in youth participatory action research and how the language of scientists can inform science education.

Maureen Callanan is Professor of Psychology at University of California, Santa Cruz. Her research takes a sociocultural developmental approach, focusing on cognitive and language development in young children. Much of her work investigates children's informal science learning in the context of family conversations and activities. Callanan was a faculty partner in the Center for Informal Learning and Schools (CILS), and a co-author of the 2009 National Academies Press volume "Learning Science in Informal Environments: People, Places and Pursuits." She also has a long-standing research partnership with Children's Discovery Museum of San Jose, where she has been co-PI on a number of NSF-funded projects focused on parent-child conversations about science.

Michael Cole's work focuses on the elaboration of a mediational theory of mind. He has conducted cross-cultural research on cognitive development, especially as it relates to the role of literacy and schooling. His recent research has been devoted to a longitudinal study of individual and organizational change within educational activities specially designed for afterschool hours. These systems link universities and local communities and allow a study of the dynamics of appropriation and use of new technologies and cultural-historical approaches to human development. According to Cole's methodology, mind is created and must be studied in communication.

Rhiannon Crain received her Ph.D. from the University of California Santa Cruz and the Center for Informal Learning and Schools. She has worked with the Exploratorium, the Monterey Bay Aquarium, and the Cornell Lab of Ornithology

Citizen Science Department to understand how informal learning opportunities can support complex conservation thinking and activity. She is interested in the evolution of science centers as learning-centered organizations and their changing role as potential contexts for public engagement with science. She has a research background in evolutionary biology and public education.

Indigo Esmonde pursues research on diversity and equity in mathematics education, with a focus on ethnography and discourse analysis. She is an Assistant Professor in mathematics education at the University of Toronto in Canada. Initially trained in mathematics at Queens University and McGill University, Indigo received her high school teaching certification and Ph.D. from the University of California, Berkeley. She has recently been awarded a Knowles Science Teaching Foundation Research Fellowship to study how secondary mathematics teachers learn to teach in culturally responsive ways.

Shelley Goldman is an Educational Anthropologist working on understanding the power of contexts to both drive and inhibit learning and success. She studies learning in both school and out of school settings, and applies findings from research to the development of teaching and learning environments. Her research and development work has two main foci: understanding the many contexts that create opportunities for rich math learning experiences, with a special emphasis on the family; and, understanding how technology can increase access to and transform learning. She is co-editor of the 2009 volume, *Educating Learning Technology Designers*. Goldman is Professor (Teaching) of Education, and by courtesy, Mechanical Engineering-Design Program, at Stanford University.

Linda Grigholm is a teacher at the Cooperative Children's Center in Seattle, Washington. She completed her M.Ed. in learning sciences at the University of Washington in 2009. She is interested in how young children organize themselves in social and play interactions and the learning opportunities that arise from those interactions.

Kris D. Gutiérrez, Professor of Social Research Methodology at UCLA, is president-elect of the American Educational Research Association. Professor Gutiérrez is a member of the faculty at the UCLA Graduate School of Education and Information Studies where she is Co-Principal Investigator of the UCLA Center for the Everyday Lives of Families in Anthropology, Director of the Education Studies Minor, and Director of the Center for the Study of Urban Literacies. Gutiérrez also serves as Associate Faculty at the University of California, San Diego Laboratory for Comparative Human Cognition. Her research interests address the relationship between literacy, culture and learning, and target how students appropriate cultural concepts.

Maura Varley Gutiérrez is interested in critical mathematics education and the intersections of gender and race at both the elementary classroom and preservice teacher levels. She completed her doctorate at the University of Arizona where she currently coordinates an NSF-funded project investigating the reform of our teacher preparation program to be a hybrid learning space including local teachers, schools, and the university. As a doctoral student, she was a fellow with the NSF-funded Center for the Mathematics Education of Latinas/os (CEMELA), working on a research project integrating community funds of knowledge in an after-school mathematics setting.

Osvaldo Jimenez is a Ph.D. candidate in learning sciences and technology design at Stanford University. He is currently a member of the Family Math team, participating in the construction of tools to gather and analyze data, as well as in the collection of data with primarily Spanish-speaking populations. His other interests lie in casual gaming and how it can be created and harnessed as a learning tool in a variety of contexts. Osvaldo has an undergraduate degree from Stanford in Computer Science.

Jennifer Jipson is an Associate Professor in the Department of Psychology and Child Development at California Polytechnic State University in San Luis Obispo. She received her B.A. in psychology and government from Smith College in Northampton, MA, and her Ph.D. in developmental psychology from the University of California at Santa Cruz. She also completed 2 years of postdoctoral training at the University of Michigan. Her research investigates how children engage with science in the context of everyday conversations with parents. She has authored numerous papers and book chapters and regularly presents her work at professional conferences.

Lena Licón Khisty is an Associate Professor in Curriculum and Instruction at the University of Illinois at Chicago. She teaches graduate and undergraduate courses in both bilingual/ESL and mathematics education. She identifies herself as a Chicana and has devoted her work to removing barriers to Latino education. Her research revolves around the application of sociohistorical activity perspectives in understanding and explaining mathematics learning processes with bilingual and Latino students, and is particularly interested in classroom interactions, communication processes, and racism in mathematics. Her published work has involved challenging conventional wisdom about Latinos by offering examples of classrooms that work for them and implications for teacher preparation.

Vivian Lim is a Ph.D. student at the University of Pennsylvania Graduate School of Education. She is broadly interested in promoting culturally relevant and social-justice oriented mathematics education. More specifically, she is interested in mathematics education that draws on the context and resources of local communities through community-based partnerships. Vivian is currently involved in several projects in which she represents the University as a resource and partner, contributing to efforts to create curriculum intended to engage students in active participation in their communities.

Molly Loomis is the Director of Education at ECHO Lake Aquarium and Science Center in Burlington, Vermont. She received her Ph.D. in science education from the University of California Santa Cruz as a fellow with the Center for Informal

Learning and Schools. Her research has focused on the history and development of interactive science centers, the organization of learning in informal settings, and family activity at interactive exhibits. She has a background in museum education, exhibit development, and program evaluation.

Megan Luce has a doctoral degree in developmental psychology from the University of California Santa Cruz. She was a research fellow with the Center for Informal Learning and Schools. Her research investigates children's cognitive development in the context of family conversation and activity. She is particularly interested in the role of parents' attitudes toward science and views on the nature of knowledge in the development of children's scientific reasoning. Megan's work in informal science learning has focused on learning research to inform the design of science exhibitions at Children's Discovery Museum of San Jose.

Caitlin K. Martin has been the project manager and a lead researcher of the Learning Ecologies Group at Stanford University since receiving her Master's degree in learning, design, and technology from Stanford in 1999. Her work involves the design, implementation, and study of environments that encourage youth from diverse communities to develop technological fluency. Specific projects include the design of a 4-year computing curriculum and corresponding professional development for government funded high schools in Bermuda, an assessment of after-school academic technology programs at community centers in Northern California, and longitudinal case studies of students participating in an urban middle school media arts program.

Lee Martin is an Assistant Professor in the School of Education at the University of California, Davis. His research looks across contexts to study how people use tools and resources to help them be adaptive thinkers in the realm of quantitative reasoning. In everyday settings, he examines the varied ways in which people assemble social, material, and intellectual resources for problem solving and learning. In school settings, he looks to find ways in which schools might better prepare students to be more resourceful and flexible in fostering their own learning.

Ray McDermott takes a broad interest in the analysis of human communication, the organization of school success and failure, and the history and use of various literacies around the world. His work includes studies of inner-city public schools, after-school classrooms, and the function of information technologies in different cultures. At present, he is working on the intellectual history of ideas like genius, intelligence, race, and capital.

Siri Mehus is a researcher with the LIFE Science of Learning Center at the University of Washington. She received her Ph.D. in communication studies from the University of Texas in 2006. She studies interaction and learning as situated in particular social and material environments (such as preschools, homes, and workplaces), employing video-based qualitative analysis of talk and embodied action.

Vera Michalchik is a Senior Social Scientist at SRI International's Center for Technology in Learning where she leads the informal learning practice. She currently heads the evaluation of the NSF Informal Science Education program, co-directs the NSF AYS Research and Evaluation Center, leads research on partnerships in the Nanoscale ISE Network, studies science and new media-focused youth development programs, conducts research with the LIFE Science of Learning Center, and coordinates worldwide evaluation for education programs funded by the Intel Foundation. Vera also does research in the areas of teacher professional development and visualizations for learning science and math. From 2006 to 2008, she served on the NSF-sponsored Committee on Learning Science in Informal Environments at the National Research Council. Vera specializes in observational, ethnographic, and video-analytic methods, and holds a Ph.D. from Stanford University, an Ed.M. from Harvard, and a B.A. from UC Berkeley.

Rodney T. Ogawa is Professor of Education and Director of the Center for Educational Research in the Interest of Underserved Students (CERIUS) at the University of California, Santa Cruz. His research has focused on how and with what consequences school organizations adapt to educational policies. Currently, his research examines how and with what consequences the social structures of educational organizations are associated with the learning contexts they present, and thus instructional interactions and educational outcomes. Rodney's most recent work has been published in *Educational Researcher*, *Review of Educational Research*, *Science Education*, and *Teachers College Record*. Rodney served as Vice President of Division A of the American Educational Research Association. He is the 2010 recipient of the Roald Campbell Award for Lifetime Achievement in the academic field of educational leadership by the University Council for Educational Administration.

Roy Pea is David Jacks Professor of the Learning Sciences and Education at Stanford University and Director of the Stanford Center for Innovations in Learning. He has published widely on such topics as distributed cognition, learning and education fostered by advanced technologies including scientific visualization, on-line communities, digital video collaboratories, and wireless handheld computers. His current studies include investigating how family mathematics works to provide occasions for learning, how video point-of-view influences learning outcomes, and working on theory for how informal and formal learning can be better understood and connected, as Co-PI of the LIFE Center funded by the NSF. He is coeditor of the 2007 volume *Video Research in the Learning Sciences*. He was co-author of the 2000 National Academy Press volume *How People Learn*. He is a Fellow of the National Academy of Education, Association for Psychological Science, The Center for Advanced Study in the Behavioral Sciences, and the American Educational Research Association.

Aria Razfar is an Assistant Professor in curriculum and instruction at the University of Illinois at Chicago. His research interests are grounded in sociocultural and critical theories of language, learning, and human development. His work examines the role of language and discourse in human development across the life-span. He has published work examining the sociocultural influence in early literacy, second language learning, mathematics education, and teacher development. He is a co-lead for the National Science Foundation's Center for Mathematics Education of Latinos/as (CEMELA). He is also the Lead Investigator for Project LSciMAct (Transforming Literacy, Science, and Math through Action Research) supported by the US Department of Education's Office of English Language Acquisition (OELA).

Suzanne Reeve teaches high school biology and chemistry in Bellevue, Washington. She received her Ph.D. in learning sciences from the University of Washington in 2009, with a dissertation focusing on young people's learning of health and nutrition across everyday life settings. Her studies included working as a research assistant for the LIFE Center and the Everyday Science and Technology research group. Prior to her Ph.D., Suzanne studied habitat ecology and conducted research on learning and teaching in undergraduate biology.

Janine T. Remillard is an Associate Professor in the Graduate School of Education at the University of Pennsylvania, specializing in mathematics education. Her research interests include mathematics teaching and learning in urban schools and communities and parents', teachers', and students' experiences with reform curriculum materials. She is P.I. of Improving Curriculum Use for Better Teaching (iCubit) and a research associate with the Center for the Study of Mathematics Curriculum. She is co-editor of *Mathematics Teachers at Work: Connecting Curriculum Materials and Classroom Instruction.*

Jennifer Rigney is a doctoral candidate in developmental psychology at UC Santa Cruz and was a research fellow with the Center for Informal Learning and Schools. Her research involves investigating children's conceptual development within meaningful social contexts, with a focus on the development of everyday theories of biology. She is also interested in how parents and children reason about food and health.

Deborah Siegel received her Ph.D. in developmental psychology from the University of California, Santa Cruz in 2010. She was also a research fellow with the NSF-funded Center for Informal Learning and Schools (CILS). Broadly, she uses mixed methodologies to study children's cognitive development within a social context. In her dissertation, she explored how preschool children learn to use artifacts in both culturally conventional and creative ways in the context of an everyday parent-child activity. She has also studied parent-child interactions within science-related tasks in a variety of informal learning contexts.

Ksenija Simić-Muller has a Ph.D. in mathematics from Carnegie Mellon University. During her postdoc at the University of Arizona, she was a fellow with the Center for Mathematics Education for Latinos/as(CEMELA). She currently teaches mathematics and mathematics education courses at Pacific Lutheran University in Tacoma, Washington. She works with future elementary and secondary teachers, and is especially interested in issues of access and equity in mathematics education, and in bringing social justice issues into courses for preservice teachers.

Reed Stevens is a Professor of Learning Sciences in the School of Education and Social Policy at Northwestern University. He is co-lead of the NSF LIFE Center (Learning in Informal and Formal Environments). Since the mid-1990s, he has pursued a program of field studies of learning and thinking in a wide range of settings that have included schools, homes, workplaces, museums, and child care centers. His work compares thinking and learning across settings and seeks to follow people learning and thinking across settings toward a goal of more complete accounts of learning. His topical focus is mathematical, scientific, and design-oriented practices.

Lara Triona is an Assistant Professor of developmental psychology at California State University, Fresno. Lara earned her Ph.D. from Carnegie Mellon University in 2004. Lara's research focuses on children's science learning in informal settings focusing on parents' influence. In prior projects she has compared student learning with physical material to a computer simulation and children's developing understanding of how to create external representations. Her research combines sociocultural and cognitive approaches to understand children's development.

Carrie Tzou is an Assistant Professor in Science Education at the University of Washington Bothell. She holds a Ph.D. in learning sciences from Northwestern University and an M.S. in teaching and learning with a concentration in science education from Vanderbilt University. Carrie's research has three major components, all connected with an interest in addressing issues of equity and social justice in science and environmental science learning: (1) ethnographic work to understand how youth and their communities are positioned through place-based education, (2) design-based research to design curricula to bring youths' out of school science and cultural practices into science and environmental science teaching, and (3) research and design of elementary and secondary preservice teacher education that explores how to orient preservice teachers to the sophisticated learning and identities that their students construct both in and out of school in order to make science more accessible to all of their students.

Gabriel Viego has a Master's degree from UIC in curriculum and instruction. He has a background in mathematics education and has taught mathematics and science for 10 years at the middle school level. He is currently a doctoral student in curriculum studies at UIC. His research interests are in lesson study as a means to facilitate mathematics teachers' professional growth in teaching ELL's. His interests also include work with CEMELA (Center for the Mathematics Education of Latinos/as) and its after-school mathematics club (for 3rd–5th grade students) that includes a research focus on the role of language and everyday mathematics practices in mathematics learning.

Eugenia Vomvoridi-Ivanović is an Assistant Professor in mathematics education at the University of South Florida. She completed her Ph.D. in curriculum and instruction, with an emphasis in mathematics education, at the University of Illinois at Chicago and was a doctoral fellow with the Center for Mathematics Education of Latinas/os (CEMELA) for 4 years. Her dissertation is entitled: "Latina/o Pre-service

Teachers' use of Language and Culture while assisting children in mathematics." Her areas of interest in research and teaching include: mathematics teacher preparation for linguistically and culturally diverse students, culturally responsive mathematics teacher preparation, and preservice teacher development in informal mathematics learning contexts.

Craig Joseph Willey is a Research Fellow with the Center for Mathematics Education of Latinas/os (CEMELA) and a doctoral candidate in mathematics education at the University of Illinois at Chicago. His research is situated at the intersection of two fields – mathematics and bilingual education – and focuses on multiple aspects of mathematics teaching and learning for Latinas/os. Specifically, he is interested in the role of teachers in creating optimal bilingual mathematics classrooms and the macro, sociopolitical structures that perpetuate differentiated mathematics learning for marginalized youth. Craig completed his undergraduate work at Butler University and taught middle grade mathematics in Denver Public Schools' Bilingual Education program before joining CEMELA.

Susie Wise is the Director of Media Operations at Stanford University's Hasso Plattner Institute of Design. Her research focuses on how people learn through design and on developing assessments for design thinking. She has an M.A. in learning, design, and technology from Stanford and worked for 3 years as a graduate research associate with youthLAB (http://youthlab.stanford.edu) where she focused on technological fluency development in informal learning contexts.

Heather Toomey Zimmerman is an Assistant Professor of Education at Pennsylvania State University. Heather has her Ph.D. from the Learning Sciences program at University of Washington, with additional degrees in Museum Studies and Science Communication. Heather's research analyzes the social-cognitive practices that people use to learn about science and technology in everyday settings. She studies how multiple learning experiences contribute to the development of knowledge, practices, and affiliations toward (or away) from science. She has published in *Science Education*, the *Journal of Museum Education*, *Museums and Social Issues*, and the *Proceedings of the International Conference of the Learning Sciences*. Her interests include parent-child interactions, designing for learning in informal institutions, technology to support learning across settings, and gender issues that intersect with STEM disciplines.

Index

B. Bevan et al. (eds.), *LOST Opportunities: Learning in Out-of-School Time*, Explorations of Educational Purpose 23, DOI 10.1007/978-94-007-4304-5,

MIX
Papier aus verantwortungsvollen Quellen
Paper from responsible sources
FSC® C105338
www.fsc.org

If you have any concerns about our products,
you can contact us on
ProductSafety@springernature.com

In case Publisher is established outside the EU,
the EU authorized representative is:
Springer Nature Customer Service Center GmbH
Europaplatz 3, 69115 Heidelberg, Germany

Printed by Libri Plureos GmbH
in Hamburg, Germany

www.ingramcontent.com/pod-product-compliance
Lightning Source LLC
LaVergne TN
LVHW020717290926
849104LV00004B/33